Elements of Structural Geology

Originally published in 1963, this classic textbook was revised fully for the 1972 edition. The author presents a comprehensive account of all topics falling within the domain of structural geology in his characteristically objective, scientific and logical manner. The book pays particular attention to definitions and the origin of terms. Geology is a global science and this book used examples and ideas from work in many countries. The book is comprehensive in scope, dealing not only with secondary structures and tectonics, but also with primary structures of secondary and igneous rocks. This was the first textbook to deal with rock material as two-phase systems rather than as solids and this approach is continued in this reissued edition by analysis of concepts such as ocean-floor spreading and plate tectonics.

Elements of Structural Geology

E. Sherborn Hills

Routledge
Taylor & Francis Group

First published in 1963 and as a second edition in 1972 by Chapman and Hall Ltd.

This edition first published in 2024 by Routledge
4 Park Square, Milton Park, Abingdon, Oxon, OX14 4RN
and by Routledge
605 Third Avenue, New York, NY 10158.

Routledge is an imprint of the Taylor & Francis Group, an informa business

© 1963, 1972 E. Sherborn Hills

ISBN 13: 978-1-032-73649-5 (hbk)
ISBN 13: 978-1-003-46536-2 (ebk)
ISBN 13: 978-1-032-73669-3 (pbk)
Book DOI 10.4324/9781003465362

ELEMENTS
OF STRUCTURAL
GEOLOGY

E. SHERBON HILLS

Research Professor of Geology in the University of Melbourne

LONDON   NEW YORK

CHAPMAN AND HALL

First published 1963 by
Methuen & Co Ltd
Second edition 1972

© *1963, 1972 E. Sherbon Hills*

First publshed as a Science Paperback 1965
Second edition 1972 published by
Chapman and Hall Ltd
11 New Fetter Lane, London EC4P 4EE
Published in the USA by
Chapman and Hall
783 Third Avenue, New York NY 10017
Reprinted 1974, 1979 and 1983

ISBN 0 412 20750 8

Filmset in Monophoto Times 11 on 12 pt by
Richard Clay (The Chaucer Press) Ltd,
Bungay, Suffolk
and printed in Great Britain by
Fletcher & Son Ltd, Norwich

Contents

Preface to Second Edition

During the ten years which have passed since the manuscript of the first edition of this book was completed, most of the topics to which the chapters are devoted have themselves been dealt with in important monographic publications, and it is increasingly difficult to condense the volume of knowledge and number of examples which may properly be grouped under Structural Geology, within one volume. I nevertheless believe that there is still a place for the general textbook, although its writing may, in future, well be a task for several collaborators.

For this edition, complete recasting of the book was not practicable, but many errors have been corrected, references have been brought up-to-date, and certain sections have been rewritten, while others have been expanded to deal with new concepts. It is not long since, in an earlier work, it was apposite to refer to gravitational tectonics and flowage as a burgeoning branch of structural geology; but it must now be said that rigid-plate tectonics and ocean-floor spreading are even more remarkable concepts, which must have fundamental effects on regional structural geology and on our understanding of global phenomena.

Additionally to the Journals mentioned in the Preface to the first edition, my thanks are due to the editors of the Journal of Geophysical Research and the Seismological Society of America, as well as to the authors concerned, for permission to publish illustrations in the revised edition.

I am indeed grateful to the many colleagues who have suggested improvements in the book, among which it has been salutary to find that what is praised by some is criticized by others. I am also again much indebted to Professor E. den Tex (Leiden), who kindly consented to revise Chapter XIII on Structural Petrology, and to Miss Cecily Finlay for her constant help during the revision of the manuscript.

University of Melbourne   E. S. H.
March 1970

Preface to First Edition

The range of fundamental topics that should be included in a general work on structural geology is indeed very wide, but, despite the condensation needed to reduce these matters to a reasonable compass, I have tried to adhere to certain standards as to completeness of coverage, style of presentation, and adequacy of illustration.

Particular attention has been paid to definitions, both for the realities of structures and for concepts and notions, since definitions, far from being dry-as-dust material for rote learning, express much of what emerges from the systematization of knowledge, and it is desirable that the student should realize the extent to which they are conceptual and thus subjective. Experimental analogues and the relevant physical or chemical principles have been freely referred to, with as much discussion of pros and cons as is possible within the limits of a textbook. It is hoped that in these ways the reader may himself be induced to consider and weigh propositions, and thus to inculcate something of the author's own abhorrence of the cult of the fashionable theory, which so often in geology involves personal loyalties rather than intellectual integrity, and certainly does nothing to raise the scientific status of the subject.

For the Chapter on Geomorphology and Structure I have, of necessity, relied upon Davisian or genetic geomorphology, partly because quantitative terrain analysis, despite the 'precise and coldly analytical' methodology claimed for it, has not yet, so far as I am aware, included sufficient of structure in its premises to make it applicable to structural geomorphology, and partly because I have found Davisian principles, modified as required by new knowledge, most warmly satisfying in the field. It may, however, readily be foreseen that morphometric studies will in due course add their contribution to structural geomorphology and morphotectonics, at scales ranging from micro- to megatectonics. For the latter, I have briefly referred to several generalizations and theories, but have emphasized the need for further precise knowledge as to the morphological and geological data. Palaeomagnetic results, although of great potential applicability in tectonics, are not discussed because the book deals mainly with the optically demonstrable features of rocks and rock masses, so that the implications of palaeoclimatology, the past distribution of animals and plants, the distribution of magma-types and such, and similar topics of significance in megatectonics are not dealt with.

Although the principles of structural geology are of general applicability, the particular needs of oilfield geology, engineering geology, and mining geology call for certain specialized field and laboratory methods, including the geometrical processing of data according to the needs of particular problems. Detailed reference to such matters would be inappropriate in a book intended to deal with fundamentals, and indeed each of these specialities is itself treated in well-known textbooks. It is, however, well to note that, unless data are correctly diagnosed in the field, their geometrical processing is valueless, and the stress in this book is accordingly on the understanding of geological structures, although reference is made throughout to certain well-known methods of representation and reconstruction of structures.

The sequence of topics in my small 'Outlines of Structural Geology' has broadly been adhered to, as it permits a gradual introduction of notions, and appears to have been acceptable to many teachers, but I have not hesitated to make repeated reference to the same notion, for example, soft-rock deformation, under several headings, as the student often fails to transfer knowledge from one compartment of his mind to another.

The Chapter on Structural Petrology was kindly written at my request by my former colleague in Melbourne, Dr E. den Tex, now Professor of Petrology in the University of Leiden, Netherlands. I am indeed grateful to him for having undertaken the difficult task of presenting this subject in a style and with a philosophy conforming generally with the rest of the book, and in so satisfying a manner in a single chapter.

I am much indebted also to Dr J. V. Harrison of Oxford for critically reading the MS. and for assistance in various other ways. It is a pleasure to acknowledge the kindness of those numerous colleagues who over the years have sent me their publications from many countries, as I have tried to draw examples and illustrations from a wide area, geology being *par excellence* a global science.

Acknowledgement is made to the under-mentioned Publishers, Journals, and Serials for permission to use previously published illustrations as a basis for line drawings specially made in a uniform style for this book. Acknowledgement to authors, and to the sources of half-tone illustrations other than the author's own photographs, are made in the captions.

Addison-Wesley Press Inc., Edward Arnold & Co., Cambridge University Press, Julius Springer Verlag, McGraw-Hill Book Co. Inc., Macmillan & Co., Ed. Masson & Cie, Oliver and Boyd, Thomas Nelson & Sons, Ltd., University of Chicago Press, John Wiley & Sons, The Williams & Wilkins Company, American Association of Petroleum Geologists, American Geophysical Union, American Journal of Science, Economic Geology, Geographical Review, Geologie en Mijnbouw, Geologische Rundschau,

Geological Society of America, Geological Society of South Africa, Geological Magazine, Geologists' Association, Geological Society of London, Journal of Geology, Koninklijke Nederlandische Akademie Van Wetenschappen, Mineralogical Magazine, Neues Jahrbuch für Geologie, etc., New Zealand Journal of Science and Technology, N.V. de Bataafsche Petroleum Maatschappig, Physical Society of London, Royal Society of London, Royal Society of New Zealand, U.S. Geological Survey, Geological Society of Australia, Geological Survey of Great Britain, Geological Survey of Victoria.

My personal thanks are due to Miss P. Carolan and Miss C. Finlay for secretarial and technical assistance.

University of Melbourne   E. S. H.
March 1961

CHAPTER I

The domain and content of structural geology

The special province of structural geology among the branches of geological science lies in the recognition, representation, and genetic interpretation of rock-structures. Without for the moment defining precisely what is meant by rock-structures, let us consider any geological cross-section showing beds in various attitudes – flat-lying or inclined, folded or faulted – perhaps in places intruded by igneous rocks or covered with lava-flows, and again in places metamorphosed. Such a section shows the shape and attitude of the various rock-units present, and is said to represent the geological structure of that part of the earth, in the plane of the section. The shapes and attitudes of the several rock-units shown, as they now exist in the earth, are the result firstly of processes connected with their original formation when their *primary structures* were developed, and secondly of all later processes whether of mechanical deformation or chemical reconstitution, that have affected them. It is usual to distinguish those effects that are connected with the later stages of the formative processes of sedimentary or igneous rocks as *penecontemporaneous*, since they are intimately connected with the conditions that obtained during the origination of the rock in question, while structures of entirely subsequent formation are regarded as *secondary*. Uplift, tilting, faulting, or folding in general give rise to secondary structures, but although it is true that structural geology is very largely concerned with these, it is also clear that one must have a very good understanding of the objects that were thus deformed, and that attention must be devoted to the primary and penecontemporaneous structures of rocks before proceeding to treat with their secondary deformation.

The necessity for this is even more obvious if we consider how structural knowledge is gained in the field. When one observes a rock exposure for the purpose of mapping, it is eventually necessary to record in the field-book some data on the geometry of the rocks, generally the dip and strike of some planar surface that one can identify in the exposure. It is therefore obvious that the real nature of such a surface must be recognized by the field-geologist. It is a common enough error to measure cleavage in slates in mistake for bedding, and other misconceptions are equally possible. In order

1

to map structural data, the geologist working in the field must be able to recognize what he is mapping, which implies a very complete knowledge of the primary structures of sedimentary and of igneous rocks, as well as of secondary structures such as jointing, cleavage, foliation, and the like.

Although too much emphasis on primary structures in a work on structural geology might be regarded as usurping the field of interest of petrology or of sedimentation, many fundamental advances in our knowledge of sedimentation have in fact been made of recent years by structural geologists who have carefully examined the primary structures of sedimentary rocks. This is true particularly of the processes of subaqueous slumping and of deposition from turbidity currents. These topics are treated along with the primary structures of sediments in Chapter II of this book, which is followed in Chapter III with a discussion of penecontemporaneous structures and of structures which, although of secondary origin, are not the result of diastrophism, but of gravitational forces acting on soft or yielding rocks at or near the earth's surface.

Before proceeding further with secondary structures, those physical principles that are particularly applicable to the interpretation of rock-deformation are dealt with in Chapters IV and V, following which discussion we are equipped to proceed to what may be regarded as the core of the subject, that is to say with joints, faults, folds, and slaty cleavage. In the study of these we shall find that we are concerned with various types of planar and linear structure-elements both megascopic and, especially in metamorphic rocks, in the crystallographic elements of their component grains.

Structure and fabric

In describing the geometry of rock-masses in terms of the various linear and planar structure-elements two distinctly different methods have been developed. The first recognizes and describes rock-structures which are actual entities or groups of entities such as folds, faults, and joints. The position of each of these is mapped, its form and other characteristics are described or illustrated, and its origin is debated. The chief methods of representation are by maps, cross-sections, block diagrams, and photographs. The second method treats with the geometry of rock-structures very largely on a statistical basis. The attitude of a great many individual structure-elements is mapped and plotted to reveal the statistical grouping of the elements – their attitude relative to geographical co-ordinates, and their mutual geometrical relationships – which together constitute the *fabric* of the rocks. Fabric studies are based on hundreds of observations recorded as notes and later transferred to *fabric diagrams*, which facilitate statistical analysis. Little may be plotted on the map as to the actual position of the many individual elements such as joint planes or dip readings which are measured. To illustrate the two methods, a house might be represented in plan and elevation by an architect's detailed drawings of the actual structures in the edifice; the fabric, however, might be said to exhibit strong maxima for vertical planes in two directions at right angles (the walls), and a third horizontal direction of planes – a flat roof, floors, and ceilings. The two methods supplement each other, but in many practical matters it is the location and attitude of actual structural entities that is in the long run essential, as in mining and in petroleum and engineering geology. Fabric analysis has been applied particularly to the microscopical and semi-microscopical features of rocks, including grain-shape, various crystallographic directions and planes (optic axes, twin planes, etc.), and small-scale structures seen in hand specimens. This is variously known as structural petrology, petrofabric analysis, or micro-tectonics. The term fabric, however, includes not only the microfabric but also the megascopic structures, and statistical treatment may usefully be applied to these latter. Accordingly the opportunity is taken in Chapter VI to give an introduction to the graphical representation of statistical data, which is of general applicability throughout the book. A more complete account of fabric analysis is given in Chapter XIII which treats with Structural Petrology and which, apart from the remarks on cleavage in Chapter X, affords the chief account of metamorphic rocks in this book. Igneous rocks are dealt with in Chapter XII. They are unique in that they preserve many structures that originated in the fluid phase and the presence of fluid also affects their space relationships in the stress–strain pattern of a region.

Geotectonics

The study of a small area inevitably leads the structural geologist to consider the 'setting' of his area in the broader framework of the surrounding region. He wishes to fit his mapped beds, folds, and faults into the pattern of folded or faulted belts of the whole country, and soon finds that he is thinking on a continental scale. While there seems no logical reason from considerations of size or area alone to divorce such broad investigations and research from structural geology, traditionally the wider aspects of continental and oceanic, and eventually of global structures, are referred to as *tectonics* or *geotectonics*. Some treatment of major fold belts, fault zones, and the like is essential in any course on structural geology and these topics are discussed in Chapter XI, but structural geology is concerned essentially with the outer layers of the globe, and detailed discussion of topics that involve the geophysics of the deep interior of the earth are accordingly omitted from this book, or mentioned only briefly as the occasion may demand.

Following Chapter XIII on Structural Petrology, much of which is concerned with micro-tectonics, the scale of our observations again changes vastly, to that of the landscape, in dealing with Morphotectonics in Chapter XIV. This is a fruitful field which no field-geologist can afford to neglect, and it is pleasing to note a revival of interest in geomorphology and its structural implications among geologists, after a long period of neglect.

The literature of structural geology

At fairly frequent intervals, one or other branch of structural geology has been given a fresh outlook by the introduction of some new concept or, perhaps more often, by the realization that some idea, previously expressed but long neglected, is, in fact, useful and stimulating. Even if at times such saltations in thought have had their opponents and have led to controversy, the lively discussions that have resulted have often stimulated further interest and research. Topics such as folding, faulting, nappe structures, cleavage-formation, the structures of igneous rocks, petrofabric analysis, gravitational sliding, or the influence of basement structures on superincumbent rocks or on younger structures, to name only a few, will amply repay historical review by the serious student.

Indeed there are so many ancillary ideas, illustrations, and facts stored in the voluminous literature of structural and regional geology that it is both profitable and pleasurable to pursue bibliographic research in these fields. The references given throughout this book are regarded as those most useful on any particular topic but are far from exhaustive. The chief textbooks in languages other than English are listed at the end of this chapter.

Brief acquaintance with the literature will immediately reveal to the student the unfortunate circumstance that there are wide variations in meaning attached to many terms in structural geology, even for some that are most fundamental and, if one recalls one's elementary instruction in geology, apparently simple, such as 'unconformity'. The following passage is taken from a recent text: 'Almost every separation plane in a stratigraphic sequence is an unconformity of some duration.' The student may be pardoned for wondering what meaning the term unconformity actually has, when it can be so used for the separation planes in a series of beds which would, by all the usual standards, be termed a conformable series. Although alternative usages are mentioned in this book, an attempt has usually been made to arrive at what appears to be the most acceptable definition, and in places new definitions are proposed. Where there has been an agreed change in nomenclature, the older usage is referred to, as with 'pitch' for the currently used 'plunge' of folds, because it will be found in the literature and on maps antedating the change.

The best geological writings contain much fine description and sound thought, and it is essential for every student to read original works for himself. It is, however, equally imperative for him to keep an open and critical mind, and also, since the eye generally sees what the mind expects it to see, to learn in the field the art and skill which De la Beche so aptly named as that of the 'Geological Observer'.

Foreign textbooks

G. D. Azhgirei, *Strukturgeologie*, 572 pp., 1963. (German translation of an outline of fundamentals of structural geology, originally published in Russian, 1956.)

V. V. Beloussov, *Basic Problems in Geotectonics*, Moscow, 1954 (in Russian). English translation, New York, 1962.

J. Goguel, *Traité de Tectonique*, Paris, 1952. English translation, San Francisco, 1962.

V. E. Khain, *General Tectonics*, Moscow, 1964 (in Russian).

K. Metz, *Lehrbuch der Tektonischen Geologie*, 2nd edition, Stuttgart, 1967.

CHAPTER II

Depositional textures and structures

BEDDING

The most obvious and characteristic feature of sedimentary rocks is *bedding* or *stratification*,[1] by which is meant both the presence of recognizably different beds or strata in a sedimentary succession, and the presence within any one bed of a fine lamination or texture which is connected with its deposition (Fig. II–1). In general each bed is fairly homogeneous and may be distin-

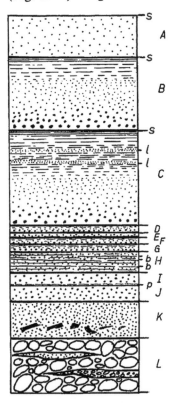

FIG. II–1. TYPES OF BEDS

Beds are bounded by separation planes (S).

A. Uniform, massive sandstone.

B. Simple graded bed, with uniform grading from grit below to shale above.

C. Complex graded bed with thin sandstone laminae (l, l) in the shale.

D, E, F, G. Thin individual bed (e.g. glacial varves).

H. Single sandstone bed with discrete bedding planes (b, b, etc.).

I. & J. Two sandstone beds separated by shale parting (p).

K. Heterogeneous bed of sandstone containing angular shale fragments.

L. Heterogeneous bed of conglomerate containing lenses of sand and gravel.

[1] W. H. Twenhofel, *Treatise on Sedimentation*, London, 2nd edn, 1932, pp. 603–756; K. Andrée, 'Wesen, Ursachen und Arten der Schichtung', *Geol. Rundsch.*, Vol. 6, 1915, pp. 351–97; F. J. Pettijohn and P. E. Potter, *Atlas and Glossary of Primary Sedimentary Structures*, Berlin, 1964; J. Bokman, 'Terminology of Stratification in Sedimentary Rocks', *Bull. Geol. Soc. Amer.*, Vol. 67, 1956, pp. 125–6; A. Lombard, *Géologie Sédimentaire*; Paris, 1956.

guished from other beds by some characteristic feature such as texture (e.g. shale, sandstone, conglomerate), composition (e.g. coal, limestone, shale), colour, hardness, or the like. Individual beds may, however, be recognized in certain sequences of uniform lithology by weathering of the *separation planes* between strata of similar lithological type or by the presence of very thin 'partings' of different rock, as for example of shale in sandstone. Rock exposures usually reveal the stratification by differences in the weathering characteristics of the beds (differential weathering).

Although many beds are homogeneous some show considerable variation, especially graded beds (in which there is a passage from coarser to finer particles towards the top); lateral gradation may also be found. Again, thin laminae or layers, differing somewhat in colour or texture, may be present without causing a bed to lose its individuality. It is the sum of its lithological features that characterizes a bed, with the implication that it was laid down under a particular set of conditions, either uniform or systematically varying. Although the decision must often be arbitrary, some very thin strata are best regarded as beds rather than as laminae within a bed. Thus in glacial varves, each annual deposit is clearly worthy of recognition as an individual bed even though its thickness is to be measured in inches or fractions of an inch, whereas sandy laminae in a graded greywacke are but parts of the whole graded unit.[1]

The upper and lower surfaces of adjacent beds combine to form the separation planes in a sedimentary succession. Since they often mark a change in conditions of sedimentation and also a time-gap they may be of almost equal significance with the beds themselves in the history of sedimentation, and top and bottom features of beds thus acquire special significance which is more fully discussed below (see p. 20). In addition, most beds possess an inner structure marked by fine laminations (the *bedding* or *stratification planes*) or a direction of ready splitting which is the *plane of bedding* (the 'bed' of quarrymen). In detrital sediments in which the component grains were washed into place by currents the lamination (or microlamination), represents successive upper surfaces of the bed as the detritus was laid down, but in rapidly deposited sands and gravels the majority of such surfaces would almost instantaneously have been buried. Indeed the preservation of fine lamination is generally an indication of rapid deposition, for with very slow accretion the chances of disturbing thin deposits by waves and currents and by organisms are greatly increased. The bedding planes may be revealed by slight differences in colour or texture and are usually best seen on weathered surfaces, but in some beds of very uniform lithology they are virtually invisible and

[1] The distinction sometimes made between lamina, layer, and bed according to thickness alone is not only entirely arbitrary but contrary to common usage whereby one speaks of a 'bed of laminated shale or limestone', a lamina being part of a bed. The term 'layer' has not gained general acceptance in the sense of a thin bed, and is best used in its everyday sense rather than as a technical term.

may then be found by splitting the rock. Fossils such as graptolites and plant remains generally lie on bedding planes, and the parallel orientation of detrital mica flakes may reveal the bedding in massive sandstones or mudstones.

Shales and laminated brown coals (*Blätterkohle*) will split into very thin sheets parallel to the bedding. The fissile nature of shales is largely due to the parallelism of micas and clay-mineral flakes, resulting partly from deposition, partly from the compaction of the deposit, and perhaps also from plastic flow parallel to the bedding under load.[1] Flaky particles such as mica generally lie flat in the stratification planes during sedimentation, even from currents, but in a slurry of mica mixed with other grains many flakes will come to rest in other attitudes. During compaction these are rotated, and the result is a tendency to parallelism.

Flat pebbles laid down in the beds of streams or on stony beaches are often piled against each other, leaning at an angle to form *edgewise conglomerates*. In streams the pebbles lean down-stream, and on beaches away from the water's edge. The resultant texture of the conglomerate is misleading as to the true bedding, but is also to be distinguished from cross-bedding, especially as the dip of the pebbles is in the opposite direction in relation to the current, from that in cross-bedding. Thus in describing bedding it is necessary to distinguish firstly between bedding or stratification planes or laminations which are individual structural entities such that each planar surface may be seen and traced, and depositional *textures*, which result from the parallel orientation of particles throughout a bed. Both are primary depositional features, and may be either parallel or inclined to the separation planes bounding beds (Fig. II–2). Both bedding planes and depositional textures are included under the term *depositional fabric*, which has been discussed by Sander and others.[2] Use may be made of the orientation of pebbles and boulders in glacial tills,[3] but the interpretation of till fabrics is complicated by glacial drag which may reorientate the pebbles.[4]

In addition, various textures, for example that due to the parallel orienta-

[1] J. V. Lewis, 'Fissility in Shales and its Relations to Petroleum', *Bull. Geol. Soc. Amer.* Vol. 35, 1924, pp. 557–90.

[2] B. Sander, *Einführung in die Gefügekunde der Geologischen Körper*, Pt II, Vienna, 1950, pp. 312 ff.; idem, 'Contributions to the Study of Depositional Fabrics', *Amer. Assoc. Petrol. Geol.*, Tulsa, 1951, 160 pp.; A. B. Vistellius, *Structural Diagrams*, Oxford, 1966, gives many examples of grain orientation studies.

[3] C. D. Holmes, 'Till Fabric', *Bull. Geol. Soc. Amer.*, Vol. 52, 1941, pp. 1299–354; R. G. West and J. J. Donner, 'The Glaciations of East Anglia and the East Midlands. A Differentiation based on Stone Orientation Measurements of the Tills', *Quart. Journ. Geol. Soc. Lond.*, Vol. 112, 1956, pp. 69–91; J. W. Glen, J. J. Donner, and R. G. West, 'On the Mechanism by which Stones in Till become Orientated', *Amer. Journ. Sci.*, Vol. 255, 1957, pp. 194–204; D. W. Harrison, 'A Clay-Till Fabric: Its Character and Origin', *Journ. Geol.*, Vol. 65, 1957, pp. 275–308. G. Lundquist, 'The orientation of block material in certain species of flow earth' in '*Glaciers and Climate*', *Geogr. Annaler*, Hft. 1–2, 1949, pp. 335–49.

[4] P. H. Banham, 'The Significance of Till Pebble Lineations . . .', *Proc. Geol. Assoc.*, Vol. 77, 1966, pp. 469–74.

tion of mica flakes, may be induced by post-depositional effects such as compaction, extrusion of connate water, and perhaps other causes. These are post-depositional fabrics, but in many instances will be very difficult to separate from true depositional fabrics.

FIG. II–2. TYPES OF BEDDING

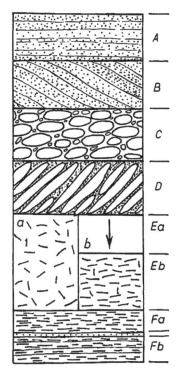

A. Sandstone with discrete bedding planes parallel to separation planes. (Parallel depositional structure.)

B. Sandstone with discrete bedding planes inclined to separation planes (cross-bedding). (Inclined depositional structure.)

C. Conglomerate with long axes of pebbles statistically parallel to separation planes. (Parallel depositional texture.)

D. Edgewise conglomerate with long axes of pebbles inclined to separation planes. (Inclined depositional texture.)

E. (a) Uncompacted mud with random orientation of mica flakes and clay particles. (Random depositional texture.)

E. (b) Compacted mudstone with flaky particles statistically parallel, and parallel with separation planes. (Parallel compactional texture.)

F. (a) Mudstone with mica flakes deposited parallel to separation planes, but lacking discrete bedding planes (parallel depositional texture, cf. C above).

F. (b) Mudstone with mica flakes deposited parallel to separation planes, and showing discrete bedding planes. A thin bed of sandstone lies between the two mudstones.

Pseudo-bedding – Structures resembling bedding are fairly common especially in metamorphic rocks. Strong cleavage or regular parallel jointing in sandstones and limestones often gives a false appearance of bedding on weathering, but the true bedding may usually be recognized by careful examination of the rock to reveal fine stratification planes or orientated fossils (Fig. II–3). In schists and gneisses, however, the original bedding may be almost obliterated by shearing and recrystallization along parallel planes (transverse schistosity, foliation, banding). In such rocks it may be difficult to identify and map original bedding. Structures closely simulating bedding are also formed in solifluxion masses and hill-creep material by differential flowage within the moving heterogeneous masses, giving rise to parallel orientation of pebbles, and layering of different grades of particles (Fig. V–22, p. 128). Colour banding and layers of concretions may form in sedimentary rocks in almost any direction and also require to be carefully distinguished from bedding.

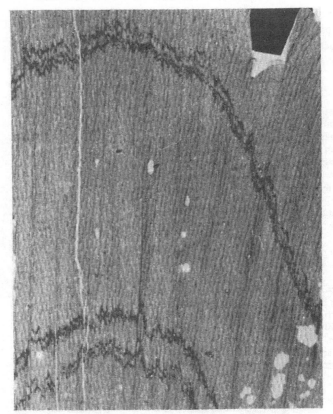

FIG. II–3. BEDDING AND CLEAVAGE IN SLATE
An anticline is shown by the carbonaceous laminae. ×7.

Facing – One of the most important practical uses of a knowledge of sedimentary structures is the determination of the stratigraphical top of strata especially in regions of complex structure where overfolding may have occurred. As many of the criteria are applicable to unfossiliferous or metamorphosed rocks, the various methods have found their most important application in Pre-Cambrian terrains. Beds are said to *face* in the direction of the stratigraphic top of the succession. Criteria for recognizing facing are mentioned in several different sections of this book, and have been treated at length by Shrock.[1]

Cross-bedding

Where the bedding planes within a bed are inclined more or less regularly to the separation planes between the beds, the arrangement is known as *cross-bedding* or *cross-stratification*[2] (Fig. II–2B).

[1] R. R. Shrock, *Sequence in Layered Rocks*, New York, 1948.
[2] The alternative terms *false bedding* and *inclined bedding* are not recommended. False

Very often the inclined bedding planes are smoothly curved, and are asymptotic to the separation plane at the base of the cross-bedded stratum (Fig. II−4). This is seen in sand and silt, but pebble beds deposited by fast-flowing streams commonly have straight cross-bedding (*torrential cross-bedding*).

A *delta* is formed by the dumping of sediment carried by a stream into a standing body of water where the delta is built out from the land, its inner

FIG. II−4. CROSS-BEDDING – TERTIARY SANDS, VICTORIA

Very regular foresets concave upwards, are truncated above, and tangential (below) to the lower separation plane of each cross-bedded stratum. Beds are about one foot thick.

structure exhibiting cross-bedding on a large scale. The finer grades of silt and clay are carried well beyond the steep front formed by the *foreset beds*, and deposited as *pro-delta clays* which are then covered by the advancing foresets to form *bottom-set beds* (Fig. II−5). The sub-aerial part of a delta is deposited by streams as a flood plain. With local exceptions at scour-channels, this consists of beds dipping very gently at an angle equal to the gradient of the

bedding might equally refer to pseudo-bedding; inclined bedding to any initial dip. *Current-bedding*, being a genetic term, should include all bedding structures due to current action, but is commonly used for the small-scale ripple-like bedding of rapidly deposited sand. The term *ripple-bedding* is here preferred for this.

For general accounts of cross-bedding, see W. O. Thompson, 'Original Structures of Beaches, Bars, and Dunes', *Bull. Geol. Soc. Amer.*, Vol. 48, 1937, pp. 723–52; W. H. Twenhofel, *Treatise on Sedimentation*, London, 2nd edn, 1932, pp. 618–23; E. D. McKee and G. W. Weir, 'Terminology for Stratification and Cross-Stratification in Sedimentary Rocks', *Bull. Geol. Soc. Amer.*, Vol. 64, 1953, pp. 381–9; H. Illies, 'Die Schrägeschichtung in fluviatilen und litoralen Sedimenten, ihre Ursachen, Messung und Auswertung', *Mitt. Geol. Staatsinst.*, Hamburg, Vol. 19, 1949, pp. 89–110; J. R. L. Allen, 'The Classification of Cross-Stratified Units, with Notes on their Origin,' *Sedimentology*, Vol. 2, 1963, pp. 93–114; K. A. W. Crook, 'Comments on J. R. L. Allen', 1963, ibid., Vol. 5, 1965, pp. 249–52.

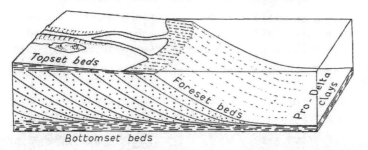

FIG. II-5. STRUCTURE OF A SIMPLE DELTA

stream in the deltaic area. The beds are known as the *topset beds*, and they sharply truncate the upper ends of the foresets. Ideally a few topset beds might be seen to turn rapidly down at a cusp to form foresets, but in general there will be actual truncation of the foresets due to scour in the wandering stream channels. A very similar arrangement of topsets and foresets is present in small sand and gravel banks deposited in river alluvia, and the clays that settle in over-deepened scour-holes in river-beds are analogous to the bottom-set beds of deltas.

Dune-bedding – Aeolian sand accumulations show a variety of cross-bedding which in regular dunes may be identified as *aeolian dune-bedding*. The sand-fall slopes are long and straight, terminating abruptly at the foot of the dune (Fig. II–6), while drift up the gentle windward slope either causes truncation of the tops of buried sand-fall slopes, or, if sand is being continually supplied, produces beds inclined in general at the angle of the slope, but with many

FIG. II-6. PLEISTOCENE CALCAREOUS AEOLIANITE, BARWON HEADS, VICTORIA

Two dunes rest on basalt (at foot of cliff). Upper dune shows bedding turning down to sand-fall slopes. Some sand-fall slopes are truncated below, others have a very short curved lower part.

local scours and small drifts of sand. Especially in calcareous dunes, certain of the bedding planes are more indurated than others, these probably representing the effects of showers of rain while the dune was building. Regular dune-bedding is well shown in Pleistocene calcareous aeolianites, but most dunes, especially in deserts, are much less regular. Wind-scoured hollows subsequently infilled with sand give rise to a heterogeneous series of cross-bedded wedges and lenses.

FIG. II–7A. COMPLETE AND TRUNCATED CURVES OF GIANT RIPPLE-BEDDING IN
PLEISTOCENE TUFFS, LAKE PURRUMBETE, VICTORIA
Ripple-crests are 3–4 ft apart. (*Photo: E.S.H.*)

Aeolian dune-bedding is of little use in determining the facing of strata because of its irregularity, and may even give false readings because of the truncation of bedding planes at the base, rather than at the top of the sand-fall slopes (Fig. II – 6).

Ripple-bedding (*current-bedding*) – The small-scale undulose bedding seen in rapidly deposited sand is built up from a series of sand ripples[1] (see Ripple-mark, p. 22). In one type the water itself takes on a wavy laminar flow and the lamination in the sand reflects this, the complete curves showing the laminae concave upwards in the troughs and concave downwards at the crests (Fig. II–7A). As a result of changes in current intensity and supply of sand, the top

[1] H. C. Sorby, 'On the application of quantitative methods to the study of the structure and history of rocks': *Quart. Journ. Geol. Soc. Lond.*, Vol. 64, 1908, pp. 171–232; A. Wood and A. J. Smith, 'The sedimentation and sedimentary history of the Aberystwyth Grits (Upper Llandoverian)': ibid., Vol. 114, 1959, pp. 163–95.

of ripple-bedding (also sometimes called current-bedding) may be truncated by local scour. Again, the down-current migration of current-ripples in sand produces a succession of miniature foresets truncated at the top and often grading downwards to finer grades of sediment, known as *ripple-drift* (Fig. II–7B).

Many other types of irregular bedding are now recognized, but most of these involve deformation of the newly deposited soft rock. They are therefore penecontemporaneous structures and are discussed below (see pp. 26–33).

It will be seen that in all examples of subaqueous cross-bedding and ripple-bedding, truncation of bedding planes which are concave upwards gives a completely reliable indication of facing.[1]

From statistical study of the dip of the cross-bedding in deltas, fluviatile deposits and ripple-marked beds, the direction of supply of sediment can be inferred, but, because of the changing position of channels, it is not to be expected that the direction of dip of the foresets in deltas will exhibit more than a general trend, from which many local deviations will exist. The variation of dip is greatest in digitate deltas such as that of the Mississippi.[2]

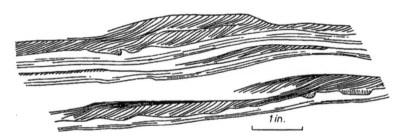

FIG. II–7B. RIPPLE-BEDDING (RIPPLE-DRIFT) IN THE ABERYSTWYTH
GRITS
The current flowed from right to left. (*After Wood and Smith, 1959*)

Initial dip – The extent to which strata deviate from the horizontal when they are deposited is their *initial dip*. Even very low initial dips, for instance in broad shallow basins, result in initial structures of importance, since a dip of only half a degree continued for 10 miles carries a bed up or down a distance of 460 ft, and when represented in sections with great exaggeration of vertical

[1] E. B. Bailey, 'New Light on Sedimentation and Tectonics', *Geol. Mag.*, Vol. 67, 1930, pp. 77–92; W. A. Cummings and R. M. Shackleton, 'The Ben Lui Recumbent Syncline (S.W. Highlands)', *Geol. Mag.*, Vol. 90, 1953, pp. 377–87.
[2] W. F. Tanner, 'Palaeogeographic Reconstruction from Cross-Bedding Studies', *Bull. Amer. Assoc. Petrol. Geol.*, Vol. 39, 1955, pp. 2471–83; P. Reiche, 'An Analysis of Cross-Lamination – the Coconino Sandstone', *Journ. Geol.*, Vol. 46, 1938, pp. 905–32; G. W. Brett, 'Cross-Bedding in the Baraboo Quartzite of Wisconsin', ibid., Vol. 63, 1955, pp. 143–8; J. Barrell, 'Criteria for the Recognition of Ancient Delta Deposits', *Bull. Geol. Soc. Amer.*, Vol. 23, 1912, pp. 377–446; H. N. Fisk *et al.*, 'Sedimentary framework of the modern Mississippi Delta', *Journ. Sed. Petrol.*, Vol. 24, 1954, pp. 76–99.

scale, gives an appearance of strong tilt. The maximum angle of initial dip depends on the angle of repose of the sediments under the conditions at the site, and is stated to be as high as 43° for sand in quiet water. The sand-fall slopes of dunes range from about 28° to 33°. Fine grades of sediment have lower angles of rest than coarse. High initial dips are seen in cross-bedded deposits, around the flanks of coral reefs, on buried escarpments and on the sides of buried hills. Compaction of sediments reduces the initial dip of current- or cross-bedding,[1] but increases the dip on the flanks of buried hills (see p. 59).

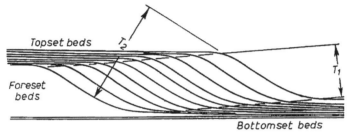

FIG. II–8. STRATIGRAPHIC EFFECTS IN LARGE DELTAS
The true thickness of deposits in the foreset beds is T_1, not T_2 as obtained by summing local thicknesses of strata.

A significant aspect of the structure of large deltas (Fig. II–8) is the effect of initial dip on the estimation of total stratigraphical thickness in the foreset beds. This may be greatly over-estimated if the structure of the delta is not recognized and due allowance made.

Graded bedding

One of the most valuable aids in determining facing is *graded bedding*, by which is meant gradation in the size of the component grains vertically within a bed. With few exceptions, the coarsest particles are at the bottom and the finest at the top, so that there is a sharp break in lithology in passing from one bed to the next. Graded bedding was for a long time best known in glacial varves, which are seasonally banded deposits formed in pro-glacial lakes. The coarser and thicker bottom of each varve represents the deposit of one spring and summer melt period, and the upper finer and thinner portion the rock-flour that settled out during winter. The grading is thus essentially due to changes in the grain-size of detritus supplied to the lake in which the varves were deposited.[2] Graded bedding is common in marine beds of the greywacke

[1] I. F. Wilson, 'Buried Topography, Initial Structures and Sedimentation in Santa Rosalia Area, Baja California, Mexico', *Bull. Amer. Assoc. Petrol. Geol.*, Vol. 32, 1948, Pt 2, pp. 1762–807.
[2] Recently Ph. H. Kuenen has suggested that the sediment-laden water reaching glacial lakes flows as a turbidity current and that the grading of varves, especially of the bottom layer, results chiefly from this, 'Mechanics of varve-formation and the action of turbidity currents', *Geol. fören. i Stockholm Forh.*, Vol. 73, 1951, pp. 69–84.

(argillaceous sandstone) lithofacies, and was formerly accounted for by the fractional settling of mixed grades of sediment carried along over still bottom waters by a superficial current, the settling being governed by Stokes' Law. E. B. Bailey, whose studies have stimulated much later thought, suggested that tidal waves (*tsunamis*) caused by submarine earthquakes might carry detritus in this way, and he showed a connection between the disturbances of sediments on submarine scarps, and the formation of graded beds from the spread-out detritus.[1] Further clarification has come through experiments and field observations by Kuenen and Migliorini, who demonstrated that a current heavily laden with detritus including clay, silt, sand, and pebbles, may flow down a very gentle slope as a *turbidity current*, and deposit its load as a graded bed.[2] Turbidity currents are now recognized as perhaps the most important single agent of marine deposition especially in bathyal waters. Because they can flow down gentle slopes and carry sand-size particles in suspension for long distances, the deposition of terrigenous sands hundreds of miles from land may thus be accounted for.

Turbidity currents are a type of density current, that is, a current which owes its properties to the possession of greater density than the surrounding fluid. Such currents preserve their individuality and maintain a sharp interface with the surrounding fluid in the absence of severe disturbances such as strong wave action in water. Salinity currents owe their higher density to salt in solution; it has been shown that in desert dust-storms the wind, laden with dust, behaves as a density current,[3] and most probably *nuées ardentes*, or glowing tuff clouds, also have many of the properties of these currents because of their high content of finely divided solid particles.

High-velocity density currents may be expected to flow over not-too-prominent elevations in their path. In aqueous turbidity currents a high density for the flow derives from admixture of detritus with the water, which admixture is continually maintained by vertical turbulence within the flow. High-density currents may have a specific gravity of about 2, and can transport particles as large as pebbles or cobbles. Experiments show that the coarser particles settle out towards the base leaving the finer at the top, but with sand–silt–clay mixtures the separation is not clean, and beds of greywacke or mudstone commonly result.

Sequences of this type deposited from turbidity currents are known as *turbidites*.[4] Turbidity currents originate in several ways. They have for long

[1] E. B. Bailey, 'New Light on Sedimentation and Tectonics', *Geol. Mag.*, Vol. 67, 1930, pp. 77–92; idem, 'Sedimentation in Relation to Tectonics', *Bull. Geol. Soc. Amer.*, Vol. 47, 1936, pp. 1713–26.

[2] Ph. H. Kuenen and C. I. Migliorini, 'Turbidity Currents as a Cause of Graded Bedding', *Journ. Geol.*, Vol. 58, 1950, pp. 91–127.

[3] R. T. Knapp, 'Étude de l'interaction entre un fluide et des particles en suspension dans les écoulements sous l'action de la pesanteur et les courants de densité', *Colloques internat., C.N.R.S., No. 35, Actions éoliennes, etc.*, Paris, 1953, pp. 153–77.

[4] A. H. Bouma, *Turbidites*: Developments in Sedimentology – 3, Amsterdam, 1964.

been recognized in engineering practice in relation to silting in reservoirs, where provision may be made to lead the detritus-laden bottom-water of floods through low-level outlet tunnels at the dam, rather than blocking the currents and allowing the silt to settle in the reservoir. This suggests that in geology also we must conceive that the load of streams entering the sea may be carried partly as turbidity currents as well as by the momentum of the river flow itself, although the lower density of fresh water must be borne in mind. Engineers, however, have recognized that density currents may flow

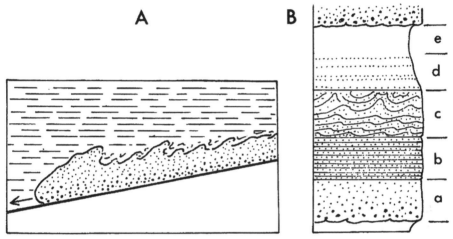

FIG. II–9. TURBIDITY CURRENT AND GRADED TURBIDITE

A. Turbidity current, bottom-flowing, with suspended load of coarse and fine particles. Behind the head, the interface with the surrounding water is unstable and some mixing takes place. (*After Knapp, 1953, and Allen, 1969*)

B. Ideal complete turbidite sandstone–mudstone sequence. a, graded unlaminated sandstone; b, sandstone with parallel lamination; c, sandstone with current ripples and convolute bedding; d, mudstone with sand laminae; e, pelagic interval with mudstone, shale, or clay. (*After Bouma, 1962, and Allen, 1969*)

not only along the bottom but also between other layers of water, or across the top of a reservoir.[1] Muddy river-water may often be seen to make a very sharp junction with clear sea-water, and it seems certain that further geological studies of sedimentation will need to take account of inter-flowing and over-flowing, as well as botton-flowing turbidity currents (Fig. II–9).

Many turbidity currents are thought to originate from submarine slumps of already deposited sediments, not fully consolidated, off the continental shelf or from submarine scarps including fault-scarps. The disturbance of thixotropic clays by submarine earthquakes, permitting them to flow, is a contributing cause. Thus slump-masses of wet clay may become modified into turbidity currents, and in Kuenen's words: 'Dependent on the relative amount

[1] See, for example, the studies of Lake Mead: C. P. Vetter, 'Sediment Problems in Lake Mead and Downstream on the Colorado River', *Trans. Amer. Geophys. Union*, Vol. 34, 1953, pp. 249–56.

FIG. II–10. CROSS-BEDDED SANDSTONE WITH INCLUDED SHALE BLOCKS,
WARRAGAMBA DAM, NEW SOUTH WALES
(*By courtesy W. R. Browne*)

of water there appear to be all transitions from mud flows to flows of water
with a high percentage of suspended sediment.[1] The slumps may carry away
with them fragmented beds that have developed some cohesion, such as stiff
clay or shale. This may account, perhaps better than scour by ordinary
currents, for many auto-breccias and auto-conglomerates, containing angular
lumps or rounded pebbles of rocks similar to those in the sediments themselves, as, for example, of angular inclusions of soft shale in sandstone (Fig.
II–10). On the other hand there is ample evidence that some such beds,
especially those containing angular fragments and slabs of soft rock such as
shale, developed almost in place either by the pulling apart of tenacious clays
in deposits that have slid a little after deposition, or by the injection of sand-

[1] Ph. H. Kuenen, 'The difference between sliding and turbidity flow', *Deep Sea Res.*, Vol. 3,
1956, pp. 134–9; R. H. Dott, Jr, 'Dynamics of Subaqueous Depositional Processes', *Amer.
Assoc. Petrol. Geologists*, Vol. 47, 1963, pp. 104–28.

water slurries along bedding planes, with consequent disruption of the shale or clay. The extent of rounding of the 'pebbles', the matching of the edges of adjoining fragments, and similar detailed evidence may assist in recognizing the origin of such breccias and conglomerates.

The occurrence of myriads of graptolites which were killed as colonies and entombed in graded greywackes thought to have been deposited by turbidity currents in Ordovician times in Victoria, indicates that the finer grades of sediment associated with such currents may affect near-surface pelagic waters.[1]

The overall gradation from coarse sand to mud in a graded bed is often complicated by the presence within the one major graded unit of a number of intervals or divisions which ideally follow the sequence of five types shown in Figure II–9B.[2] A bed showing less than the standard five is described as *truncated*, this being largely a function of the distance from the source of the current.[3] The coarser laminae, higher in the bed than the level normal for their grain size, may be ascribed to the presence of secondary waves, pulsations, or upward-moving eddies in the turbidity current. Pulsations might be caused by successive small slumps, derived by fretting of the main slump-scar from which the mass originated, but, since turbulence is an essential feature of turbidity currents, lamination of the graded bed may also be ascribed to eddies carrying up sand-size particles. Whatever their cause, the current-bedded sand layers may afford valuable confirmation of the use of graded bedding in defining facing.

In normal grading the base of the bed is coarser grained, but rare instances are known of the reverse arrangement, especially in fluviatile beds or other environments where the supply of detritus governs the gradation. Inverse grading is also seen in some chemical deposits, especially gypsum and anhydrite.[4] A succession of graded beds in a series of regularly stratified rocks, however, affords a most reliable indication of facing,[5]

[1] E. S. Hills and D. E. Thomas, 'Turbidity Currents and the Graptolitic Facies in Victoria', *Journ. Geol. Soc. Aust.*, Vol. 1, 1954, pp. 119–33.

[2] A. H. Bouma, *Sedimentology of Some Flysch Deposits*, Amsterdam, 1962.

[3] R. G. Walker, 'Turbidite Sedimentary Structures and their Relationship to Proximal and Distal Depositional Environments', *Journ. Sed. Petrol.*, Vol. 37, 1967, pp. 25–43.

[4] L. Oguiben, 'Secondary Gypsum of the Sulphur Series, Sicily, and the so-called Integration', *Journ. Sed. Petrol.*, Vol. 27, 1957, pp. 64–79.

[5] For further information on graded beds and turbidity currents consult Ph. H. Kuenen, 'Properties of Turbidity Currents of High Density', *Soc. Econ. Geol. & Min.*, Spec. Pub. No. 2, 1951, pp. 14–33; Ph. H. Kuenen and H. W. Menard, 'Turbidity Currents, Graded and Non-graded Deposits', *Journ. Sed. Petrol.*, Vol. 22, 1952, pp. 83–96; Ph. H. Kuenen, 'Significant Features of Graded Bedding', *Bull. Amer. Assoc. Petrol. Geol.*, Vol. 37, 1953, pp. 1044–66; F. P. H. Kopstein, *Graded Bedding of the Harlech Dome*, 'sGravenhage, 1954; M. L. Natland and Ph. H. Kuenen, 'Sedimentary History of the Ventura Basin, California, and the Action of Turbidity Currents', *Soc. Econ. Geol. & Min.*, Spec. Pub. No. 2, 1951, pp. 76–107; J. R. L. Allen, 'Some Recent Advances in the Physics of Sedimentation,' *Proc. Geol. Assoc.*, Vol. 80, 1969, pp. 1–42. For an excellent review, critical of the over-enthusiastic application of the turbidite concept, see G. J. van der Lingen, 'The Turbidite Problem', *New Zealand Journ. Geol. Geophys.*, Vol. 12, 1969, pp. 7–50.

which may be observed even when the rocks have undergone considerable metamorphism.[1]

TOP AND BOTTOM FEATURES OF STRATA

Structures of the top of beds

Mud cracks (*sun cracks*) – Mud cracks form on the surface of clayey sediments when these are exposed to a drying atmosphere for some time. If the mud-layer is thin, it may both crack and flake off producing clay pellets, but

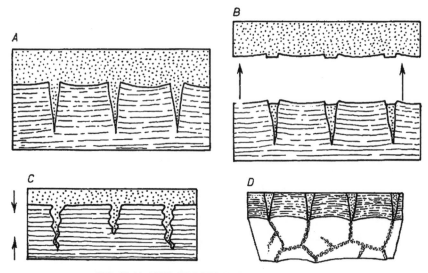

FIG. II–11. MUD CRACKS AND THEIR FILLINGS

A. Mud cracks filled and covered with sand.
B. Sandstone separated from mudstone, showing broken casts of cracks on base of sandstone (compare Fig. II–12A).
C. Compaction of mud causes crumpling of sand-fillings.
D. Tips of sand-fillings show as polygons on *lower* surface of mud (compare Fig. II–12B).

in deep mud the cracks penetrate to considerable depths, tapering downwards. They are preserved by becoming filled with sand or with mud of different texture, but only if the new sediment is spread out sufficiently quickly, before the underlying cracked mud swells again on wetting. In weathered exposures of mud-cracked strata, the clayey bed often disintegrates, leaving casts of the cracks on the lower surface of an overlying sandstone (Fig. II–11B, 12A) but the lower tips of the wedge-shaped sand-fillings may also appear, surrounded by shale (Fig. II–11D, 12B). Because of compaction of the surrounding mud,

---

[1] See H. H. Read, 'Dalradian Rocks of the Banffshire Coast', *Geol. Mag.*, Vol. 73, 1936, pp. 468–76, for an account of the metamorphism of graded beds.

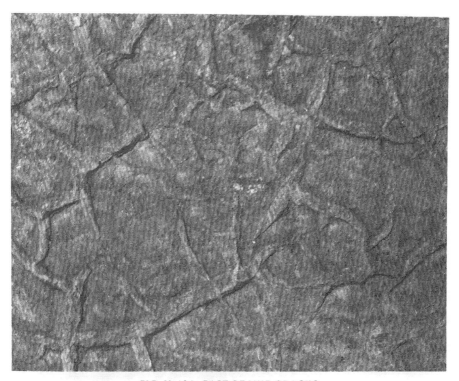

FIG. II–12A. CAST OF MUD CRACKS

Shown on lower surface of sandstone slab, Grampian Mountains, Victoria. $\times \frac{1}{6}$. Bed faces away from the observer (cf. Fig. II–11B).

FIG. II–12B. FILLED MUD CRACKS FROM BELOW

Lower surface of mudstone shows tips of sand-fillings in mud cracks. Bed 'faces' away from the observer (cf. Fig. II–11D).

the sand-fillings of mud cracks are squeezed and folded in a vertical plane (Fig. II–11C).[1]

Rain prints – Scattered raindrops falling on soft mud make shallow pits, which are obliterated if the rain continues and floods the surface, but if the isolated prints can be identified they indicate facing as shown in Fig. II–13. A slightly raised rim may assist in their recognition.[2]

Pit-and-mound – This is formed by air or other gases escaping through unconsolidated sediments. It may be seen in sand as the swash of waves

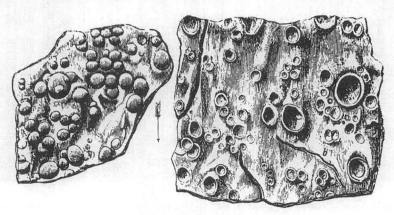

FIG. II–13. RAIN PRINTS

Rain prints (*right*) and cast of rain prints (*left*), in green shale, Cape Breton, Nova Scotia.
The arrow shows direction of impact of drops and hence wind direction. (From Pepper, *Playbook of Metals*, London, 1861, after Lyell, *Elements of Geology*, London, 4th edn, 1855.)

moves up a beach, forcing out entrapped air, and is also found in mud from which bubbles of CO_2 or methane escape. Pit-and-mound ideally resembles a miniature mud-volcano, with a tubular pipe and encircling rim, but the pipe tends to collapse. Some examples are only a millimetre or two across.

Ripple-mark – Ripple-mark[3] is formed on sediments, particularly sand, in which the grains are separate and free to move in water or air. The movements generating ripples may be currents, or oscillations due to wave action at a depth shallower than wave-base; thus there are two general types, *current*

---

[1] I. Hessland, 'Studies in the lithogenesis of the Cambrian and basal Ordovician of the Böda Hamm sequence of strata', *Bull. Geol. Inst. Univ. Uppsala*, Vol. 55, 1955, pp. 35–109; R. Teichmuller, 'Sedimentation und Setzung im Ruhrkarbon', *Neues Jahrb. f. Geol. u. Pal.*, *Monatshefte*, 1955, pp. 146–68.

[2] K. Kaiser, 'Ausbildung und Erhaltung von Regentropfen-Eindrücken', *Geol.. Inst. Univ. Köln*, Bd. 13, 1967, pp. 143–56.

[3] For comprehensive accounts of ripple-mark, see E. M. Kindle, 'Recent and Fossil Ripple-mark', *Geol. Surv. Canada*, Mus. Bull. No. 25, 1917, 56 pp.; W. H. Bucher, 'On Ripples and Related Sedimentary Surface Forms and their Palaeogeographic Interpretation', *Amer. Journ. Sci.*, 4th Ser., Vol. 47, 1919, pp. 149–210, 241–69; J. R. L. Allen, op. cit., 1969.

ripple-mark and *wave* (or *oscillation*) *ripple-mark* (Fig. II–14). Simple current ripples formed under water are asymmetrical ridges with intervening rounded troughs, which move slowly downstream and resemble miniature sand-dunes. They form only with a certain range of current-velocity, and with faster currents are replaced first with giant ripples or dunes, then with a planar surface and finally by low-amplitude anti-dunes in which the bedding is obscure. The steeper lee-slopes of current ripples, which face down-current, are not simply sand-fall slopes as in dunes, but are affected by eddies located in the troughs, which tend to lift the grains and thus to effect a separation of the lighter from the heavier particles, which latter are concentrated in the

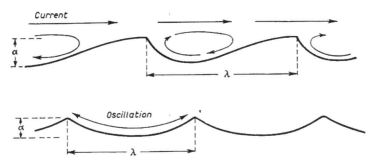

FIG. II–14. CROSS-SECTIONS OF RIPPLE-MARK
A. Current ripples; asymmetrical with sharp crests. Current flowed from left to right. Eddies form in troughs.
B. Wave (oscillation) ripples; symmetrical with sharp crests (compare Fig. II–15). α = amplitude. λ = wave length.

troughs. The ripples, if well formed, have sharp crests and rounded troughs, which enables them to be used for determining facing,[1] and again their internal structure, with small-scale ripple-bedding, may also be used for this purpose.

Although common in shallow water, current ripples may form at any depth and have been photographed at 790 ft in the sea. They are also common in sand laminae in beds deposited by turbidity currents and exhibiting graded bedding. Aeolian sand ripples are in general longer and of smaller amplitude than sub-aqueous ripples.

Wave ripples are formed by oscillatory movements of the water on a sandy bottom as waves pass. The water is theoretically affected only to a depth

[1] Despite repeated statements to the contrary in some standard texts, it is true that current ripples may be used to determine facing if the sharp crests are preserved. They have, for instance, been so used in mapping the Proterozoic quartzites of the Yampi Sound iron-ore field in Western Australia. The shape of well-formed current ripples is admirably reproduced by H. W. Menard, 'Current-Ripple Profiles and their Development', *Journ. Geol.*, Vol. 58, 1950, pp. 152–5; see also H. C. Sorby, 'On the Application of Quantitative Methods to the Study of the Structure and History of Rocks', *Quart. Journ. Geol. Soc. Lond.*, Vol. 64, 1908, pp. 171–233; J. R. L. Allen, *Current Ripples*, Amsterdam, 1968.

known as wave-base, and water deeper than about 100 fathoms should not be affected by even the largest waves of surface origin.[1] In cross-section wave ripples are symmetrical, with sharp crests and rounded troughs which give a ready reading of the facing of strata (Fig. II–15). However, there are many modified forms of both wave and current ripples, such as linguoid current ripples, interference ripples, and partially destroyed ripples with flat or rounded crests, from which facing is rarely determinable, except that such markings are often confined to the top surface of thin beds with flat bases, which gives some guidance.

FIG. II–15. WAVE (OR OSCILLATION) RIPPLE-MARK

Cathedral Sandstone, Chapel Hill, Taggerty, Victoria. This photograph is of the upper surface of a bed; but owing to the common optical illusion of inversion of relief, it may appear as a cast.

The form and inner structure of ripple-mark and also of current-bedding may be used to estimate subsequent deformation and plastic flow of beds (see also p. 124), but certain secondary structures, with shearing and lensing of thin beds, produce pseudo-ripple-mark (Fig. II–16) and the distinction between primary and secondary structures may in fact be difficult.[2]

In beds composed of a vertical succession of climbing ripples in which the

[1] H. W. Menard, however, reports possible oscillation ripples at 4500 ft: *Journ. Sed. Petrol.*, Vol. 22, 1952, pp. 3–9.

[2] E. Ingerson, 'Fabric Criteria for Distinguishing Pseudo-Ripple Marks from Ripple Marks', *Bull. Geol. Soc. Amer.*, Vol. 51, 1940, pp. 557–70.

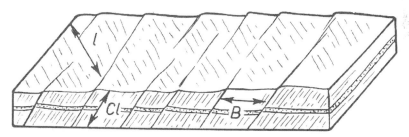

FIG. II-16. PSEUDO-RIPPLE-MARK

Sandy slate shows cleavage (Cl.) intersecting bedding (B). Micro-faulting on a limited number of curving cleavage planes produces the pseudo-ripples. A strong lineation (l) is also visible. Ordovician, Bendigo, Victoria. $\times \frac{1}{4}$.

position of crests and troughs gradually shifts from top to bottom of the bed, lines of troughs, ripple-fronts, and crests may give apparent cross-bedding (*ripple-drift cross-lamination*) in the reverse direction to the real cross-bedding (Fig. II-17). The effect is heightened if grains of different size or colour occupy the troughs, and if platy particles lie parallel on the ripple-fronts.[1]

[1] See E. D. McKee, 'Some Types of Bedding in the Colorado River Delta', *Journ. Geol.*, Vol. 47, 1939, pp. 64–81; A. V. Jopling and R. G. Walker, 'Morphology and Origin of Ripple-Drift Cross-Lamination . . .', *Journ. Sed. Petrol.*, Vol. 38, 1968, pp. 971–84; R. G. Walker, 'Distinctive Types of Ripple-Drift Cross-Lamination', *Sedimentology*, Vol. 2, 1963, pp. 173–88. A similar feature in oscillation ripple-mark is mentioned by G. K. Gilbert, 'Ripple marks and Cross Bedding', *Bull. Geol. Soc. Amer.*, Vol. 10, 1899, pp. 135–40.

FIG. II–17. APPARENT CROSS-BEDDING DUE TO CLIMBING RIPPLES IN SANDSTONE
At the base, sinusoidal ripple-lamination, passing upwards to ripple-drift cross-lamination.
Current flowed from left to right. ×⅓. (*After F. J. Pettijohn and P. E. Potter, op. cit., 1964, pl. 39*)

SOLE-MARKINGS[1]

A variety of small-scale structures for which several explanations have been proposed, occurs in the lower few inches of sandstones resting on argillaceous beds or on coal. In considering the origin of these and of certain related structures, it is significant that they occur at sandstone contacts with underlying originally plastic material while the top of the sandstone remains flat and undisturbed. This suggests that appropriate physical properties of the rocks are fundamental for the development of the structures. In general they exhibit downward projecting lobes of sandstone interspaced with upward projections of argillite, or coal where this is involved. Several factors are concerned in the origin of sole-markings, and as these may operate together, there are many complex variations in the details of the structures formed.

Load-cast – In its simplest form, the structure known as *load-cast* demonstrates the results of the deposition of beds, generally of sand or greywacke, above plastic material such as wet clay or peat, which latter penetrates upwards into the superincumbent beds (Fig. II–18). This is demonstrated by the breaking-through of the stratification of the sandstone, the upturning of its bedding planes adjacent to injections, and by the filamentous prolongation of injections for several inches into the sandstone. These inferences are confirmed by experiments designed to reproduce the features of salt domes, which are also believed to originate by the upward injection of light mobile

[1] The term *sole-markings*, being non-committal as to genesis but descriptive of the important stratigraphical aspect of the structures to be discussed, may be used for a variety of features of different origin.

beds under load (see p. 279). The shape of certain load-casts so much resembles, in miniature, that of salt domes that a similar origin is undoubted.[1] The injections rise vertically, their location being in places determined by the crests of ripples on an underlying sandstone (Fig. II–18A).

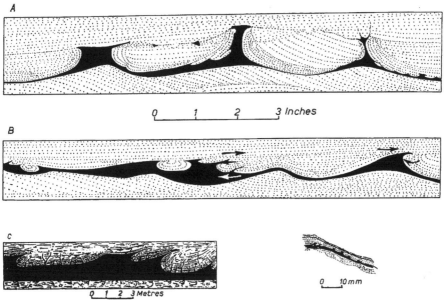

FIG. II–18. LOAD-CAST IN SANDSTONE

A. Injections of mudstone (black) into sandstone above crests of underlying ripples. No lateral sliding. Silurian, Studley Park, Victoria.

B. Lobes and asymmetrical mudstone injections into sandstone. Lateral sliding in sense of arrows. Silurian, Studley Park, Victoria, adjacent to A.

C. Asymmetrical lobes in sandstone overlying coal, Humboldt Mine, Westphalia. (*After Kukuk*)

D. Minute injections of type shown in A. Ordovician, Castlemaine, Victoria.

The presence of simple, cuspate injections alongside the miniature 'salt domes' demonstrates that these, too, may originate directly by loading. Many cuspate injections are, however, asymmetrical, pointing to slight lateral movement of the sand during load-casting. Where the amount of displacement is small, it may be attributed simply to local readjustments in response to

---

[1] L. L. Nettleton, 'Fluid Mechanics of Salt Domes', *Bull. Amer. Assoc. Petrol. Geol.*, Vol. 18, 1934, pp. 1175–204; M. B. Dobrin, 'Re-creating Geological History with Models', *Journ. App. Phys.*, Vol. 10, 1939, pp. 360–71; T. J. Parker and A. N. McDowell, 'Model Studies of Salt Dome Tectonics', *Bull. Amer. Assoc. Petrol. Geol.*, Vol. 39, 1955, pp. 2384–470. The origin of load-cast was recognized several years ago: see P. Kukuk, 'Bemerkungswerte Einzelerscheinungen der Gasflammkohlenschichten in der Lippemulde', *Glückauf*, Vol. 56, 1920, pp. 805–10, 829–35. See also O. Stutzer and A. C. Noé, *Geology of Coal*, Chicago, 1940, Fig. 122. Scott Simpson, 'Das Devon der Südost-Eifel zwischen Nette und Alf', *Abh. Senck. Naturf. Gesell.*, Abh. 447, 1940, pp. 1–67; E. S. Hills, 'The Silurian Rocks of the Studley Park District', *Proc. Roy. Soc. Vict.*, Vol. 53, 1941, pp. 167–91.

uneven flow of the argillite, as shown in Fig. II–18B, but stronger asymmetry may indicate some sliding of the sand over the clay, and such movement passes in some instances into actual differential flowage, as discussed below.

Injections are found in some places where it is clear that the thickness of the overlying beds deposited at the time of injection was no more than a few inches. This is seen, for example, in sections of abandoned slime-dumps of mines (Fig. II–19), and is also found in the experiments on salt-dome formation. In slime-dumps the mud may squeeze through the overlying sand and

FIG. II–19. SEDIMENTARY STRUCTURES IN TAILINGS DUMP OF MINE

A, B. Upwellings of soft fine-grained silt through sand, the latter with enrolled edges. Height of injections about 3 inches. Note flat tops of beds and lateral spread of fine silt at *A*. Eaglehawk, Bendigo district, Victoria. (*Photo: D. E. Thomas*)

appear at its upper surface before being covered by later deposits, so that the effect may also be thought of as involving rather the sinking of the sand through the lighter mud than the squeezing-up of mud above its own original level, which is the basic principle of load-casting.

The settlement of sand into clay is best demonstrated where isolated lenticular sand-masses were deposited, which occurs on a small scale in incomplete sand-ripples. As the lenses settle into the clay their thin edges are turned up and the base becomes rounded, the shape eventually having close analogies with the glass meteorites known as Australites, which are moulded by aerodynamic forces affecting the liquid glass in its passage through the atmosphere.

The resulting inrolled sand-balls have been termed *pseudo-nodules*.[1] (Fig. II–20.)

Transposition disturbances – Load-casts at sharply defined separation planes between sandstones and argillaceous beds, and complex flow-folding or break-

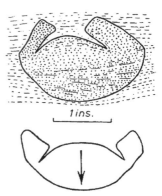

FIG. II–20. PSEUDO-NODULE OF SANDSTONE IN MUDSTONE

Ordovician, Castlemaine, Victoria.
Below, section of glass meteorite
(Australite) showing flanges
formed by motion in direction of arrow.

ing up of laminae within a bed point to disturbance of the bedding by relative movements among the strata, which must have been in a very wet and mobile condition, but nevertheless originally well-bedded before the displacements occurred (Figs. II–21, 22). Although such features are best known from successions of graded beds believed to have been deposited by turbidity

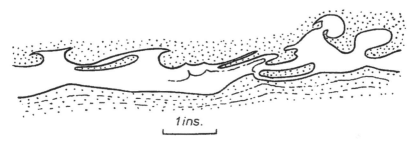

FIG. II–21. TRANSPOSITION STRUCTURES IN ORDOVICIAN SANDSTONES (STIPPLED) AND SHALE (WHITE)
Chewton, Victoria.

currents, their absence from many such beds suggests that they are not inherently caused by residual flow due to the current itself. There are, however, many possible causes that might lead to subsequent internal sliding adjustments within a set of beds, especially at the contacts of argillaceous deposits. These include the expression of water from underlying clays due to

[1] P. Macar and P. Antun, 'Pseudo-nodules et glissements sous-aquatique dans l'Emsien inférieur de l'Œsling', *Bull. Soc. Géol. Belg.*, Vol. 73, 1950, pp. 121–50. For the form of Australites, see G. Baker, 'Nirranda Strewnfield Australites, South-east of Warrnambool, Western Victoria', *Mem. Nat. Mus. Melb.*, No. 20, 1956, pp. 1–114.

compaction under load, the effects of differential loading which would squeeze beds towards areas of thinner deposition, and the effects of differential compaction in inducing dips that were lacking when deposition occurred, and which might cause some sliding of the beds. The clear evidence that these structures form after the beds were deposited in a well-stratified condition strongly suggests that thixotropy of the argillaceous beds may be involved, but sandstones in which the pores are filled with water are also liable to

FIG. II-22. TRANSPOSITION STRUCTURES, ORDOVICIAN, VICTORIA

Sandstone with sole markings rests on slate (shale). In the sandstone, below the tape-measure, slabs of tenaceous but plastic shale disrupted first by 'necking' (note narrowed ends), then by flowage of sand from left to right. Above the tape measure, paler sand is disrupted by soft-rock flowage. The whole demonstrates renewed but limited hydro-plastic flow after deposition of the full sequence of beds. Hargreaves St., Castlemaine. (*Photo: D. E. Thomas*)

flow. Sorby's prescient conclusion that the beds involved in sole-markings illustrated by him in 1908 had the 'consistency of thick cream', may be recalled.[1]

Many of the readjustments leading to sole-markings may commence as soon as the bottom sediment of a graded bed has settled out, and may thus be virtually contemporaneous with deposition, but as the involutions and intermingling here discussed are not regarded as due directly to current action but rather to hydro-plastic or fluid flow after deposition, they are termed

[1] H. C. Sorby, 'On the Application of Quantitative Methods to the Study of the Structure and History of Rocks', *Quart. Journ. Geol. Soc. Lond.*, Vol. 64, 1908, pp. 171–233, Plate 14.

transpositional structures.[1] Examples of transposition such as that of the argillaceous bed in Fig. II–22 that has been first pulled into segments, and later moved so that the segments overlap, show that movements of this type may go on after some consolidation has occurred, while others, more complex, must belong to an earlier stage of compaction.

Convolute-bedding – In this structure the lamination planes in sandstones are thrown into deep convolutions which consist of rounded synclines with intervening cuspate or flat-topped anticlines. The overlying bed truncates the convolutions, and this, combined with the increase in amplitude of the anticlines upwards, affords a guide to facing (Fig. II–23A, B). Convolute bedding is formed during deposition, by forces due to the current and by loading.[2]

Groove-cast – Straight, parallel grooves in the top of an argillaceous bed overlain by sandstone are preserved as casts in the sandstone base. Observed grooves range from isolated large examples to parallel series and even to two sets crossing each other. Possible causes may include the dragging of floating vegetation or ice over the sea- or lake-bed, which might be postulated where plant remains or glacigene deposits are associated, but the drag of boulders at the base of a turbidity current accounts for many examples.

Flute-cast – Sole-markings having a linguoid appearance and somewhat resembling scales have been termed 'flow-cast' but are now known as 'flute-cast', and are ascribed to the filling of fluted scour-marks on the top of the argillaceous bed, formed by a turbidity current that deposited the sand. Each tongue in the array has one deeper end which is pointed or slightly rounded, becoming broader and shallower distally (Fig. II–24). Many examples have a tripartite form with a narrow lateral fluting on either side of the larger median structure. Plotting the long axes of flute-casts generally shows that this corresponds statistically with the direction of turbidity current flow in the sedimentary trough, the deeper ends pointing towards the source of the current. The effects of load casting on flute-casts is to cause the deeper end (the bulbous end of the cast) to be partially surrounded or overhung by the argillite. The casts are commonly arranged in a definite pattern, either in lines oblique to the long axes of the tongues, or in two such lines, giving a rhomboidal appearance. A variety of sole-markings is now identified including tool-marks made by objects caught up in the turbidity currents.[3] It is difficult to see what

[1] It is appropriate at this point to mention that movements of sediments after deposition range from slight internal hydro-plastic flowage to the displacement of large masses by sliding or slumping (see pp. 61–8), and on a grander scale, gravitational tectonics (see pp. 345–50). Original sediments may be redistributed especially by sliding or slumping off the continental slope or the outer edge of the continental shelf.

[2] E. ten Haaf, 'Significance of Convolute Lamination', *Geol. en Mijnb.*, N.S., Jhg. 18, Nr. 6, 1956, pp. 188–94; S. Dzulynski, 'New Data on Experimental Production of Sedimentary Structures', *Journ. Sed. Petrol.*, Vol. 35, 1965, pp. 196–212; R. H. Dott, Jr, and J. K. Howard, 'Convolute Lamination in Non-Graded Sequences', *Journ. Geol.*, Vol. 70, 1962, pp. 114–21.

[3] R. R. Shrock, *Sequence in Layered Rocks*, 1948, New York. Ph. H. Kuenen, 'Sole

FIG. II–23. CONVOLUTE BEDDING

A. Silurian sandstone, Studley Park, Victoria. Note the gradual development upwards from planar bedding, of the anticlines, the overturning of these, and the break-up of bedding in overlying sand, in the troughs. The top is an undeformed sandstone deposited later. ×⅓.

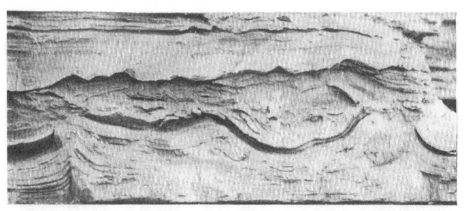

B. Sand deposited in very shallow water in the tailings dump of a mine, and showing convolute bedding similar to A. Eaglehawk, Bendigo district, Victoria. ×⅓. (*Photo: D. E. Thomas*)

C. Ordovician sandstone, Castlemaine, Victoria. Convolutions are further developed with troughs deeply depressed, becoming isoclinal. ×½.

FIG. II–24. FLUTE CAST ON BOTTOM SURFACE OF SANDSTONE
Casts revealed by weathering away of underlying shale (slate). Ordovician, Chewton, Victoria.
Current flowed from top left corner, in the direction of fluting. (*Photo: E.S.H.*)

attitude the bedding in the underlying argillaceous bed takes in linguoid casts since these are not determinable as such until the sandstone base is exposed by weathering, and their precise origin has not yet been satisfactorily explained.

In considering the various notions that have been propounded to account for sole-markings of all kinds, the extremely small scale of some examples (Fig. II–18D), the occurrence of isolated examples, and the involvement of beds such as coal (at the stage of peat) must be borne in mind. Superficial deposits in slime-dumps and in Pleistocene beds serve to indicate the variety of conditions under which the structures may form, which also includes periglacial soil movements.[1]

Facing from fossils

Tracks, trails, castings, and burrows, as well as fossil organisms themselves, may be used to determine facing. For tracks, trails, and castings, detailed

Markings of Graded Graywacke Beds', *Journ. Geol.*, Vol. 65, 1957, pp. 231–58 (with Bibliography); S. Dzulynski and E. K. Walton, *Sedimentary Features of Flysch and Greywackes*, Amsterdam, 1965.

[1] K. O. Emery, 'Contorted Pleistocene Strata at Newport Beach, California', *Journ. Sed. Petrol.*, Vol. 20, 1950, pp. 111–15. See also A. Jahn, 'Some Periglacial Problems in Poland', *Lodz Tow. Nauk., Soc. Scient. Lodz.*, Vol. 3, Sect., 3. Biul. Peryglacjalny Nr. 4, 1956, pp. 169–83.

knowledge of the possible form and structure of a wide range of such traces of life is an absolute necessity, since, for example, worm-castings may be confused with the sedimentary fillings of depressed tracks and tracks or imprints may include both depressions and elevations on a bedding plane.[1]

The burrows of boring Annelid worms, preserved in the top of a bed that has been exposed for some time as the sea-bed, have various characteristic shapes by which they may be recognized. Many such borings in sandstones have been considered to be 'pipe stem' concretions. The open ends of the tubes face upwards, and a mound of debris is built up at one opening[2] (Fig. II–25A).

Actual fossil organisms indicate facing either by their preservation in the position of growth of the animal or by the parallel arrangement of shells after death, by waves and currents. For the interpretation of fossils in their growth

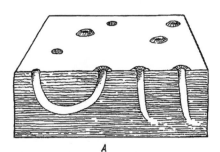 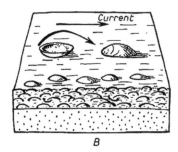

FIG. II–25. FACING FROM FOSSILS

A. Worm burrows and mounds.
B. Concavo-convex shells turned convex side up by waves of translation or by currents.

position considerable palaeontological knowledge is required, details of which may be found in standard works.[3] The late Pre-Cambrian calcareous alga *Collenia* is, however, particularly important because of its widespread occurrence in ancient, otherwise unfossiliferous rocks. Its normal pattern of growth is in colonies with rounded and markedly convex tops, which give a good indication of facing (Fig. II–26).[4]

[1] O. Abel, *Vorzeitliche Lebensspuren*, Jena, 1935; J. Lessertisseur, 'Traces fossiles d'activité animale et leur signification paléobiologique', *Soc. Géol. France Mem.*, No. 74, 1955, 150 pp.; A. Seilacher, 'Studien zur Palichnologie', Nos. 1 and 2, *Neues Jahrb. Geol. u. Pal., Abh.*, Bd. 96, 1953, pp. 421–52; Bd. 98, pp. 87–124. Idem, 'Die geologische Bedeutung fossiler Lebensspuren', *Zeitschr. Deutsch. Geol. Gesellsch.*, Bd. 105, 1953, pp. 214–27.

[2] H. Faul, 'Fossil Burrows, from the Pre-Cambrian Ajibik quartzite of Michigan', *Nature*, Vol. 164, 1949, p. 32; R. Richter, 'Die Fossile Fahrten und Bauten der Würmer', *Pal. Zeitsch.*, Vol. 9, 1927, pp. 193–235.

[3] See especially R. R. Shrock and W. H. Twenhofel, *Principles of Invertebrate Palaeontology*, New York, 1953.

[4] For good illustrations, see R. B. Young, 'Further Notes on Algal Structures in the Dolomite Series', *Trans. Geol. Soc. S. Africa*, Vol. 43, 1940, pp. 17–22.

Shells redistributed after the death of the animal are oriented by the action of currents and waves, so that concavo-convex valves come to lie with their convex surfaces upwards (Fig. II–25B) in which position they offer less resistance to the flow of water over them than they do with their concave surfaces upwards. A spurious appearance of such orientation may be

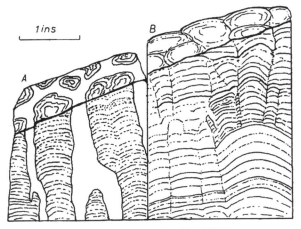

FIG. II–26. COLLENIA COLONIES

A. Separate colonies showing domical growth layers.
B. Contiguous colonies, with growth layers resembling bedding, Macdonnell Ranges, Central Australia.

produced with bivalves if the valves remain in contact and the rock shows only convex surfaces when weathered on a bedding plane; but with isolated valves the indication of facing is good if sufficient shells are present.

It has also been noted that internal moulds formed by the entry into the shells of fine-grained mud from the surrounding sediments, are not always fully developed, but may only partially fill the shell cavities. In such instances, the flat upper surfaces of the partial moulds are parallel with the original bedding planes, and face the stratigraphical upper surface.[1]

Marker beds and key horizons

In mapping stratified rocks, means must be sought not only of identifying the various strata at any one exposure but also of recognizing the same bed when it appears again in another place. Skilled mappers have long been noted for their ability to piece together almost intuitively the criteria that identify a bed or formation, but it is true that identification is possible only if a bed possesses some unique characteristic of colour, texture, mineral or fossil constituents or some unique combination of these properties. Some formations consist of many beds which, although clearly distinguishable at any one exposure, are so

[1] J. S. Cullison, 'Origin of Composite and Incomplete Internal Moulds and their Possible Use as Criteria of Structure', *Bull. Geol. Soc. Amer.*, Vol. 49, 1938, pp. 981–8.

similar to each other that no one bed can be identified elsewhere. Of such a type are the Flysch deposits on the northern margin of the European Alps, and the rapidly alternating sandstones and mudstones of greywacke type, commonly found in marine geosynclinal zones. Among rocks of this kind any distinctive bed that may occur is of particular value in mapping, and is known as a *marker bed* or *key horizon*. The distinguishing feature may be some

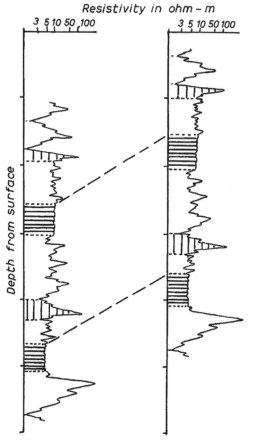

FIG. II–27A. ELECTRICAL WELL-LOGGING

Resistivities in ohm/metres are plotted against depth. Correlation depends on absolute values of resistivity and the arrangement of values in sequence, like a spectrum.

pecularity of colour, the presence of particular types of concretions or fossils, cone-in-cone structure, and so on, but it is most important to test any bed regarded as a marker to decide that it is so in fact, for repetition of similar conditions is always to be expected, leading to repetition of a similar type of bed. The test is generally made by trial and error.

Use of lithological series – In mapping and perhaps more especially in the correlation of bore logs and vertical sections, structural use may be made of the arrangement of beds of different lithologies and thicknesses in a definite sequence. This is the basis of the correlation of glacial varves and of the method of bore-hole correlation based on electrical resistivities of strata, since the resistivity is itself a function of the porosity, mineralogy, and water-bearing properties of the bed (Fig. II–27A). The resultant geophysical logs are characterized by the resistivity values, the thicknesses of beds having certain resistivities, and the serial order in which the beds are arranged, by which data, correlation may be effected. The emission of gamma-rays resulting from the decay of radio-active elements in rocks may be used in a similar way to obtain gamma-ray logs of bore-holes.[1]

On the other hand the repetition as a whole, of a sequence or group of lithologies may lead to misconceptions in correlation and structural interpretation. The best-known example of such repetition is in the *cyclothems* of the Carboniferous coal measures of the Northern Hemisphere. The sequence differs somewhat in different regions, and even within the one coal province, but an ideal sequence may be illustrated from the Yoredale Series in the West Riding of Yorkshire.[2]

SEDIMENTARY RHYTHMS IN THE YOREDALE SERIES

Eroded Limestone surface

Limestone
- 5. Algal phase or chert bed
- 4. Shale with 'modified limestone' fauna
- 3. Limestone with 'normal' fauna (corals, brachiopods)
- 2. Pseudo-breccias
- 1. Coral phase

Break

Sandstone
- Sandstone or coaly-shale
- Coal
- Fireclay or ganister
- Cross-bedded sandstone

Shale
- Unfossiliferous ferruginous shale
- Calcareous shale with 'normal shale' fauna
- Limestone-conglomerate

Break

Limestone of preceding rhythm

[1] D. S. Parasnis, *Principles of Applied Geophysics*, London and New York, 1962, pp. 157–8.
[2] R. G. Hudson, 'On the Rhythmic Succession of the Yoredale Series in Wensleydale', *Proc. Yorks. Geol. Soc.*, N.S., Vol. 20, 1923–4, pp. 125–35; L. J. Wills, *Physiographical Evolution of Britain*, London, 1929, pp. 317–19. See also J. M. Weller, 'Argument for Diastrophic Control of Late Palaeozoic Cyclothems', *Bull. Amer. Assoc. Petrol. Geol.*, Vol. 40, 1956, pp. 17–50; idem, 'Cyclothems and larger sedimentary cycles of the Pennsylvanian', *Journ. Geol.*, Vol. 66, 1958, pp. 195–207; W. Schwarzacher, 'An Application of Statistical Time-Series Analysis of a Limestone–Shale Sequence', *Journ. Geol.*, Vol. 72, 1964, pp. 195–213; J. R. L. Allen, 'Studies in Fluviatile Sedimentation: Six Cyclothems from the Lower Old Red Sandstone, Anglo-Welsh Basin', *Sedimentology*, Vol. 3, 1964, pp. 163–98; A. E. Trueman, Pres. Address, *Quart. Journ. Geol. Soc. Lond.*, Vol. 102, 1946, pp. xlix–xciii.

FIG. II–27B. AERIAL PHOTOGRAPH, FLINDERS RANGES, SOUTH AUSTRALIA
Showing continuity of individual strata. (*Photo: R.A.A.F.: reproduced by permission*)

The presence of numerous rhythms, which are also found in many sequences other than the coal measures,[1] may lead to error in the identification of formations and thus to erroneous structural interpretation, especially if careful palaeontological work is not combined with lithological and structural mapping or borehole correlation.[2]

Shape of strata and formations

Although any one bed may locally preserve a uniform thickness, it must eventually terminate either by lateral thinning (*lensing* or *wedging out*) or by passing gradually into a bed of different lithology. The rate of variation in thickness may be very rapid, or again the same bed may be traced with virtually constant thickness over many miles[3] (Fig. II–27B).

[1] J. Barrell, 'Rhythms and the Measurement of Geologic Time', *Bull. Geol. Soc. Amer.*, Vol. 28, 1917, pp. 745–904; T. Kimura, 'Thickness Distribution of Sandstone Beds and Cyclic Sedimentation in the Turbidite Sequences at Two Localities in Japan', *Bull. Earthquake Res. Inst. Tokyo Univ.*, Vol. 44, 1966, pp. 561–607; A. Hallam, 'Eustatic Control of Major Cyclic Changes in Jurassic Sedimentation', *Geol. Mag.*, Vol. 100, 1963, 444–50.

[2] A useful guide is afforded by the series 'Rhythm in Sedimentation'. Papers by twelve authors, in *Internat. Geol. Congr., Rept. 18th Session, Great Britain, 1948*, London, 1950, Part IV, Section C, p. 99.

[3] The same remarks, also those that follow, apply as well to a formation as to individual beds.

Channel deposits, laid down in former stream-beds, are strongly lenticular in cross-section, filling the former channel, but are persistent in a winding pattern along their length. They are generally composed of sands and gravels and are of economic importance in certain uranium occurrences, in placer deposits especially of gold and tin, and in some oilfields (Fig. II−28A).

Shoestring sands are somewhat similar long sand stringers, but they originate as off-shore bars and are found only in marine successions. With a migrating shoreline they occur in step-like succession, and are straight or simply curved rather than wandering like channel deposits (Fig. II−28B).

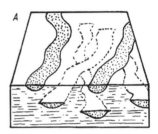

FIG. II-28. CHANNEL DEPOSITS AND SHOESTRING
SANDS

A. Channel deposits of sand and gravel interbedded in fluviatile silts.

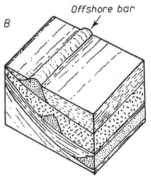

B. Shoestring sands formed by burial of off-shore bars
   along a subsiding coast as sea-level rises across a plain.

The *splitting* of strata or formations is particularly important in coal basins, where it is common to find that a thick seam, followed laterally, divides into two or more smaller seams interdigitating with sandstone or clays. This is interpreted as indicating more rapid sinking of the coal-swamp where the splitting occurs, compared with quiet subsidence where the seam is thick. Infilled scour-channels in the original peat also cause a 'split' within an otherwise continuous seam (Fig. II−29).

Strongly wedge-shaped formations are formed where deposition takes place against steep slopes such as the flanks of coral reefs, around volcanoes, and against fault-scarps.

Reefs − Reefs which are known as coral reefs while living in present-day seas are termed *bioherms* or simply *reefs* in geological examples because corals form only a part of their fossil content. Their geology has a particular interest because of the increasing importance of the reef facies in oil-geology. Reefs are ridges or mounds rising from the sea-floor to the surface, and containing an association of corals, algae, bryozoa, sponges, molluscs, fishes, and a host of other groups, in which, however, the corals seem to be the organisms that determine the location of the reefs and afford the habitat suitable for the other forms. There is a fairly definite control of reef growth by depth and temperature in relation to the common reef-building corals, which, today, require warm tropical or sub-tropical water about 22°C (72°F); they prefer depths shallower than 100 ft and are not profuse below 150 ft;[1] their growth is also inhibited by turbidity of the water and by reduced or increased salinity.

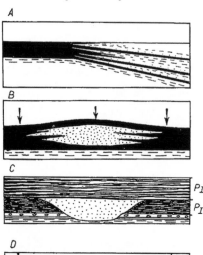

FIG. II-29. FORMATION OF SPLIT COAL SEAMS

A. According to Bowman (vide Lyell, Manual of Elementary Geology, London, 1855). Floor of swamp sinks more rapidly at right. Stable area at left has continuing peat deposition.

B. River deposits of sand in peat mark course of stream through peat-bog. Arrows show compaction.

C. Washout in peat I is filled with sand, covered by younger peat II.

D. Compaction of peat to form coal seams C_1, C_2, split by the channel-fill deposits. (After Raistrick and Marshall, Geology of Coal and Coal Seams, London, 1939)

Modern reefs build up to much greater thicknesses than 150 ft, as for instance at Bikini Atoll, where 2500 ft of reef-rock, all of shallow-water origin, was penetrated by boring.[2] This indicates slow subsidence, as a slow rise of sea-level of this amount is unlikely. Both factors are presumably involved to account for thick reefs, such as the Permian El Capitan Reef of the Guadalupe Mountains region of Texas and New Mexico,[3] which not only attained a maximum vertical thickness of about 1000 ft of reef-rock exclusive

[1] Certain solitary corals and others that do not form large reef-masses are more tolerant, growing in colder and deeper waters. Reefs of a branching coral Lophelia have been built at depths of 600–2000 ft along the Norwegian coast. It is uncertain under what environmental conditions some of the older reefs in the geological succession were formed, since the genera are not closely allied to living forms.

[2] F. P. Shepard, 'Evidence of World-Wide Submergence', Journ. Marine Res., Vol. 7, 1948, pp. 66–78; J. A. Steers, The Unstable Earth, London, 1931, gives a good account of older borings, including Funafuti.

[3] N. D. Newell et al., The Permian Reef Complex of the Guadalupe Mountains Region, Texas and New Mexico, San Francisco, 1953; J. W. Wells, 'Note on Mississippian and Permian Reef Suites', Journ. Geol., Vol. 60, 1952, pp. 97–8.

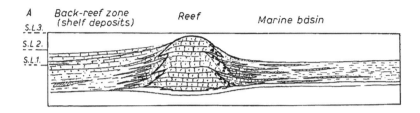

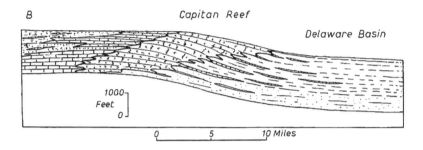

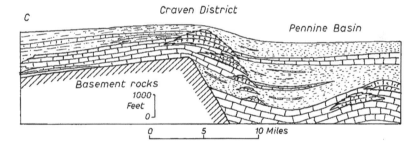

FIG. II–30. REEFS IN STAGNANT SEAS

A. Diagram showing reef with three main stages of upward growth accompanying slow rises of sea-level (SL. 1, 2, 3). The reef grew near land in a mediterranean marine basin (not an oceanic basin).

Back-reef zone: lies between reef and land-mass and has shelf deposits of dolomitized reef detritus, calcarenite, biostromal limestones, and dolomites, with lagoonal deposits, terrigenous clastics, and evaporites nearer the land.

Reef rock: structureless biohermal limestone, with marginal reef talus (breccia) due to wave action chiefly on basin side.

Basin deposits: reef talus passes down to calcarenite, thin limestone and black sandy limestone with interbedded tongues of breccia near reef. Quartz sandstones and siltstones appear in main basin, derived from marginal land-masses.

B. Upper Permian Capitan Reef, Texas and New Mexico. The reef migrates basinwards on a marginal flexure.

C. Upper Carboniferous Reefs, Craven District, Northern England. Reef knolls migrated on and near marginal fault and flexure.

(B. and C. semi-diagrammatic after J. W. Wells, *Journ. Geol.*, Vol. 60, 1952, pp. 97–8, Fig. 1.)

of talus but migrated laterally on the flank of an active flexure within the basin of marine deposition (Fig. II – 30) the maximum lateral extent being six times the vertical thickness.

A typical reef-mound has a core, massive and more or less structureless, flanked by talus slopes consisting of blocks detached by wave action, and surrounded farther out with fine detrital limestone extending laterally for a considerable distance, even for several miles. Within these flanking deposits, especially those close to the reef-core, many effects due to submarine slumping may be found, such as have been described for the Capitan Reef. These include injections of calcareous muds between partially consolidated beds, as well as the normal, superficial slides. A fringing reef built on the edge of a land-mass is strongly asymmetrical; barrier reefs have shallow sea on the landward side, in which wave action is reduced, and a great drop on the seaward side: atolls are roughly symmetrical, although modified by the prevailing winds and currents. The beds covering a reef are affected by differential compaction and an anticline or dome will normally develop over the massive core. The weight of large reefs causes sagging of strata on which they are built up. In the Capitan Reef this causes a basinward tilting of deposits laid down in the back-reef lagoon.[1]

Stratigraphical breaks

Interpretation of the tectonic history of a region is based very largely upon the mutual relationships of strata, and the recognition of breaks of various kinds in the sedimentary succession.

Diastems – The top of any bed was exposed as part of the floor of the sea, lake- or river-bed on which deposition took place, or for a time formed part of the land-surface in the case of aeolian deposits. Such surfaces represent a certain period of time, of the order of minutes or hours in the case of torrential deposition of sands and gravels, but in other environments of the order of thousands of years. The time so represented is a *diastem*.[2]

Although the tendency is perhaps natural to minimize the importance of the diastems and to regard the beds as representing the major part of the time involved in deposition, a simple calculation indicates that for a series of thick-graded beds deposited from turbidity currents, the deposition of each bed

[1] F. A. Bather, 'Reef Structures in Gotland', *Proc. Geol. Assoc. Lond.*, Vol. 25, 1914, pp. 225–8; N. D. Newell *et al.*, *The Permian Reef Complex of the Guadalupe Mountains Region, Texas and New Mexico*, San Francisco, 1953, p. 129.

[2] The term diastem was coined by J. Barrell in his classic study 'Rhythms and the Measurement of Geologic Time', *Bull. Geol. Soc. Amer.*, Vol. 28, 1917, pp. 745–904. He states (p. 794) that diastems may be recognized especially by sharp surfaces separating beds, and again (p. 797) 'the light lines of separation [in his Figure 5] are the bedding planes, or diastems'. There can therefore be no doubt as to his intention, and recent departures from this usage are not favoured. For larger intervals represented by missing fossil zones the correct term is *non-sequence* (q.v. below).

would involve only weeks or months, and that, correspondingly, the diastems may represent thousands of years.[1] One of the most striking evidences of rapid deposition in a non-marine environment is afforded by the burial in erect position of growth, of large trees of the Carboniferous coal measures of Europe and Great Britain, which must have rotted had the enclosing deposits of sandstone been slowly laid down (Fig. II–31). In the Ruhr the sandstones were deposited chiefly from turbidity currents.[2]

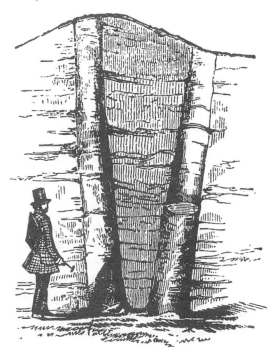

FIG. II–31. STANDING COALIFIED TRUNKS (SIGILLARIA)

Very rapid deposition of surrounding sands is indicated, Cwm Llech, Glamorgan. (From De la Beche, *The Geological Observer*, London, 1853)

On the other hand, during the slow accumulation of a foraminiferal limestone the diastems may be relatively unimportant, a virtually continuous record of sedimentation being represented.[3]

At long diastems, features such as solution pipes, borer holes, worm burrows, scour channels, and animal tracks may be expected on, or in the top of, the lower bed; phosphatic and manganese nodules, fossils representing benthonic life and also forms whose remains accumulated on the sea- or lake-bed after death, may be preserved in the bottom of the overlying bed; on land, fossil soils, root zones, and remains of land animals occur at diastems and

[1] E. S. Hills and D. E. Thomas, 'Turbidity Currents and the Graptolitic Facies in Victoria', *Journ. Geol. Soc. Aust.*, Vol. 1, 1954, pp. 119–33.
[2] R. Teichmuller, 'Sedimentation und Setzung im Ruhrkarbon', *N.J.B. f. Geol. u. Pal., Monatshefte*, 1955, pp. 146–68.
[3] D. G. Moore, 'Rate of Deposition shown by relative Abundance of Foraminifera', *Bull. Geol. Soc. Amer.*, Vol. 66, 1955, pp. 1594–600.

are a feature of most aeolianites. All these may aid in the determination of facing.

Diastems are the smallest breaks recognizable in the deposition of a sedimentary series. In general it is to be expected that no marked palaeontological break will be found across diastems, but in dealing with very numerous fossils such as graptolites, it is in fact found that the different 'bands', of which there may be some five to a biostratigraphical zone, have distinctly different fossil assemblages. Where a depositional break is long enough for a zone to be lacking the break is termed a *non-sequence*.

Disconformity – At a disconformity younger beds overlie distinctly older, the beds are approximately parallel, but an erosional interval is represented at the junction. If the older beds have been folded such relationships may occur locally whereas an angular unconformity will be found elsewhere, but to be of significance disconformities should be regional and they are more likely to occur on relatively stable crustal blocks which have been subjected to vertical movements, than in active geosynclinal zones. Major disconformities are rare, for with close study an angular unconformity, however slight, will generally be found instead of regional parallelism.

Unconformity – An unconformity, or more properly an *angular unconformity*, is represented where younger beds overlie older rocks which have been tilted or folded and subsequently eroded before the overlying beds were deposited. Differences of dip, or in both strike and dip, obtain between the two groups of strata. Some unconformities represent many millions of years of geological time, in which case there is often a notable difference in the compaction, induration, or degree of metamorphism of the two groups of rocks, for instance where normal sandstones and shales overlie slates or schists. The term *non-conformity* may be used for a buried erosion surface separating an intrusive complex from overlying sediments, the essential concept being that prolonged erosion must have occurred to expose the intrusives before burial, although dip and strike relationships are not involved. There is, however, little objection to the extension of the term unconformity to cover such cases.

Major unconformities extend over a wide area and represent the effects of important earth movements followed by prolonged erosion and later deposition. As such they have been used in the subdivision of the geological column, and it has been argued that the periods of diastrophism represented by unconformities might afford an objective basis for correlation if diastrophism were synchronous over wide areas of the globe.[1]

Important contributions to the study of the regional distribution of oil have

---

[1] T. C. Chamberlin, 'Diastrophism as the ultimate basis for correlation', *Journ. Geol.*, Vol. 17, 1909, pp. 685–93. H. Stille concludes that diatrophic episodes are virtually synchronous over wide areas – see his *Grundfragen der Vergleichenden Tektonik*, Berlin, 1924; but it is doubtful if precise age-equivalence can be demonstrated.

been based by Levorsen on widespread unconformities and their broad geological implications.[1]

On the other hand local unconformities between rocks that do not differ greatly in age have little tectonic significance beyond indicating instability in a region of overall sedimentation. In one square mile of Cambrian rocks at Comley, Shropshire, seven local unconformities have been mapped, none of which is taken to imply an important orogenic epoch.[2]

The evaluation of the time-gap or *hiatus* represented by an unconformity and its significance with respect to erosion and non-deposition is not easy. For instance if Tertiary beds rest on Silurian, it may be that deposition originally continued through Devonian times, the Devonian beds being subsequently eroded away leaving only the Silurian. Deposition may again have occurred at any time between Devonian and Tertiary, say during the Permian, and erosion again may have removed the rocks so deposited, before the Tertiary deposition occurred. Regional mapping will assist in clarifying the picture, but invariably much will remain hypothetical about any major unconformity.

Recognition of uncomformities – In cliff sections unconformities are in many cases clearly recognizable, but some difficulties may arise, for instance where slump-masses of bedded rocks are seen resting on the truncated ends of penecontemporaneous strata along a slide plane. The absence of weathering, and of types of sediments in the overlying beds characteristic of transgression will be noted, and both palaeontological evidence and wider mapping will reveal the true relationships. Contemporaneous scour-channels seen at local exposures might also be taken to indicate an unconformity which would not be confirmed by regional mapping or by palaeontological study. Indeed the recognition, evaluation, and regional plotting of an unconformity requires the fullest study by structural mapping, sub-surface exploration, and biostratigraphy. Apparently simple criteria such as weathering of the basement, presence of a conglomerate, and the like may be misconstrued, for instance where in surface exposure beds are preferentially weathered beneath a fault, or where conglomerates are in fact intra-formational rather than basal.

The presence of pebbles of older rocks in a conglomerate is no indication of an unconformity, since the area of erosion from which the pebbles came may have been continuously a land-mass, bordering the basin of sedimentation which received the conglomerates. Unless deposition occurs on the erosion surface an unconformity is not represented there, but merely a region of non-deposition. The difficulties of recognizing unconformities

---

[1] A. I. Levorsen, 'Studies in Paleogeology', *Bull. Amer. Assoc. Petrol. Geol.*, Vol. 17, 1933, pp. 1107–32.

[2] E. S. Cobbold, 'The Stratigraphy and Geological Structure of the Cambrian Area of Comley (Shropshire)', *Quart. Journ. Geol. Soc. Lond.*, Vol. 83, 1927, pp. 551–73.

are greatly increased in strongly folded rocks, especially if strike-faulting is prevalent.

Overlap and off-lap – Important structural relationships arise where a sloping basement is subject to transgression by the sea accompanied by deposition of younger rocks unconformably on the basement. As the shoreline advances deposition extends farther on to the basement so that successively deposited beds *overlap* each other and *overstep* farther on to the basement. Occurring with an advancing shoreline this is known as *transgressive overlap. Off-lap* is the reverse of overlap and is formed in marine beds with a retreating shoreline (Fig. II–32). More particularly, these relationships may be related to the cycle of sedimentation.[1]

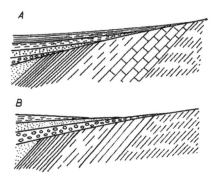

FIG. II–32. UNCONFORMABLE OVERLAP AND OFF-LAP

A. Overlaps formed with rising sea-level.
B. Off-lap formed with falling sea-level.

In many basins of sedimentation, retreats and advances of the sea alternate cyclically; thus, should regression of the shoreline follow a previous advance, the area of continental deposits extends over the marine beds. Interdigitating wedges of marine and continental beds result, each pair representing one cycle of sedimentation. On an advancing shoreline a basal conglomerate may be expected in the marine beds (*gravier d'immersion*), while with recession a second conglomerate is formed (*gravier d'émersion*) (Fig. II–33A). When considered also in relation to the kinds of sediments deposited at a given time, the cyclic sedimentation is more complex than has been indicated. At any one time, with stationary sea-level, clastic deposits will normally range from coarse to fine seawards.[2] Successive advances of the strandline thus result in belts or zones in which one facies dominates, and a given formation of conglomerate, sandstone, or clay ranges in time from point to point normal to the shore. This is known as *facies-ranging* or *diachronism* (Fig. II–33B). An instance is afforded by the Liassic sands of southern England, in which the

---

[1] L. D. Stamp, 'Cycles of Sedimentation in the Eocene Strata of the Anglo-Franco-Belgian Basin', *Geol. Mag.*, Vol. 58, 1921, pp. 108–14, 146–57, 194–200; F. F. Sabins, Jr, Symmetry, Stratigraphy, and Petrography of Cyclic Cretaceous Deposits in San Juan Basin', *Bull. Amer. Assoc. Petrol. Geol.*, Vol. 48, 1964, pp. 292–316.

[2] This is a generalization which is often, in nature, quite incorrect, but may be used to illustrate the point at issue.

sandy facies ranges through nine ammonoid zones in passing from Cheltenham to Dorset.[1]

Sedimentation and tectonics — The recognition that sedimentary or lithofacies of certain types are normally deposited in particular environments that

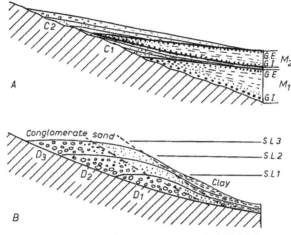

FIG. II–33. FACIES – TIME RELATIONSHIPS

A. Cycles of Sedimentation. Marine wedges M1, 2, show diachronism. GI, basal conglomerate (*gravier d'immersion*) formed with advancing sea-level (compare *A* above); GE, conglomerate formed with retreating sea (*gravier d'émersion*). M1 is followed by continental wedge C1, which spreads seawards as the coastline retreats. A second cycle involves M2 and C2. *Gravier d'émersion* may be spread partly on land, from rejuvenated streams, partly in the sea. (*After Stamp, 1921*)

B. Diachronism or facies-ranging: overlapping deposits on an advancing shoreline, are formed with successive sea-levels 1, 2, 3; D1, 2, 3, are corresponding deposits, varying from coarse clastic to fine grained. Litho-facies are conglomerate, sand, shale, and clay, which cut time lines. Any one facies ranges in time. (After Boswell, *quoted in Wills*, Physiographical Evolution of Britain)

are determined by the tectonic setting of realms of sedimentation is a principle of wide application in regional tectonic analysis.[2] In addition, the statistical treatment of mapped directions deriving from observations on crossbedding, current ripples, flute casts, and other data permits inferences to be

[1] L. J. Wills, *Physiographical Evolution of Britain*, London, 1929, gives an excellent account based on the original work of Buckman, Richardson, and Boswell.

[2] See especially P. D. Krynine, 'A Critique of Geotectonic Elements', *Trans. Amer. Geophys. Union*, Vol. 32, 1951, pp. 743–8; W. C. Krumbein and L. L. Sloss, *Stratigraphy and Sedimentation*, San Francisco, 1951; E. C. Dapples, W. C. Krumbein, and L. L. Sloss, 'Petrographic and Lithologic Attributes of Sandstones', *Journ. Geol.*, Vol. 61, 1953, pp. 291–317; 'Sedimentary Facies in Geologic History', *Geol. Soc. Amer.*, Mem. No. 39, 1949; A. Lombard, *Géologie Sédimentaire*, 'Les Séries Marines', Paris, 1956.

drawn as to palaeocurrent directions in sedimentary troughs.[1] There are, however, inherent complexities in the relationships of palaeocurrents, deposition and palaeoslopes which necessitate careful evaluation of the significance of primary current indicators in sediments.[2]

[1] W. Niehoff, 'Die Primär Gerichteten Sediment-strukturen, insbesondere die Schrägschichtung, im Koblenzquartzit im Mittelrhein', *Geol. Rundsch*, Vol. 47, No. 1, 1958, pp. 252–321 (with Bibliography); P. E. Potter and F. J. Pettijohn, *Palaeocurrents and Basin Analysis*, Berlin, 1963; B. R. Pelletier, 'Palaeocurrents in the Triassic of Northwestern British Columbia', in *Primary Sedimentary Structures and their Hydrodynamic Interpretation, Soc. Econ. Palaeont. and Mineralogists*, Spec. Pub. No. 12, 1965, pp. 233–43.

[2] R. C. Selley, 'A Classification of Palaeocurrent Models,' *Journ. Geol.*, Vol. 76, 1968, pp. 99–110.

CHAPTER III

Non-diastrophic structures

The structures discussed in this chapter are formed chiefly by gravitational forces, which are involved in the compaction of sediments, in the slumping of unconsolidated deposits, and in landslides and related effects. In Chapter XI gravity is further discussed as a factor in diastrophism, but the structures now to be described are generally recognized as *non-diastrophic*. This view is taken because these structures do not necessarily involve deformation of the deeper layers of the crust of the earth and thus do not imply the transmission of major earth-forces through the rocks themselves or the basement on which they rest. They may for the most part be explained by the yielding of inadequately supported rocks at or near the surface, and are susceptible to study by the established principles of civil engineering and soil mechanics in relation to earth-masses. Certain structures of a different origin, for instance those due to chemical changes, especially hydration, and to glacial action, are also for convenience dealt with here.

Compaction and diagenesis

After sediments are deposited they are subject to compaction which affects different deposits to various degrees. The chief factors influencing compaction are the constitution of the deposit, the load to which it is subjected on burial, and the length of time that has elapsed since its deposition; but the process is also affected by the geological environment of a rock, which may be such as to inhibit compaction. This applies for instance to an artesian aquifer in which the grains are prevented from assuming close-packing by the pressure of water in the pores. Vibrations set up by earthquakes, stresses due to diastrophism, increase of temperature, and chemical changes either in the original sedimentary grains themselves or in the deposition of cementing minerals around and between the grains, also affect the process and modify the original sediment. The complex of changes whereby an unconsolidated sediment is gradually transformed into a more or less coherent 'rock' is known as *diagenesis*.[1] Compaction implies chiefly the rearrangement of the grains in

[1] G. D. de Segonzac, 'The Birth and Development of the Concept of Diagenesis', *Earth Science Rev.*, Vol. 3, 1968, pp. 153–201; F. J. Turner, *Metamorphic Petrology*, New York, 1968, p. 263 et seq., discusses the transition from diagenesis to metamorphism, which affects rocks very deeply buried in some geosynclinal troughs.

closer packing accompanied by the expulsion of fluid from between the grains, and especially the loss of connate water entrapped in subaqueous deposits.[1] The term also includes, however, any actual loss of substance from chemical changes such as are involved in coalification, or in the loss of inter-layer water from the crystal lattices of certain clays. Compaction results in a decrease of porosity, and a corresponding increase in density of the sediment. Measurements made on samples obtained from bores in the Mid-Continent and the Gulf Coast regions of the U.S.A. show progressive compaction with increasing thickness of cover, but it appears that full compaction requires a considerable time, as certain Miocene shales at 15,000 ft are still incom-pletely compacted, and in general older rocks at the same depth have higher specific gravities than younger.[2] Fig. III—1 shows the maximum possible

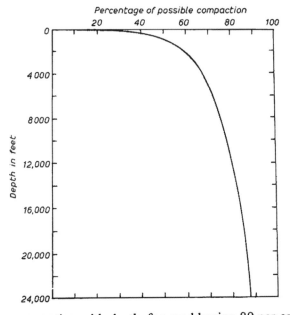

FIG. III-1. CURVE SHOWING PERCENTAGE OF TOTAL POSSIBLE COMPACTION OF MUD WITH DEPTH

(*After Dickinson, 1953*)

compaction with depth, for mud having 80 per cent porosity and a density of 1·4 as deposited. Broadly speaking, the compaction of sediments is best considered in relation to the properties of two-phase (solid-fluid) systems, especially as by so doing the role of the fluid phase will not be overlooked.[3]

Aeolian deposits, in which the fluid phase is air, afford a special case.

[1] O. T. Jones, 'The Compaction of Muddy Sediments', *Quart. Journ. Geol. Soc. Lond.*, Vol. 100, 1944, pp. 137–60.

[2] L. F. Athy, 'Density, Porosity and Compaction of Sedimentary Rocks', *Bull. Amer. Assoc. Petrol. Geol.*, Vol. 14, 1930, pp. 1–24. G. Dickinson, 'Geological Aspects of Abnormal Reservoir Pressures in Gulf Coast Louisiana', ibid., Vol. 37, 1953, pp. 410–32, discusses the earlier literature and recent results.

[3] K. Magara, 'Compaction and Migration of Fluids in Miocene Mudstone, Nagaoka Plain, Japan', *Bull. Amer. Assoc. Petrol. Geol.*, Vol. 52, 1968, pp. 2466–501.

Calcareous aeolianites are commonly lithified to form dune-rock with virtually no compaction. The sand-fall slopes are preserved at angles of rest similar to those in modern dunes, but the grains are cemented with small amounts of secondary calcite. Dune-rock strong enough to be used for domestic buildings (Fig. III−2) may thus have a porosity as high as 55 per

FIG. III–2. QUARRY IN PLEISTOCENE DUNE LIMESTONE (AEOLIANITE)
Warrnambool, Victoria, showing dip of sand-fall slopes of dune. Footprints of a large bird (emu) are seen on the trapezoidal block, photographed separately at the right.

cent and a permeability so great that a beaker of water may be emptied through the stone.

Sedimentary clays contain a high proportion of water when deposited, and may be reduced to as little as a quarter of their original thickness by compaction. Water is held in clays in several ways, and compaction is correspondingly complex. 'Free' water occupies the voids between clay mineral particles or aggregates, and is readily removed by pressure. Around each clay particle there are unsatisfied chemical bonds which polarize and attract water-molecules to form water-films, the thickness of which varies inversely as the pressure, for pressures up to 600 lb per square inch.[1] At higher pressures the clay particles themselves begin to deform but it seems most likely that some film would remain even at very high pressures. Some clay minerals, notably montmorillonite, also take up 'inter-layer' water as sheets interleaved in the crystal lattice. When a montmorillonite lattice absorbs such water it expands

[1] F. H. Norton and A. L. Johnson, 'Fundamental Study of Clay: V, Nature of Water Film in Plastic Clay'. *Journ. Amer. Ceram. Soc.*, Vol. 27, 1944, pp. 77–80. R. E. Grim, *Clay Mineralogy*, New York, 1953.

and exerts pressure, and likewise if this pressure is exceeded, inter-layer water may be expressed from the lattice. Again, temperatures below 100°C cause the expulsion of some inter-layer water from halloysite, and other clay minerals also lose inter-layer water at low temperatures such as will be encountered at quite shallow depths in the earth. Clay compaction is also affected by the state of aggregation of the particles, which are rarely freely dispersed but form linked groups or aggregates, thus affecting the porosity and more especially the permeability of the clay. The state of aggregation of clays is greatly influenced by the presence of certain cations in solution (particularly Na·, K·, and Ca··) and in general, clays deposited in fresh water will be more highly dispersed than those deposited in saline waters, which tend to flocculate dispersed clay particles. Also the several clay minerals have markedly different properties with regard to dispersion and water absorption. In general, the expulsion of water and compaction of clay is a reversible process, so that clay compacted in depth might be expected to swell again on removal of load by erosion. However, at depths between 6000 and 10,000 ft it seems likely that montmorillonite progressively loses water layers from its crystal lattice and is diagenetically transformed to illite.[1] The growth of illite is probably the main cause of the transformation of clay to mudstone or shale, some contribution to the lithification process being also made by iron and manganese oxides and calcium carbonate.

Sands and detrital limestones compact much less than clays. Many detrital limestones are slowly deposited and contain grains of various sizes. As a consequence the packing is close and the voids between large grains are partly filled with fine detritus so that little compaction is possible. The same applies to certain sands, but those that are deposited from heavily laden slurries or turbidity currents have a very open texture at the moment of settling and contain much connate water, which is immediately reduced, perhaps within a period of hours, as the deposit settles down and normal slow compaction will then follow. However, the first settlement will almost certainly give rise to greater reduction in thickness than will subsequent compaction. The presence of silt and clay in an impure sandstone leads to greater eventual compaction and to a great reduction in the size of the inter-granular pore-spaces, which in turn reduces the permeability of the rock to fluids such as water or oil. Close-packing of quartz and calcium carbonate is accompanied also by some fracturing of grains and by solution and redeposition. Solution occurs at points of strong stress concentration, and redeposition where the stress is low in the voids[2] (Fig. III–3A, B). The effects of solution at points where pressure is

[1] M. C. Powers, 'Fluid-Release Mechanisms in Compacting Marine Mudrocks and their Importance in Oil Exploration', *Bull. Amer. Assoc. Petrol. Geol.*, Vol. 51, 1967, pp. 1240–54.
[2] For studies in the diagenesis of sands, see W. D. Lowry, 'Factors in Loss of Porosity of Quartzose Sandstones of Virginia', *Bull. Amer. Assoc. Petrol. Geol.*, Vol. 40, 1956, pp. 489–500. M. T. Heald, 'Cementation of Simpson and St Peter Sandstones', *Journ. Geol.*, Vol. 64, 1956, pp. 16–30.

concentrated among grains in contact with each other is best shown by *dimpled pebbles* in conglomerate. Sorby[1] and later workers[2] have demonstrated that the dimples are formed by solution, according to Riecke's Principle, in a saturated solution of the material of the pebbles (calcite or silica chiefly) (Fig. III–3C).

In acid tuffs (ignimbrites) deposited from *nuées ardentes* compaction is very considerable. This is shown by the progressive increase in specific

FIG. III–3. SOLUTION AND REDEPOSITION IN COMPACTION

A. Quartz grain showing outgrowths of secondary quartz.
B. Quartz grain showing small dimples due to solution at pressure-points.
C. Conglomerate (Nagelfluh, Switzerland), showing limestone pebbles (L) dimpled by pressure from dolomite pebble (D).
(A. & B. *after Lowry, 1956;* C. *after Trurnit, 1968*)

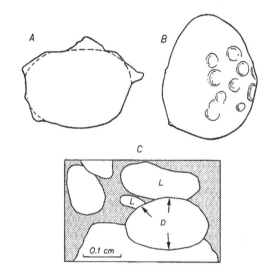

gravity of New Zealand ignimbrites (from 1·67 to 2·37) with a corresponding decrease in porosity (from 28·6 per cent to 3·4 per cent)[3] from top to bottom of a deposit, and also by the flexing of so-called 'flow layers' around quartz phenocrysts (Fig. III–4B).

A coal-seam, owing to the loss of substance in the process of coalification as well as to the packing of the plant detritus, may finally represent only 15 to 20 per cent of the thickness of the original peat deposit.[4] It should be made clear that, as applied to coal, compaction refers to processes that go on after the burial of the original peat, not to the stage of biochemical decay in the open peat-bog. The reduction in thickness of solid coalified tree-trunks embedded in coal, sand, or clay is in some instances not great, but thin

[1] H. C. Sorby, 'Ueber Kalkstein-Geschiebe mit Eindrücken', *Neues Jahrb. f. Min., etc.,* Vol. 34, 1863, pp. 801–7.
[2] Ph. H. Kuenen, 'Pitted Pebbles', *Leidsch. Geol. Med.,* Dl. 13, 1942, pp. 189–201; R. Trurnit, 'Pressure Solution Phenomena in Detrital Rocks', *Sediment. Geol.,* Vol. 2, 1968, pp. 89–114.
[3] P. Marshall, 'Acid Rocks of the Taupo-Rotorua Volcanic District', *Trans. Roy. Soc. New Zealand,* Vol. 64, 1935, pp. 1–44. H. E. Enlows, 'Welded Tuffs at Chiricahua National Monument, Arizona', *Bull. Geol. Soc. Amer.,* Vol. 66, 1955, pp. 1215–46.
[4] R. Teichmuller, 'Sedimentation und Setzung im Ruhrkarbon', *Neues Jahrb. f. Geol. u. Pal. Monatshefte,* Jhg. 1955, pp. 146–68.

sections of softer plant tissues often show concertina-like compression of the cell-walls.

Structures formed about resistant inclusions demonstrate clearly the effects of compaction in a variety of rocks. This is shown by the squeezing together

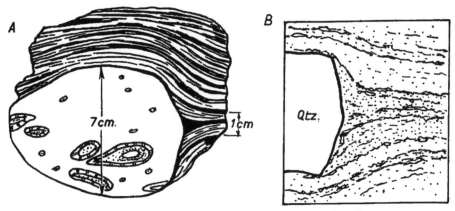

FIG. III–4. COMPACTION ABOUT INCLUSIONS

A. Compaction of coal about pebble of dolomitic limestone. (*After Teichmüller, 1955*)
B. Compaction in ignimbrite about quartz phenocryst, as seen in thin section.

of bedding planes on either side of the inclusion, which results in flexing of the bedding above, and in some rocks also below the inclusion. Fig. III–4A shows the effect about dolomite pebbles in coal. The compaction of sandstones around standing tree-trunks in roof-rock of coal-measures (Fig. III–5) shows not only the reduction in thickness of the sands but also the much

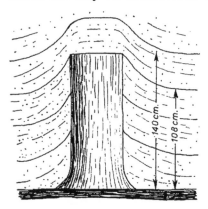

FIG. III–5. COMPACTION OF RAPIDLY DEPOSITED SANDS ABOUT STANDING TREE-TRUNK IN RUHR COAL MEASURES

(*After Teichmüller, 1955*)

smaller compaction of an object such as a tree compared with that of a detrital peat-mass. Bedding planes are flexed about large boulders in shale or sandstone (Fig. III–6). In the example illustrated the boulder occurs in a bed of mixed pebbles and sandy silt which was most probably deposited from a slurry or rock-flow. Compaction of the finer-grained portions left the large granite boulders projecting upwards and the overlying strata were flexed accordingly. In Fig. III–7 an underlying bed of clay was squeezed by the weight of the boulder and reduced in thickness beneath it.

This flexing of compacted strata about resistant boulders resembles in some ways the effects produced where boulders are dropped from floating ice

FIG. III–6. COMPACTION IN SANDY SILTS OVER GRANITE BOULDER IN PERMIAN GLACIO-LACUSTRINE BEDS, COIMADAI, VICTORIA
(*Photo: E.S.H.*)

or vegetation onto soft sediments and are then covered by later deposits, or again where large volcanic bombs or ejected blocks are projected into bedded tuffs. The origin is clear where the boulder actually breaks through the underlying beds, which is generally found to occur with ejected blocks from volcanoes (Fig. III–8) and also with pebbles or boulders dropped from

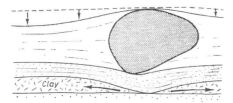

FIG. III–7. COMPACTION IN SANDY SILTS ABOVE, AND LATERAL SQUEEZING OF CLAY BENEATH A GRANITE BOULDER IN PERMIAN GLACIO-LACUSTRINE BEDS, COIMADAI, VICTORIA
(*Photo: E.S.H.*)

floating ice (Fig. III–9). It is said that such boulders often come to rest with the long axis vertical.

Syngenetic concretions that formed on a sea- or lake-floor also behave as resistant inclusions, and strata are flexed above them as a result of compaction.[1]

[1] W. A. Richardson, 'The Relative Age of Concretions', *Geol. Mag.*, Vol. 58, 1921, pp. 114–24.

FIG. III-8. EJECTED BLOCK OF GRANITE INDENTING CAINOZOIC BASALT-TUFF, LAKE BURRUMBEET, NEAR BALLARAT, VICTORIA

(Photo: E.S.H.)

FIG. III-9. PEBBLE DROPPED FROM FLOATING ICE ON TO VARVED CLAYS, SEAHAM GLACIALS (UPPER CARBONIFEROUS), NEW SOUTH WALES

Natural size

The use of the flexing of bedding about boulders, pebbles, or concretions to determine facing is liable to misinterpretation unless a clear breakthrough of the underlying beds is to be made out. If not, it is in general most likely that the flexing of bedding will be greater above rather than below a boulder, but much depends on the actual origin of the deposit.[1]

Elastic expansion

A bed containing fluid under pressure greater than that due to the load of overlying rocks tends to expand, the grains being forced apart. This is

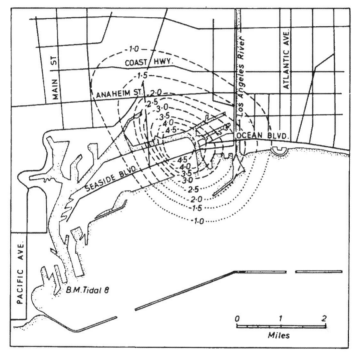

FIG. III–10. SUBSIDENCE (IN FEET) IN THE LONG BEACH HARBOUR
AREA, CALIFORNIA, BETWEEN 1933 AND 1946
(*From Gilluly and Grant, 1949*)

demonstrated on piercing the bed by a bore which penetrates the covering impervious strata and yields flows of oil or water. Subsidence of the area about the flowing bore may then take place. The effect is strikingly shown at Long Beach Harbour, California, where subsidence of the Wilmington oilfield between 1933 and 1946 attained a maximum of 29 feet in an elliptical bowl-shaped depression 20 square miles in area (Fig. III–10). Stabilization has

[1] W. H. Collins, 'The Geology of the Gowganda Mining Division', *Mem. Geol. Surv. Canada*, No. 33, 1913, p. 84, Fig. 1.

been achieved by repressuring the field through water-injection wells.[1] The phenomenon is, however, now recognized as a general one affecting particularly artesian basins, in which the first flows are chiefly the result of the expression of water from the expanded aquifer, which subsides until the grains are close-packed, at which stage the steady state of flow is attained, this being much less than the original flush-flow.

Clearly the notion of compaction must be modified to include the possible effects of fluids under pressure in the pores of rocks that are confined by impervious strata. It is also interesting to speculate on the results that might be expected were strong diastrophism to affect an artesian basin, or, indeed, any sediments in which the voids are filled with water, oil, or gas. The phenomena associated with earthquakes in regions of deep alluvium certainly indicate that notable readjustments of the packing of sediments occur such that objects embedded in the ground may be forced up, or in other places may sink, while jets of water containing sand and mud are squirted up cracks.

Sand volcanoes ranging from one to fifty feet in diameter are associated with sub-aqueous slumps in the Carboniferous of County Clare, Ireland,[2] and, like many other phenomena, point to readjustments connected with the expression of water from the voids of unconsolidated rocks.

The injection of clastic material by such means is one of the mechanisms giving rise to *clastic dykes*. These injections take place mainly because of the elastic expansion of the fluids contained in the pores of rocks which are under considerable load. When a crack occurs in the impervious enclosing strata, the fluids expand to fill the newly created void, carrying the clastic fragments with them.[3] As elastic expansion is the chief cause of the dykes, their injection may equally well take place either upwards or downwards according to the location of newly formed fissures, and many examples do appear to have been introduced from above. Examples formed simply by the falling of overlying rocks down cracks are also reported as clastic dykes, although the term is not particularly appropriate to these. Some groups of clastic dykes are formed in relation to local structures (Fig. III–11), others to regional tectonic strain:[4] others again appear to exhibit no such correlation and these have been regarded as injections connected with earthquakes.[5]

[1] J. Gilluly and U. S. Grant, 'Subsidence in the Long Beach Harbour Area, California', *Bull. Geol. Soc. Amer.*, Vol. 60, 1949; pp. 461–530; M. N. Mayuga and D. R. Allen, 'Long Beach Subsidence', *Abstr. N. Amer. Geol.*, July, 1967, p. 921.

[2] W. D. Gill and Ph. H. Kuenen, 'Sand Volcanoes on Slumps in the Carboniferous of County Clare, Ireland', *Quart. Journ. Geol. Soc. Lond.*, Vol. 113, 1958, pp. 441–60.

[3] M. S. Walton, Jr, and R. B. O'Sullivan, 'The Intrusive Mechanics of a Clastic Dike', *Amer. Journ. Sci.*, Vol. 248, 1950, pp. 1–21. R. R. Shrock, *Sequence in Layered Rocks*, New York, 1948, gives the best general account of clastic dykes.

[4] D. Richter, 'On the New Red Sandstone Neptunian Dykes of the Tor Bay Area (Devonshire)', *Proc. Geol. Assoc.*, Vol. 77, 1966, pp. 173–86.

[5] H. Terashima, 'On the Paleogene Sandstone Dikes in the Muroto District, Shikoku Island, Southwest Japan', *Mem. Fac. Sci. Kyoto Univ., Ser. Geol. & Min.*, Vol. 34, 1967, pp. 23–34.

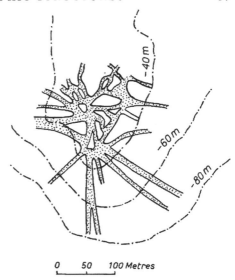

FIG. III–11. PLAN OF CLASTIC (CLAY)
DYKES AT THE CREST AND NOSE OF
AN ANTICLINE, DÖHLEN, SAXONY

Structure contours on coal-bed, in metres
below datum plane. (*After Hausse, Erläut.
geol. Spezialkarte Kön. Sachsen, Leipzig,
1892*)

Compaction folds

The results of compaction are important structurally in forming folds over
buried hills, over buried but sometimes still active fault-scarps and folds, and
over inter-bedded lenses of sand, reef-rock, or other little-compactable rocks.
Compaction folds are of particular interest in oil geology, notably in Kansas
and Oklahoma.

With buried hills, structures in the overlying rocks may form simply from
the compaction of the smaller thicknesses that exist over the hills as against
the deeper sections in the buried valleys. Compaction causes the beds to
reflect the buried topography, although with reduced slopes[1] (Fig. III–12A).
This simple picture is complicated towards the base of the succession, where
the formations will have initial dips on the flanks of the hills; they will, however, thicken in the valleys and wedge out on the hill-slopes, so that the higher
beds of the sequence will be deposited with originally horizontal bedding.

Earth-movements affecting the basement also form structures in the covering rocks. Thus if the buried hills are anticlines or uplifted fault-blocks on
which movements are renewed after burial, the overlying sediments will also
be affected and the total folding of the covering rocks is then due both to
differential compaction and to continuing diastrophism. The Mid-Continent
oilfield of the U.S.A. is a classic region for compaction folds formed over buried
hills,[2] but even there the effects of renewed folding in the basement may play a

---

[1] Owing to the greater load in the troughs, compaction would, in fact, be a little greater there
than in the thinner sections over the hills.

[2] M. G. Mehl, 'The Influence of Differential Compaction on the Attitude of Bedded Rock',
Science, Vol. 1, 1920, p. 520. S. Powers, 'Reflected Buried Hills and their Importance in
Petroleum Geology', *Econ. Geol.*, Vol. 17, 1922, pp. 233–59.

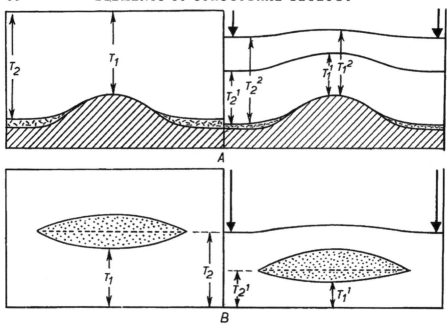

FIG. III–12. COMPACTION OVER BURIED HILLS AND SAND LENSES

A. Compaction of original thicknesses T_1, T_2 to one-half ($T_1{}^1, T_2{}^1$), or alternatively to three-quarters. Note also effects in the lower bed due to deposition with initial dips on the flanks of the buried hill.

B. Compaction of original thickness T_1, T_2 to one-half ($T_1{}^1, T_2{}^1$) about an incompactible sand-lens.

significant part in determining the geometry of overlying structures,[1] especially in forming anticlines that are of larger amplitude than could be formed by compaction. The same may apply to faults in the basement, and again, distributed shearing movements on innumerable small faults, many of which might appear to be only joints, may permit warping even of massive crystalline basement rocks, with corresponding structures being formed in the over-mass.[2]

Where less compactible formations are interbedded as lenses in other rocks, folds result from *differential compaction*, the sum total of compaction in the sections where the buried lenses exist being less than in neighbouring sections where they are absent (Fig. III–12B). Differential compaction may also give rise to normal faults where large changes in total compaction of adjacent sediment masses occur.[3] A further account of folds due to compaction will be found in the discussion of supratenuous folds (see p. 258).

[1] S. Powers, 'Structural Geology of North-eastern Oklahoma', *Journ. Geol.*, Vol. 39, 1931, pp. 117–32.

[2] F. S. Hudson, 'Folding of Unmetamorphosed Strata superjacent to massive Basement Rocks', *Bull. Amer. Assoc. Petrol. Geol.*, Vol. 39, 1955, pp. 2038–52.

[3] R. E. Carver, 'Differential Compaction as a Cause of Regional Contemporaneous Faults', *Bull. Amer. Assoc. Petrol. Geol.*, Vol. 52, 1968, pp. 414–19.

Subaqueous slumping and gliding

Subaqueous slumping refers to the displacement *en masse* of soft or partially consolidated sediments in marine or lacustrine basins of sedimentation, and includes a rather wide variety of phenomena. In July 1887, the front of the delta at the head of Lake Zug in Switzerland slipped into the lake, carrying away some of the buildings of the town of Zug. The soft deltaic sediments were transformed into a subaqueous mud-flow which could be traced on the lake-bed for over 1000 metres from the origin, with a width of 150–250 metres and a thickness of 1–4 metres (Fig. III–13). The catastrophe was

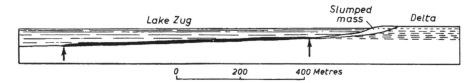

FIG. III–13. SLUMP OF DELTA-FRONT, LAKE ZUG, SWITZERLAND, JULY 5, 1887
The slump-mass spread out on the lake floor to a distance of 1020 metres, between the arrows.
(*After A. Heim, 1908*)

described by Heim,[1] who pointed out that deposits formed by the slumping of previously deposited beds should exhibit folding, faulting, and brecciation due to the movement, and would be found, after further sedimentation, as a layer of contorted rocks lying between undeformed strata, and thus exhibiting *intraformational contortion*. Heim interpreted examples from the Flysch as due to subaqueous slumping, and subsequently subaqueous slumping has been widely reported in rocks of all ages, especially to explain intraformationally contorted beds showing complex folding (*slip-bedding*), thrusting, and brecciation (*slump-breccias*). The assumption is that sediments initially deposited on a sloping platform such as the continental shelf or slope, or on submarine scarps, especially fault-scarps, may be dislodged and slide from their original position due to excess loading, earthquake shocks, or a reduction of support as with lowering of water-level in a lake.

The effects of slumping vary according to the physical properties of the displaced mass and the mode of origin of the slide. Many features of landslides, rock-flows, mud-flows, and the like, seen on land,[2] may be reproduced in subaqueous slides at their point of origin. At the base there will usually be a slip-surface that is concave upwards, and the mass may come to rest while still partly resting on this, and partly on the sediments down-slope. These

[1] A. Heim, 'Über rezente und fossile subaquatische Rutschungen und deren lithologische Bedeutung', *Neues Jahrb. f. Min.*, etc., Jhg. 1908, Band II, pp. 136–57.
[2] C. F. S. Sharpe, *Landslides and Related Phenomena*, Columbia University Press, 1938.

effects are known on the flanks of submarine scour-channels,[1] in inclined soft rocks,[2] and around reef complexes.[3]

A slide affecting consolidated rocks would produce a mass of very large brecciated blocks, which might not move far under some conditions, but which, if mixed with mud, could on the other hand spread out over a considerable area. Examples in the Ordovician and Kimmeridgian of Scotland contain blocks up to 100 ft long,[4] and much recent work on slumping has been concerned rather with the formation of breccia beds and slip-sheets of the same general type as landslides than with intraformational folding. Slices of rock ranging up to more than a kilometre in length, that have slipped onto younger sediments, have produced stratigraphic puzzles, especially in Venezuela, and similar features are also seen in the *exotic blocks* of the Alps[5] and in the Tertiary of Nevada.[6] In the Alps the effects are seen mainly in the Flysch, a succession of rapidly alternating marine micaceous shales and calcareous sandstones. In the so-called Wildflysch, conglomerates, breccias, large slabs of fossiliferous limestone and also of igneous rocks are included, some of which are hundreds of yards long. Formerly regarded as due simply to erosion of a rocky coast, it is now recognized that the exotic blocks represent the results of slumping and attendant turbidity-current deposition, some of which went on from the front of contemporaneously advancing nappes.

In Venezuela prominent peaks are known as 'morros' some of which are formed of Cretaceous and Tertiary limestone massifs lying among soft shales. Some such are reefs stratigraphically *in situ* (Fig. III–14), but others are blocks detached by submarine slides. Near Bucarito, slabs of Cretaceous limestone more than 100 metres thick and a kilometre long are incorporated in slipped masses and now lie among Paleocene trough sediments. There are indications that sliding occurred repeatedly. Associated boulder beds have been spread out by turbidity currents for 30 km from the source. Very similar

[1] J. C. Brindley and W. D. Gill, Summer Field Meeting in Southern Ireland, 1957, *Proc. Geol. Assoc.*, Vol. 69, 1958, pp. 244–61; E. G. Williams, A. L. Guber, and A. M. Johnson, 'Rotational Slumping and the Recognition of Disconformities', *Journ. Geol.*, Vol. 73, 1965, pp. 534–47.

[2] M. G. Laird, 'Rotational Slumps and Slump Scars in Silurian Rocks, Western Ireland', *Sedimentology*, Vol. 10, 1968, pp. 111–20.

[3] N. D. Newell et al., *The Permian Reef Complex of the Gaudalupe Mountains Region, Texas and New Mexico*, San Francisco, 1953.

[4] S. M. K. Henderson, 'Ordovician Submarine Disturbances in the Girvan District', *Trans. Roy. Soc. Edinburgh*, Vol. 58, Pt 2, 1935, pp. 487–509.

[5] O. Renz, R. Lakeman, and E. van der Meulen, 'Submarine Sliding in Western Venezuela', *Bull. Amer. Assoc. Petrol. Geol.*, Vol. 39, 1955, pp. 2053–67. M. Lugeon, 'Sur l'origine des blocs exotiques du Flysch préalpin', *Eclog. Geol. Helvet.*, Vol. 14, 1916, pp. 217–21. *Idem*, 'Hommage à Aug. Buxtorf et digression sur la nappe de Morcles', *Verh. Naturf. Ges. Basel*, Vol. 58, 1947, pp. 108–31.

[6] E. M. Moores, 'Mio-Pliocene Sediments, Gravity Slides, and Their Tectonic Significance, East-Central Nevada,' *Journ. Geol.*, Vol. 76, 1968, pp. 88–98.

FIG. III–14. OBLIQUE AIR PHOTOGRAPH OF THE MORROS DE SAN JUAN, STATE OF
GUÁRICO, VENEZUELA

The morros are Paleocene reef limestones interfingering with the Guárico formation. (*Photo by
courtesy Dr G. Heyl*)

phenomena are also reported from Trinidad.[1] The view has also been
expressed that these morros are tectonically disrupted masses that have been
forced into anomalous positions by squeezing among the surrounding shales,
as a pip may be squeezed between the fingers.[2]

A distinguishing feature of most slump-breccias is the presence of only a
few rock-types among the brecciated fragments, these representing disrupted
beds (Fig. II–9). If the disrupted beds were coherent but only partially
consolidated when the slump occurred, the fragments will show the effects of
squeezing or stretching due to subsequent pressures and movements in the
surrounding detritus (Fig. III–15). Disrupted fragments of shale or coherent
sands in slump-masses tend to become rounded, forming conglomerates
which, like the breccias above-mentioned, are recognizable as *autoclastic*
partly because of their uniform lithology, and partly from evidence of the

[1] H. G. Kugler, 'Jurassic to Recent Sedimentary Environments in Trinidad', *Bull. Assoc.
Suisse Géol. et Eng. Pétrol.*, Vol. 20, 1953, pp. 27–60; idem, 'Trinidad' in *Handbook of South
American Geology, Geol. Soc. Amer.*, Mem. 65, 1956.

[2] J. V. Harrison, personal communication. See also W. H. Bucher, 'Geologic Structure and
Orogenic History of Venezuela': *Geol. Soc. Amer.*, Mem. No. 49, 1952, pp. 53–5, for an
earlier interpretation.

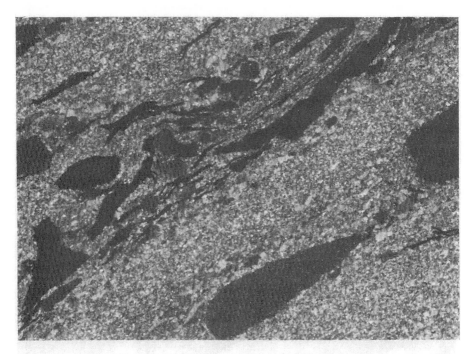

FIG. III–15. BRECCIA WITH SHALE FRAGMENTS INCORPORATED IN FELSPATHIC
SANDSTONE, JURASSIC, WONTHAGGI, VICTORIA

Note the indentations and projections at the edges of shale fragments (left-hand side) and the
evidence for stretching and tearing of the thinner lumps. Natural size.

FIG. III–16. PINCHED PEBBLE

Sandstone pebble from slump-conglomerate, Lilydale, Victoria,
showing shear fractures and 'pinching' indicating deformation
in the soft-rock condition. Natural size.

FIG. III–17. SLUMP FOLDING AND SLIDING

A. Slump-folds in Upper Carboniferous glacial varves, Seaham, New South Wales. Natural size.

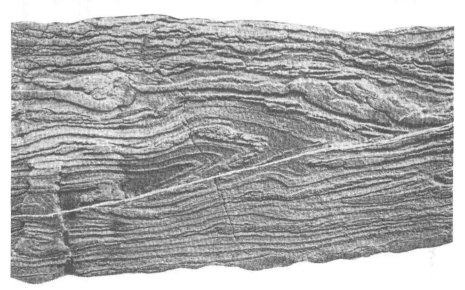

B. Slide-plane and associated folding, Silurian sandstones and mudstones, Seymour, Victoria. $\times \frac{1}{2}$.

softness of the 'pebbles' at the time of deposition, shown by shearing, stretching, and indentation during the final stages of movement of the mass (Fig. III–16).

In the example described by Heim from Lake Zug, the slumped mass spread out so far that it must have moved as a mud-flow or slurry, pointing to the complete disintegration of the deltaic deposits with consequent loss of original structures such as bedding. Although such mud-flows may incorporate fragments of brecciated beds, it is not to be expected that they will exhibit

FIG.   III–18.   RECUMBENT   FOLDS   FORMED   IN   SUBAQUEOUS   SLUMP-MASS,
CARBONIFEROUS
Fisher Street Bay, Co. Clare, Ireland. (*Photo: W. D. Gill*)

folding, but rather that resedimentation would produce new bedding structures. Numerous examples of intra-formational folding and thrusting to be
found in sedimentary rocks imply, in fact, less severe disruption and perhaps
shorter distances of translation of the slump-masses. Highly plastic beds,
especially such as entrap much connate water and contain clay materials, may
undergo limited flowage or gliding without actual disruption or tearing away
of the sliding mass, due to a variety of causes including tilting of the basin,
unequal loading due to sporadic deposition of later beds, or inequalities in the
compaction and settlement of earlier deposits. The part towards which gliding
goes on is thickened and folds are formed with or without curved thrust
planes (Fig. III–17). The hinder part is on the contrary stretched and thinned,
as may be seen by the breaking or *pull-apart* of slightly harder beds included in
the gliding mass. In examples of this type the extent of movement is limited to
some few feet, or even inches. Even in true slump-masses that have undergone
considerable translation, however, internal cohesion may be retained despite
the readjustments that are required to form large and numerous folds, such as
the recumbent overfolds in the Carboniferous of Co. Clare (Fig. III–18).[1]

It must be pointed out that many supposed examples of disturbed bedding

[1] J. C. Brindley and W. D. Gill, Summer Field Meeting in Southern Ireland, 1957, *Proc.
Geol. Assoc.*, Vol. 69, 1958, pp. 244–61.

interpreted as the results of slumping (slip-bedding) may now be subject to reinterpretation as convolute-bedding, basal deformation due to load-cast, and similar primary or near-primary features of sediments, especially in beds deposited by turbidity currents. A connection with slumping may, nevertheless, be hypothecated in relation to the origin of the turbidity currents, many of which are believed to derive from slumping at the edge of the continental shelf. Because of the thixotropy of clays, disturbances may cause a slump-mass to change in part into the fluid condition, when it flows rather than slides, and eventually it may disperse and intermingle with the surrounding water, transforming to a turbidity current which is capable of carrying along the brecciated or rounded fragments of beds that do not disperse but remain coherent (Figs. II−9; III−15). Under such conditions too, thin layers of sand may be torn apart and the fragments rolled into sand-balls (*balling-up* of sandstones).[1] Thus in many cases it seems that graded beds deposited from turbidity currents originate from residimentation, which accounts for the occurrence of shallow-water types of sediment and fossils, intermingled with deeper-water sediments.

A thixotropic mixture of clay with water changes from the gel condition to a fluid (sol) on agitation, and reverts to a gel on standing. Tremors due to earthquakes may perhaps cause the gel–sol change to occur.[2] Again, a slump-mass advancing into soft sediments may churn them up and originate a turbidity current.

An example of the possible association of an earthquake, slump, and turbidity current was afforded in the Grand Banks in 1929.[3] All the submarine cables lying down-slope from the epicentre of the earthquake were broken in succession from north to south, and geophysical evidence and sediment cores since obtained suggest that the earthquake activated submarine slumping. Some slump material appears to have remained intact, but part may have been transformed into a turbidity current which, if it caused the cable breaks, would have a velocity of about 30 km per hour.[4] The slope over which the current spread after leaving the continental slope itself is only one-thirtieth of a degree. Similar events have also been reported off the coast of Algeria,[5] but verification of the postulated effects is required.

[1] Ph. H. Kuenen, 'Slumping in the Carboniferous Rocks of Pembrokeshire', *Quart. Journ. Geol. Soc. Lond.*, Vol. 104, 1948, pp. 365–85.

[2] Clays containing montmorillonite (i.e. bentonitic clays) are notably thixotropic, but most natural clays and shales exhibit the property in some degree. See P. G. H. Boswell, 'The Thixotropy of Certain Sedimentary Rocks', *Sci. Prog.*, Vol. 36, No. 143, 1948, pp. 412–22.

[3] B. C. Heezen and M. Ewing. 'Turbidity Currents and Submarine Slumps, and the 1929 Grand Banks Earthquake', *Amer. Journ. Sci.*, Vol. 250, 1952, pp. 849–73; Ph. H. Kuenen, 'Estimated Size of the Grand Banks Turbidity Current', ibid., pp. 874–84.

[4] F. P. Shepard, 'Deep-Sea Sands,' *Int. Geol. Congr. 21st Session Norden*, Pt XXIII, pp. 26–43, Copenhagen, 1960.

[5] B. C. Heezen and M. Ewing, 'Orléansville Earthquake and Turbidity Currents', *Bull. Amer. Assoc. Petrol. Geol.*, Vol. 39, 1955, pp. 2505–14.

Intraformational folding and faulting due to slumping may simulate tectonic drag-folding and shearing, and distinguishing criteria then require careful evaluation in relation to the geometry of the folds, and the probable directions of slumping at the time of deposition,[1] inferred from palaeo-geographic considerations.

Effects in mobile beds

Salt tectonics – Because of their ready plastic yielding and low specific gravity (Sylvine 1·9, Halite 2·2 – note however, Anhydrite, with S.G. 2·9 equal to that of basic rocks) salt deposits are peculiarly liable to deformation by gravitative forces. Salt-domes are discussed elsewhere (Chapter IX), but it may be noted here that the uprise of salt-masses through surrounding heavier sediments is believed to be due mainly to hydrostatic forces arising from the greater gravitative pull on the surrounding rocks. Our concern now is to remark on the surficial and near-surface structures of salt deposits, induced by hydration and by free flowage under gravity. Anhydrite beds affected by ground-water are changed to gypsum, and the increased volume and high plasticity of this mineral give rise to complex folding and minor thrusting over large tracts of country, resembling and often confused with slump-structures. A notable example is the Lower Fars of Persia. Triassic gypsum, anhydrite, and salt-beds in the internal zones of the French Alps are excess-ively thinned or thickened, in places giving chaotic mixtures of gypsum and *cargneules*,[2] while tectonically displaced masses of other rocks may virtually float in the plastic gypsum as *Klippen*. When combined with super-ficial sliding, breakaways from plateaux, and the like, the above-described effects make structural mapping a matter of great difficulty.

'Wants' and 'stone-rolls' in coal – Coal-seams are found in places to be reduced in thickness and locally disturbed by bulges of shale from the floor, which appear to force their way into the coal and are known in Australia as *stone-rolls*. The term 'want' or 'nip' refers in general to places where a seam is locally thinned, with rock in place of coal; some wants are due to the filling with sand of scour-channels in the seam, but many result from squeezing of the coal by shale bulges either from below, or, less commonly, from above.

[1] See especially the following accounts of the Silurian rocks of Wales: O. T. Jones, 'On the Sliding or Slumping of Submarine Sediments in Denbighshire, North Wales, during the Ludlow Period', *Quart. Journ. Geol. Soc. Lond.*, Vol. 93, 1937, pp. 241–83; idem, 'The Geology of the Colwyn Bay District: A Study of Submarine Slumping during the Salopian Period', ibid., Vol. 95, 1940, pp. 335–82. See also M. R. Gregory, 'Sedimentary Features and Penecontemporaneous Slumping in the Waitemata Group, Whangaparaoa Peninsula, North Auckland, New Zealand', *New Zealand Journ. Geol. Geophys.*, Vol. 12, 1969, pp. 248–82.

[2] *Cargneules* are somewhat cavernous, calcareous rocks, probably originally dolomite and gypsum, in which the dolomite is dissolved away and the gypsum replaced by calcite. See M. Gignoux, *Stratigraphic Geology*, San Francisco, 1955, 'The Alpine Triassic', pp. 280–304 and note 15.

Woolnough[1] ascribed stone-rolls to swelling of the shale on hydration, but their considerable linear horizontal extension suggests some other origin, as yet not clear, but perhaps associated with minor tectonic disturbances in part occurring during deposition, which leads to truncation of structures in the stone-roll by the overlying rocks (Fig. III—19A), and partly after deposition,

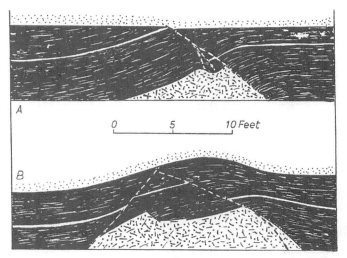

FIG. III–19. SECTIONS 90 FT APART ALONG THE SAME STONE ROLL, WONGAWILLI COLLIERY, NEW SOUTH WALES
(*By courtesy Australian Iron & Steel Ltd. and B. L. Taylor*)

as in places where the roof-rock is also involved in the deformations (Fig. III—19B). Flowage of shale is clearly indicated by the disturbed bedding, and minor faults are present on the flanks of the bulges, which are polished and slickensided, and also in the coal near by.[2]

Effects of volume changes

Enterolithic structures — Very regular close folding in some examples of intraformational contortion in salt deposits, especially in gypsum, are known as *enterolithic* from their resemblance to intestinal convolutions. Examples such as that shown in Fig. III—20 are thought to have resulted from volume changes, especially an increase in volume on hydration as with anhydrite yielding gypsum.

Expansion folds and thrusts — Superficial folds and thrusts due to the expansion of clay and shale on hydration during weathering are wide-spread in flat-lying Permian and Triassic beds in New South Wales, but they do not exhibit

[1] W. G. Woolnough, 'Pseudo-Tectonic Structures', *Bull. Amer. Assoc. Petrol. Geol.*, Vol. 17, 1933, pp. 1098–106.
[2] A. B. Edwards *et al.*, 'The Geology of the Wonthaggi Coalfield, Victoria', *Proc. Aust. Inst. Min. Met.*, N.S., No. 134, 1944, pp. 1–54.

the chaotic effects seen in beds containing salt or gypsum. Excellent examples have been exposed in recent engineering works (Fig. III – 21).

Swelling of clay soils – The volume changes that occur in the wetting or drying of rock and soil containing clay are often considerable. In heavy clay soils, lateral pressures developed after heavy rain or flooding are such that fence-posts may be forced out of the ground, and buildings founded on posts badly damaged. Likewise the swelling of clays causes a rise of ground-level, up to 9 in. having been measured in New South Wales.[1]

FIG. III–20. ENTEROLITHIC STRUCTURE IN GYPSUM, HILLBOROUGH, NOVA SCOTIA

(From Twenhofel, *Treatise on Sedimentation*)

Jointing due to weathering – Naturally, as all rocks are affected by weathering either directly, or through the influence of surrounding rocks, there is much differential movement in the weathered zone, which results in jointing. The various types are more fully discussed in Chapter IV, but it may be pointed out in the present context that the examination of most mine-dumps readily demonstrates the absence, in the deeper rocks, of much of the jointing that is common at the surface.

In *spheroidal weathering*, boulders develop circumferential cracks as a

[1] See also the figures quoted for an example in England: 'House Foundations on Shrinkable Clays', *Watford Building Research Station*, Digest No. 3, 1949 (Anon.).

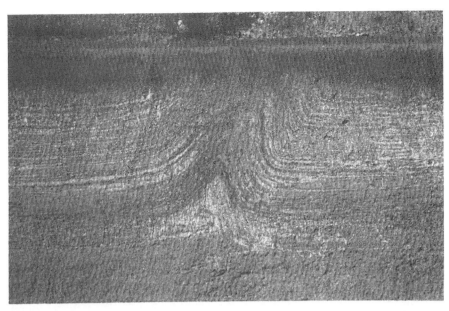

FIG. III–21. EXPANSION FOLD, DUE TO HYDRATION OF SHALES ON WEATHERING AND UNLOADING BY EROSION

Wianamatta Shale, Triassic, near Warragamba, New South Wales. (*Photo by courtesy D. G. Moye*)

FIG. III–22. SPHEROIDAL WEATHERING OF GRANITE BOULDER IN COLLUVIUM

Snowy Mountains area, New South Wales (*Photo: E. S. H.*)

result of expansion of the outer layers due to the growth of hydrated minerals during chemical weathering. In basalts the effect may extend deeply into boulders, but in granites, which are much less porous, the effect is limited to the outer layers (Fig. III–22), and in granites at the surface these flake off forming *exfoliation sheets*.

Structures in frozen ground – Many minor structures are formed in ground liable to freezing in regions of severe climate. Cryoturbate involution of clay and sand, stone polygons, stone stripes, and the like are, however, so rare except in superficial Pleistocene and Recent deposits and soils that discussion is not necessary here [1] except to note the possible occurrences of examples in older formations, and to suggest careful examination to avoid confusion with structures of different origin.

Superficial structures related to topography

Camber; valley bulges – In the Northampton Ironstone Field of England a network of valleys appears to follow synclinal troughs in the resistant flat-bedded rocks capping plateaux. Previously regarded as an example of rivers following structural troughs, it has been demonstrated by Hollingworth and Taylor that the apparent synclines, and other related structures, actually formed as a result of the erosion of the valleys, which follow a tectonic network probably along major joints. [2]

The geological succession includes Upper Liassic clays overlain successively by the Northampton Sand, the Estuarine Bed, and the Great Oolite (Fig. III–23). When erosion reaches the clays, these are squeezed out into the valleys by the weight of the nearby superincumbent rocks, or again a series of small faults and gashes may be produced in overlying hard rocks by outward flowage of an underlying clay. The faulting produces *dip and fault* structures: the gashes become filled with detritus and are called *gulls*, the hard rocks of the plateau-edges turning down towards the valleys as the clay flows out, forming a *camber*. In places the clay bulges beneath the valleys, with complex minor folds due to flowage in these *valley bulges*, and upturning of the overlying beds on the flanks simulating anticlinal limbs.

Similar structural relationships are to be sought where geological conditions analogous to those in the Northampton Ironstone area exist, and indeed valley bulges are closely allied to diapiric injections of mobile beds, which are more fully discussed in a latter section (Chapter IX).

On a much smaller scale, but having important implications in engineering

[1] See the excellent account by K. Bryan, 'Cryopedology', *Amer. Journ. Sci.*, Vol. 244, 1946, pp. 622–42.

[2] S. E. Hollingworth, J. H. Taylor, and G. A. Kellaway, 'Large-scale Superficial Structures in the Northampton Ironstone Field', *Quart. Journ. Geol. Soc. Lond.*, Vol. 100, 1944, pp. 1–44. S. E. Hollingworth and J. H. Taylor, 'An Outline of the Geology of the Kettering District', *Proc. Geol. Assoc. Lond.*, Vol. 57, 1946, pp. 204–33.

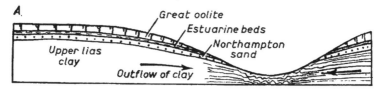

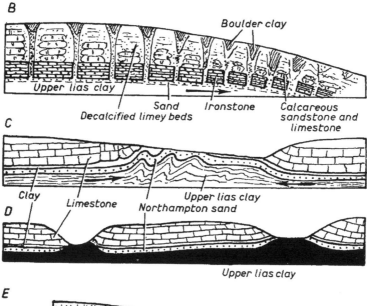

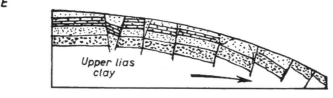

FIG. III-23. LARGE-SCALE SUPERFICIAL STRUCTURES IN
NORTHAMPTONSHIRE

A. Cambering; B. Gulls; C. and D. Valley Bulges in Liassic Clays; E. Dip and
Fault structures. (*After Hollingworth* et al., *1944*)

geology and one's attitude of mind towards movements in rocks, are the
effects of man's activities in excavation, dam construction, and the like.

Where, at a dam-site, considerable volumes of rock are removed from the
walls of a valley to afford sound abutments, the reduction of lateral pressure
on the remaining rock results in measurable elastic expansion towards the
valley, and also in sliding of sandstone on shale, resembling the movements
due to natural causes discussed above.[1] On the other hand, the load of a great

[1] A. Stuckey, 'Foundation Problems in Water Storage Dams', *Bull. Tech. Suisse Romande*,
Vol. 80, 1954, pp. 317 ff., 329 ff.

reservoir depresses the country beneath and adjacent to the basin, the maximum subsidence measured at the Boulder Dam (Lake Mead) being about 7 in.[1]

The flowage of rocks in mines and bore-holes (wells) demonstrates the ready deformability of normal sedimentary rocks such as shales, argillaceous sandstones, and greywackes on removal of lateral support or superincumbent load. The collapse of the roof of a worked-out coal-seam by mine subsidence is readily understandable, but, in addition, the effect of upward 'heaving' or 'creep' of the floor is very common, leading even to the complete closure of the space formerly occupied by the coal (Fig. III–24). Similarly in drilling, great difficulty may be experienced in soft shales due to the continual flow into the well. This is countered by the use of heavy drilling muds which exert lateral pressure on the rock, preventing its movement into the well.

Hill-creep – On steep hillsides, soil and weathered rock creep slowly down-slope towards valleys, this slow flowage of the mantle of weathering products

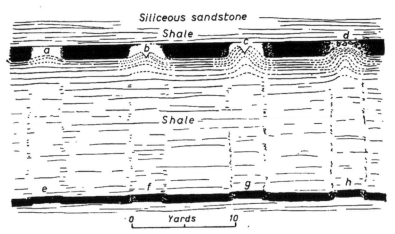

FIG. III–24. SECTION AT WALLSEND, NEWCASTLE, ENGLAND, SHOWING 'CREEPS'

Length of section 174 ft. The upper, or Main coal, now worked out, was 630 ft below ground surface, and 6 ft 6 in. thick. (*After J. Buddle, from Lyell, Elements of Geology, 6th edn, 1865*)

being an important factor in the wasting of hill-slopes during dissection of the countryside. The structural effects are significant chiefly in mapping, for as a result of creep the ends of strata turn more or less strongly downhill in the weathered zone, so that dip-readings may be quite unreliable (Fig. III–25). The depth affected ranges up to 20 ft or more in soft rocks.

Landslides – Landslides are essentially small faults formed in inadequately supported rock-masses at the earth's surface, under the action of gravity. On

[1] J. M. Raphael, 'Crustal Disturbance in the Lake Mead Area', *U.S. Dept. of Interior*, Eng. Monog. No. 21, 1954.

great escarpments the size of blocks breaking away under gravity may be so large that they may themselves be regarded as fault-blocks rather than as landslide-masses, and the collapse of the unsupported edges of upthrusts is a corollary of the ramp-valley hypothesis for the origin of major fault-troughs such as those of East Africa (see Chapter VII). Commonly, however, the structural geologist is concerned with landslides in engineering geology either

FIG. III–25. HILL-CREEP IN SILURIAN MUDSTONES AND SANDSTONES, STUDLEY PARK, VICTORIA
(*Photo: E.S.H.*)

in the prediction of the probability of slides occurring under the particular structural–geological conditions at a site, or in suggesting ways, related rather to geological factors than to engineering practice, whereby slides already initiated may be stabilized.

The fundamental geometry of sliding earth-masses is best seen in homogeneous rocks of moderate strength, in which failure is governed by a stress-distribution that is not greatly affected by differences in lithology or by the presence of weak structural surfaces such as bedding, joints, or faults. In general, failure occurs on a curved slip-surface which is nearly vertical at the head of the slide and nearly horizontal at its toe (Fig. III–26). As the slide moves beyond the intersection of the slip-surface with the ground at the toe, the moving mass is disrupted and if sufficiently mobile it may take on the features of a rock-flow. This usually requires the presence of water mixed with the earth and rock.

This basic geometry of landslides is affected by pre-existing geological structures. Thus if bedding planes dip down-slope at a lesser angle than the slip-surface, a slide may take place on the bedding, before an excavation has reached such a depth or steepness that sliding would be expected. Similarly, master joints or faults inclined towards an excavation render the rock less stable. Discussion of the many varieties of rock-failure by sliding, flow, and the like is, however, not intended here, but further information may be sought from standard works.[1]

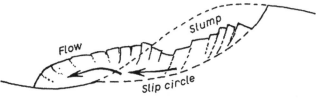

FIG. III-26. LANDSLIDE WITH MOVEMENT ON A CURVED
RUPTURE SURFACE (SLIP-CIRCLE)

Gravitational tectonics — The extent to which folding and faulting in zones of strong rock-deformation may be a result of the sliding under gravity of very great rock-masses on a scale matching mountain-building has of recent years become a matter of lively discussion. The topic is further treated in Chapter XI, but it may be said that *gravitational tectonics*, as the production of major structures by gravitational forces has been termed, is now a well-established concept and one that may well be applicable, as is claimed, to a wide range of structures in particular regions, without discarding the notions of lateral compression or active vertical forces in orogeny.

Deformation by glaciers

Continental ice-sheets, which at one or other period of geological time have existed in all continents, attain thicknesses of the order of 10,000 ft, and in soft rocks such as shale, soft sandstone, or coal they produce important deformation by drag beneath the ice, or by push in front of an advancing ice-sheet. The effects are chiefly seen in folding and thrusting of the under-mass, with thrust planes curving in lystric surfaces. Till and out-wash are affected during readvance of the ice, but certain of the complex structures in till are

[1] C. F. S. Sharpe, *Landslides and Related Phenomena*, Columbia University Press, 1938. A. Heim, *Über Bergstürze*, Zürich, 1882. L. Bendel, *Ingenieur geologie*, Teil II, Vienna, 1948.

inherited from the englacial debris, which was frozen and subject to shearing movements within the glacier.[1]

The folding of Tertiary brown coals by a Pleistocene ice-sheet advancing from Scandinavia over the North European Plain affords a fine example of glacial tectonics,[2] in which the pattern of the terminal moraines closely matches the arcuate folded zone (Fig. III–27). *Ice-thrust ridges* are topographic features resembling moraines (Ger. *Stauchmoränen*), which are in fact formed by the push of glacial lobes on soft rocks affected by permafrost.[3]

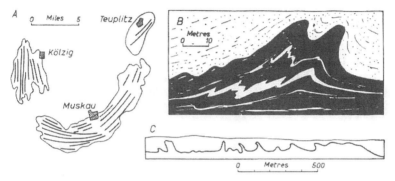

FIG. III–27. FOLDS DUE TO GLACIAL DRAG, SOUTH-EAST GERMANY

A. Brown coalfield of Muskau, showing fold axes in arcuate belt formerly occupied by a glacial lobe moving south. (*After Klein*)
B. Folding of Tertiary brown coal, sand, and clay, Borna-Gnandorf, due to glacial drag (*left to right*). (*After Etzold from Stutzer and Noé, 1940*)
C. Folding of Tertiary brown coal by glacial ice moving from right to left. Murker mine near Drebkau. (*After Russwurm from Stutzer and Noé, 1940*)

In glacial lakes, grounded ice-floes plough into soft sediments, causing severe local disturbance which is increased as the ice melts and the sides of the scours slump in. Similar effects are perhaps less frequent in marine beds because of the winds which normally blow strongly away from ice-caps, and because of the greater depths of water.

[1] G. Slater, 'Studies in Glacial Tectonics', *Proc. Geol. Assoc. Lond.*, Vol. 37, 1926, pp. 392–400; ibid., Vol. 38, 1927, pp. 157–216; idem, 'Structure of the Mud Buttes and Tit Hills in Alberta', *Bull. Geol. Soc. Amer.*, Vol. 38, 1927, pp. 721–30. J. Challinor, 'A Remarkable Example of Superficial Folding due to Glacial Drag, near Aberystwyth', *Geol. Mag.*, Vol. 84, 1947, pp. 270–2.

[2] G. Klein, *Die Deutsche Braunkohle-Industrie*, Bd 1, Tl. 1, Halle, 3rd edn, 1927 (see pp. 78 ff.); O. Stutzer and A. C. Noé, *Geology of Coal*, Chicago, 1940; D. de Waard, 'Direction of the Pleistocene Ice-Flow as determined by Measurements on Folds', *Geol. en Mijnb.*, N.S. Jhg. 14, Nr. 2, 1952, pp. 44–5.

[3] W. O. Kupsch, 'Ice-Thrust Ridges in Western Canada', *Journ. Geol.* Vol. 70, 1962, pp. 582–94.

CHAPTER IV

Physics of deformation

Most strongly folded rocks are, as now observed, hard, brittle solids which it would not be possible to bend by any simple means. It seems obvious, therefore, that at the time of folding they must have possessed very different physical properties, such that they could be sharply bent and flexed without rupturing – that is, they were plastic. The assumption might be shaken a little by the knowledge that slabs of marble such as tombstones and mantelpieces, supported only at their ends, have been observed to sag over a period of years, suggesting that very small forces acting for a very long time can cause permanent deformation of this brittle rock. Although one might retaliate that the marble had, in fact, been repeatedly heated by the sun or the fire it would remain true that the rock did yield under low stress and under no great severity of physical conditions as to temperature and pressure. The slow deformation or *creep* of rocks under low stress will be discussed later, but whatever weight may be attached to this as a factor in the formation of rock-structures, one may be quite certain that the yielding of a great many rocks took place under conditions such that the deformation was analogous to the plastic yielding either of crystalline materials such as metals, or of complexes such as clays. It is known, for instance, that the microscopic structures of crystalline schists are closely analogous to those of plastically deformed ductile metals. It is therefore admissible to draw analogies between the mechanics of deformation of such rocks with the deformation of metals, both as to the effects in individual crystals and for the crystal aggregate as a whole. Further, inferences may be made by reference to experimental data on rocks and on metals, as to what may have been the physical conditions at the time of deformation of rocks in nature. In experiments, however, it is not yet possible to reproduce exactly the physical and chemical conditions that hold in nature, the time factor, in particular, being vastly discrepant. For instance, an observed geological deformation might, given time, have been caused under less severe conditions of temperature and pressure than the experimental analogues would suggest.

    This chapter will discuss such topics, but it is well to point out before proceeding that the problem is not, in fact, to account for plastic yielding of a rock in the condition in which it is now observed, but rather of the original rock-matter at the time of deformation. A folded quartzite, for instance, may well have been a sandstone, and perhaps not even a hard sandstone, when it was folded, and have been changed to quartzite at some later time. The present condition of a rock is the result of a long sequence of events which for sediments includes deposition, compaction, diagenesis, the processes of deformation themselves, and all subsequent changes. Ideally, we have to account for the petrological and structural features formed at each stage in the rock's history. Rock-structures, moreover, include the effects both of plastic yielding as in folding, and of fracturing as in the formation of joints and faults.

Model experiments reproducing geological structures in miniature have been made by many investigators from the time when James Hall squeezed layers of clay between wooden blocks and produced folds, but recent experimentation of this type has been more soundly based on the theory of scale models than was possible in earlier years (see p. 136). The aim is generally to gain an insight into regional stresses and the displacement of crustal blocks involved in forming major structures.

There remain, however, many factors known to affect rock-deformation which it is very difficult, or even impossible, to evaluate in physical terms on present knowledge. Firstly, most rocks are heterogeneous as to their constitution and texture, including as they do many minerals in a variety of grain-shapes, sizes, and mutual arrangements, and these minerals are chiefly silicates about which not much is known from laboratory investigations, as to their deformability. Secondly, the prevalence of chemical changes during recrystallization of rocks, and the influence of mobile components, chiefly water, introduce effects that are rarely matched in laboratory experiments.

The information available from direct observation of rocks is, therefore, of considerable interest to all students of the physics of solids.

The materials of the crust

The materials of the crust include solids, liquids, and gases. The chief *solid* constituents are the crystalline mineral grains of rocks, among which quartz, silicate minerals, calcite, and dolomite are the commonest. In addition, igneous glass and amorphous minerals such as opaline silica are present in some rocks, and the skeletal parts of organisms preserved as fossils must also be considered as components of rocks.

Special consideration should be given to the igneous glasses and to the peat-coal series. Despite the great economic importance of coal, very little is known about it as a deformable rock, and the extensive chemical reconstitution and loss of substance during coalification introduces complicating factors. As to glassy rocks, no particular attention has been directed to their mode of yielding or fracturing, although thick and extensive flows of obsidian or glassy rhyolite are important members of many strongly deformed rock-groups, especially in the Pre-Cambrian and the Palaeozoic.

Liquids are significant because flow-structures are formed in igneous rocks during the fluid stage and again because water is present in the voids of almost all rocks and has an important effect on their physical properties and on recrystallization and reconstitution. Oil is generally considered in relation to the retention or migration of the petroleum itself rather than from the point of view of its possible effects on rock-structures (see, however, elastic expansion and settlement of rocks, p. 57). *Gases* include air, volcanic exhalations, and the various natural gases, but apart from their effects in volcanic phenomena (Chapter XII) they will not be further treated herein.

The terms solid, liquid, and gas, first used to distinguish substances according to their physical properties, are now recognized as describing also the states of molecular aggregation of matter, so that an element or compound may exist in the solid, liquid, or gaseous state. The solid state is typified by the crystalline condition, and it is recognized that glass and pitch are not typical solids, although in some of their physical properties they may resemble them. Again, the physical properties of polycrystalline aggregates are not simply those of single crystals, but are greatly affected by grain-boundary effects and further by the presence of fluid in the voids between crystal grains. This is very clear for clays, for which the water-content of a mixture is the factor most affecting its physical properties. It is equally true, however, of sand-size particles.

Granular materials containing interstitial fluid are *two-phase systems*. The individual grains are solid; the mixture may have the physical properties of a solid or of a liquid, or some special properties not typical of either, according to the strength of adherence of the grains, and the content of fluid. Among rocks, recently deposited sediments in the so-called *soft-rock stage* are examples of two-phase systems, but so also are igneous rocks at any stage after crystallization has begun and before the residual magma is completely crystallized. At an early stage of crystallization the liquid properties predominate, as in porphyritic lavas and dykes; but in later stages when a crystal mesh has developed the mass may fracture like a solid. In metamorphism, too, a fluid phase is often present, affecting both recrystallization and the formation of structures such as cleavage, augen, and pressure fringes, which are discussed in later Chapters.

FORCE, STRESS, AND STRAIN

Force

A *force* may be said to be a push or a pull.[1] As such it requires to be specified both as to magnitude and direction, so that force may be treated as a *vector*

[1] H. Margenau, W. W. Watson, and C. G. Montgomery, *Physics Principles and Applications*, New York, 2nd edn, 1953.

quantity, and the *composition* of a number of forces to find a resultant force, or the *resolution* of a force into components acting in different directions, may be carried out by vector analysis according to the parallelogram and polygon rules (Fig. IV–1). In such analysis the forces are conceived to act on a 'particle', that is, on some indefinitely small portion of a body, nothing being implied as to the effects *within* the particle, which is virtually a point. The resultant of two equal and opposite forces acting on the same particle is the vector sum, which is zero, and the particle will therefore remain at rest. However, two equal and opposite forces acting on a real object which is not perfectly rigid have the important effect of shortening or extending the object

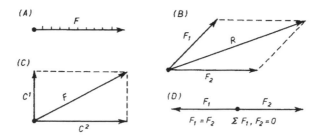

FIG. IV–1. FORCE-VECTORS

A. Vector representing a force F of magnitude 10 force-units, acting on a particle P in the direction of the line in the sense of the arrow.

B. Composition of forces F_1 and F_2 by the parallelogram rule, giving the resultant force R.

C. Resolution of a force F by the parallelogram rule into two components C_1 and C_2 acting along chosen directions.

D. Vector sum of equal forces F_1 and F_2 acting on the particle P in opposite senses in the same straight line is zero.

according to whether the forces are directed towards or away from each other. It must be realized that a statement that 'the vector sum of the forces is zero' does not imply that no force is acting, but rather that forces *are* acting, which are balanced and produce no motion of a particle.[1]

A distinction may be drawn between *surface forces*, which act at the contact of two bodies tending to move relatively to each other, and *body forces*, such as gravity. In general, body forces are proportional to the mass of the body on which they act, and are determined as to magnitude and direction by the position of the body in a *field of force*. Gravity, because it is all-pervasive and continuously acting, is one of the most important body forces concerned in geological phenomena. Although it is the earth's field that dominates, there are also earth tides (as well as oceanic and atmospheric

---

[1] The symbols for the forces in the vector equation should be printed in **Clarendon type** to indicate that they represent 'force vectors' rather than simply 'forces'. See K. E. Bullen, *An Introduction to the Theory of Dynamics*, London, 1948, p. 92.

tides) which are due to gravitational pulls of the moon and the sun on the body of the earth, and which are possibly involved in certain tectonic effects.

Units of force are determined on the basis of Newton's Second Law, so that (in absolute units) unit force produces unit acceleration in unit mass; but in engineering the unit usually adopted is the *pound force* which is the practical unit in the English system – that is, the gravitational pull on a mass of one pound at the earth's surface (1 pound force = 1 pound mass $\times g$; $g = 32$ ft/sec.²).

Stress and strain [1]

Stress – When forces are applied to the external surface of a body, they set up internal forces within the body, which is then said to be in a state of *stress*, with a balanced internal action and reaction between adjacent parts. Stress is

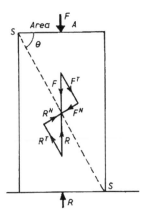

FIG. IV–2. STRESSES ON A SURFACE SS DUE TO AN APPLIED FORCE F ACTING ON A BLOCK OF MATERIAL OF END AREA A.

R = resistance, equal and opposite to F. For further details, see text.

specified as the *intensity* of these internal forces, that is, the force acting on unit area in any chosen direction, the unit area enclosing the point at which the stress is to be considered.[2]

It will be seen in Fig. IV–2 that the resolutes of a force F on a plane SS within a body, inclined to the force, are F^N normal to the plane and F^T parallel to it. The *normal stress* or *traction* across the plane SS is the intensity of F^N;

---

[1] The following works are recommended for additional reading: W. H. Bucher, 'The Mechanical Interpretation of Joints, Part II', *Journ. Geo.*, Vol. 29, 1921, pp. 1–28; M. K. Hubbert, 'Mechanical Basis for Certain Familiar Geologic Structures', *Bull. Geol. Soc. Amer.*, Vol. 62, 1951, pp. 355–72; W. Hafner, 'Stress Distributions and Faulting', ibid., pp. 373–98; C. O. Swanson, 'Notes on Stress, Strain, and Joints', *Journ. Geol.*, Vol. 35, 1927, pp. 193–223. J. C. Jaeger, *Elasticity, Fracture and Flow*, London, 1956.

[2] This is so if the stress is homogeneously distributed within the body, but because in the general case the stress is in fact variable from point to point, it is necessary to specify stress as the limiting value of the ratio of force (δF) to area (δA) as the area is indefinitely reduced $\left(\text{stress} = \lim \dfrac{\delta F}{\delta A} \right)$. The problem is similar to determining the scale at a point on a map in which the scale is variable from point to point.

the intensity of F^T is the *tangential* or *shear stress* in the plane SS. Normal stresses are usually represented by the symbol σ and tangential stresses by τ. Considering the magnitude of the stress at a point in SS, it will be seen firstly that the force F has already been distributed over a greater area than the end section of the body, $\left(SS = \dfrac{A}{\cos \theta} \right)$; and again, $F^N = F \cos \theta$. Thus the normal stress is given by

$$\sigma = \frac{F \cos \theta}{A/(\cos \theta)} = \frac{F}{A} \cos^2 \theta = \frac{F}{A} \left(\frac{1 + \cos 2\theta}{2} \right)$$

from trigonometry. Similarly the tangential stress is given by

$$\tau = \frac{F}{A} \cdot \sin \theta \cdot \cos \theta;$$

but

$$\sin \theta \cos \theta = \frac{\sin 2\theta}{2},$$

therefore

$$\tau = \frac{F}{A} \cdot \frac{\sin 2\theta}{2}.{}^{1}$$

It will be seen from this illustration that stress-analysis cannot be carried out by the use of vectors, as with forces. In mathematical treatment it requires the use of tensors, for which reason stress-components are referred to by some authors as components of the stress 'tensor'.

If a body is stationary under a system of forces, it is possible to choose at any point three mutually perpendicular planes, intersecting in the point and so orientated that the resultant stresses on the planes are wholly normal stresses or tractions. The three lines along which the planes intersect are known as the *principal axes of stress* and the stresses acting along them are the *principal stresses* at the point considered. In the general case the principal stresses are unequal, so we have maximum principal stress (σ_1), the mean principal stress (σ_2) and the minimum principal stress (σ_3) at the point considered. $\sigma_1 - \sigma_3$ is the *stress-difference*, which may be regarded as the agent causing deformation at the point.[2] Each principal axis of stress represents two equal and opposite stresses, which are compressive when directed towards each other and tensile when directed away[3] (Fig. IV–3). Surfaces in any direction other than that of

---

[1] Equations similar to these are of great use in the study of stressed bodies by means of Mohr circles (see pp. 139–42).

[2] The stress-difference is the algebraic difference between σ_1 and σ_3.

[3] In elastic theory it is usual to regard tensions as positive and compressions as negative, but this practice is generally reversed in dealing with plastic flow and fracture in geological and engineering work.

the three orthogonal planes mentioned will be subject to shearing stresses as well as to normal stresses, and the shear-stresses are a maximum in two (orthogonal) directions at 45° between the axes of maximum and minimum stress.

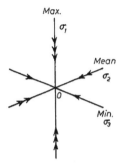

FIG. IV–3. PRINCIPAL STRESSES (COMPRESSIONAL) ON A CUBICAL ELEMENT ENCLOSING POINT O WITHIN A BODY

Stress distribution (fields of stress) — When surface forces are applied to a body, the resultant stresses within the body will in general vary in direction and intensity from point to point. Figure IV−4A shows the stress field in one plane, resulting from pressure applied to a small part of the surface of a large

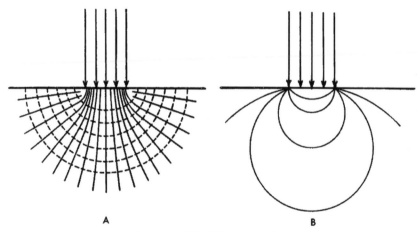

A   B

FIG. IV–4. STRESS TRAJECTORIES

A. Stress trajectories (isostatics) in a body subjected to pressure over a small area as shown by the arrows. Full lines, maximum stress (compressive); broken lines, minimum stress.

B. Distribution of constant maximum shear-stresses (isochromatics) under the same conditions of pressure as in A.

object. It may be intuitively understood that the full lines represent directions in which the compressive stress is distributed within the body, and these are indeed curves drawn to include the axes of maximum principal compressive stress at successive points. The broken curves have been drawn to include the axes of minimum principal stress, and together the two orthogonal families of curves constitute the *stress trajectories* (or isostatics) in the plane considered. Stresses in the third dimension are not dealt with. Other curves may be drawn to represent additional information, and those for the loci of points of equal maximum shear-stress (Fig. IV–4B) are significant in geology, especially in relation to faulting.

It happens that the distribution of maximum shear-stress is readily studied experimentally using the photoelastic effect, by which certain transparent isotropic materials become anistropic under strain. A pattern of coloured fringes is obtained when the strained material is viewed between crossed

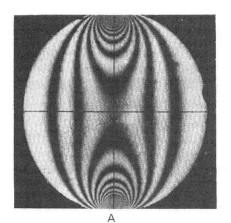
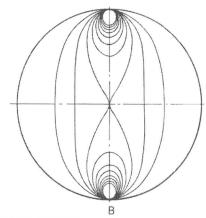

A   B

FIG. IV–5. PHOTOELASTIC EFFECT

A. Stress pattern (isochromatics = loci of equal maximum shear-stress) in a thin disc subject to compression along axis emerging at A. (*Photo by courtesy of Royal Melbourne Institute of Technology*)

B. Theoretical shear-stress pattern computed for same test. Compare Fig. IV–4B with the top of the specimen in Fig. 5.

polarizers using white light. Lines of equal maximum shear-stress then appear as bands or fringes of the same colour (isochromatics), or, in monochromatic light, as a series of light and dark bands. The close correspondence between such patterns and the calculated maximum shear-stress trajectories is demonstrated in Fig. IV–5. Photoelastic tests also yield additional information about stress fields, and they are susceptible to quantitative analysis, which has led to their increasing use especially in engineering applications of structural geology.[1]

Strain – Strain is the measure of the deformation of a body. It involves the displacement of the parts relative to each other, that is, a change in shape or in volume, or both. Changes in shape are called *distortion*, and in volume

[1] Useful reference works are A. J. Durelli, E. A. Phillips, and C. H. Tsao, *Introduction to the Theoretical and Experimental Analysis of Stress and Strain*, New York, 1958; M. M. Frocht, *Photoelasticity*, 2 vols, New York, 1941–8.

dilatation, which is either positive (in expansion) or negative (in shrinking). Strain may be measured by reference to the change in length of a line in the strained body (*extension*) and by the change in angle between two lines (*shear strain*). Strain results when a body is subjected to stress by external surface forces, by body forces, or by other causes such as changes in the heat-balance within the body. During the period of application of stress, stress and strain are in fact inseparable, and although we normally regard strain as caused by stress, stress-transmission takes place simultaneously with the propagation of strain in the body.

Most of the basic physical concepts of stress and strain are derived from the study of elastic deformations, that is, in general, small strains from which the body recovers and returns to its original configuration on removal of the stress. Especially with ductile or plastic materials, if the stress be continually increased, a point is reached at which the material suffers permanent deformation and is said to have *yielded*. On removal of stress there is then a small elastic recovery, but the permanent, plastic deformation remains.[1]

Homogeneity of strain

A body is said to be homogeneously strained if every portion of it suffers a strain of the same type, magnitude, and direction of displacement. With *homogeneous strains* an originally regular geometrical pattern is changed into

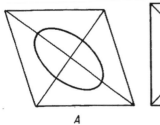 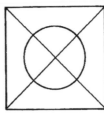 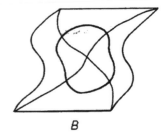

A   B

FIG. IV-6

A. Homogeneous strain of a square and circle.
B. Heterogeneous (inhomogeneous) strain of a square and circle.

another regular geometrical pattern in which similar parts remain similarly oriented throughout. In the general case, parallel straight lines remain straight and parallel, a square becomes a rhombus, and a circle an ellipse. If the strain is not the same in all the parts, it is said to be *inhomogeneous* (Fig. IV−6). However, many inhomogeneous strains may be considered as the

[1] Much confusion of thought would be obviated if the term *strain* were limited to elastic deformations, plastic yielding being referred to as permanent *deformation*. One important point is that plastic deformation produces inhomogeneities in the material, which immediately leads to great complexity in the internal stress–strain distribution.

summation of homogeneous strains affecting indefinitely small elements of the body, but deviating continuously in direction, magnitude, or type from element to adjacent element in the body. A fold such as is illustrated in Fig. IV−7 results from inhomogeneous deformation, but any one small sector shows virtually homogeneous strain.

On the other hand, the broad pattern of folds in plan and section, while

FIG. IV−7. BENDING OF A SLAB
The small circles are deformed approximately homogeneously to ellipses, but the total strain is inhomogeneous. (*After Sander*)

demonstrating inhomogeneous deformation, does exhibit a regular variation in the deformation which produces the fold-pattern, and this is, in fact, one of our main interests in geology.

The strain ellipsoid

Consider a spherical element of a homogeneous elastic body which is homogeneously strained below the elastic limit. The original sphere may remain spherical, merely changing its size by dilatation; or it may become an ellipsoid of revolution with two of its co-ordinate diameters remaining equal and the third becoming longer or shorter; or again, each diameter may change relative to the other two, giving a triaxial ellipsoid. The axes of the resulting *strain ellipsoid* are the *principal strain axes*, and are labelled respectively the maximum (A), mean (B), and minimum (C) strain axes (Fig. IV−8). In ideal elastic isotropic substances the principal axes of stress and the principal axes of strain are parallel at any point.

Many geological examples of strain approximate to *plane strain*, that is, all movements occur in parallel planes, the mean axis of the strain ellipsoid remaining constant in length. Discussion may then be simplified by considering the strain in these parallel planes only. They contain the greatest and least axes of strain, and the cross-section of the strain ellipsoid in this direction is called a *strain ellipse*. In experiments, ellipses may be formed from the deformation of circles impressed on a test specimen (Fig. IV−16).

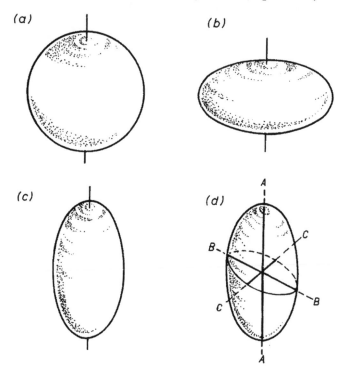

FIG. IV−8. THE STRAIN ELLIPSOID

(*a*) Original sphere; (*b*) Biaxial ellipsoid (oblate); (*c*) Biaxial ellipsoid (prolate); (*d*) Triaxial ellipsoid.

The strain ellipsoid has been used in structural geology in relation both to small-scale features such as cleavage and the deformation of pebbles, oöids, and fossils, and to the structural analysis of regions many square miles in extent, for which the pattern of faults, folds, and joints is known. In all such geological examples the direct relationship between the principal axes of strain and of stress which obtains in elastic strains affecting isotropic materials does not hold, since the rocks are in general strongly anisotropic and the strains large and permanent rather than small and elastic. Nevertheless, it is often possible to infer the likely stress directions from details of the deformation of objects whose original shape and attitude are known. On the other hand, the shape of inferred strain ellipsoids may indeed give a false impression as to the magnitude and direction of the stress, and this must be guarded against. Not only may the length of a 'stretched pebble' be far in excess of any possible elastic strain, but if there is any rolling action, the elongation may actually be parallel to the axis of rotation rather than normal to the applied pressure, as with rolling a ball of clay along a table-top.[1] Again, close examination may reveal that the shape of deformed pebbles and fossils is due rather to apparent than to actual compression or stretching, as discussed below (see p. 128), but where reasonably reliable data are available as to the geometry of deformation it is useful to attempt quantitative analysis and description, using the strain ellipsoid (or the strain ellipse if plane strain only is discussed),[2] or again the Mohr diagram for strain as expounded by Brace.[3]

Such quantitative studies afford a preliminary basis for inferences as to stress, but remembering that stress is an internal effect, a clear distinction must be drawn between the stress-distribution and the operative forces, which are what the geologist is seeking to infer.

Detailed study of local geology is required to permit the interpretation of the origin and causes of structures, and from the synthesis of such local studies, inferences as to the applied forces and their mode of application may ultimately be made on a regional basis. This is perhaps the ultimate aim of structural geology in the sphere of natural philosophy, and one that preserves the essential role of the field-geologist in geological science.

Rotational and irrotational (pure) strains

If an object be homogeneously deformed as in Fig. IV–9, in such a way that the principal axes of strain do not change their position in space during deformation, but merely alter in length, strain is *irrotational*, and is said to be a *pure strain*. Fig. IV–9, for example, illustrates *pure shear*, a strain in which the extension along one principal axis is so related to the shortening on the other that an inscribed rhombus becomes after deformation a congruent rhombus, the obtuse and acute angles being interchanged.

In what is known as *simple shear*, on the other hand, the deformation is such as might geometrically be accounted for by the sliding of infinitely close parallel planes, each being displaced an equal amount relative to neighbouring planes, and in the same sense. This may be illustrated by the deformation of a

[1] H. W. Fairbairn has discussed the possible effects of rotation: 'Elongation in Deformed Rocks', *Journ. Geol.*, Vol. 44, 1936, pp. 670–80. See also D. Flinn, 'A Tectonic Analysis of the Muness Phyllite, Block of Unst and Uyea, Shetland': *Geol. Mag.*, Vol. 89, 1952, pp. 263–72.

[2] J. H. Howard, 'The Use of Transformation Constants in Finite Homogeneous Strain Analysis,' *Amer. Journ. Sci.*, Vol. 266, 1968, pp. 497–506.

[3] W. F. Brace, 'Analysis of Large Two Dimensional Strain in Deformed Rocks', *Internat. Geol. Congr.*, Rept. 21st Session, Norden, 1960, Pt 18, pp. 261–9; idem, 'Mohr Construction in the Analysis of Large Geologic Strain', *Bull. Geol. Soc. Amer.*, Vol. 72, 1961, pp. 1059–80.

square into a rhomb, whereby the vertical sides are rotated through an angle, whose tangent is a measure of the shear-strain (Fig. IV−10). In a strain of this type the maximum and minimum axes of strain rotate as the deformation

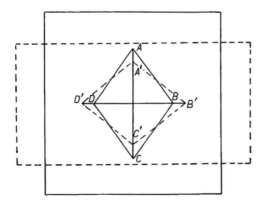

FIG. IV–9. PURE SHEAR (IRROTATIONAL STRAIN)

Note that the rhombus *ABCD* is transformed into a congruent rhombus *A'B'C'D'*, the acute and obtuse angles being interchanged. The strain axes are not rotated.

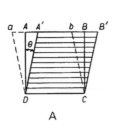

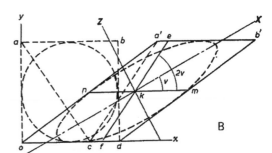

FIG. IV–10A. SIMPLE SHEAR-STRAIN

The square *ABCD*, sheared through θ, becomes the rhombus *A'B'C'D'*. There exists the congruent rhombus *abCD*, the corresponding angles being interchanged.

FIG. IV–10B. SIMPLE SHEAR AND THE STRAIN ELLIPSE

The major axis of the ellipse *X* is neither the diagonal nor the vertical diameter of the original circle. *nm* and *a'c* and lines parallel to them remain unchanged in length and are directions of maximum shear-strain (*ef* = *a'c*; *nm* = *ab*). (After *Gurevitch, Acad. Sci., U.S.S.R.*, No. 31, 1955)

proceeds, and the total strain therefore involves that pure strain which is represented by the relative lengths of the strain axes at any time, together with the necessary rotation of these axes (Fig. IV−10). The rotation of the strain axes is thus an essential part of the strain, which is therefore termed a *rotational strain*. An approximation to the conditions of simple shear may be produced experimentally by the application of a force tangential to the top of a thin strip fixed at its base, or by enclosing a thin sheet of a substance between friction-plates which are slid over each other under pressure, the essential being to prevent the body as a whole from rotating while applying the tangential forces. A convenient example is seen in the action of scissors or shears from which latter the word 'shear' as a technical term has been adopted.[1] 'Rigidity' implies the resistance of a body to shear, and the Modulus of Rigidity is the ratio of shear-stress to shear-strain in simple shear.

Since, in elastic deformation, a rotational strain is equivalent to an irrotational strain combined with rotation, it cannot be inferred by observation of the end-product how the strain actually came about. However, a significant difference is found in plastic deformation, due to the development of inhomogeneities in the material during the plastic yielding (see p. 96), which influence later straining. For example, shear-fractures formed under end-on compression (which results in pure strain) ideally develop in two comple-

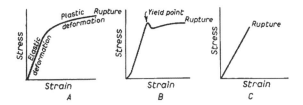

FIG. IV–11. STRESS–STRAIN CURVES

In each case the curve terminates at the point of rupture
(fracture) of the material.
A. Ductile substance; B. Mild steel, with yield-point; C.
Brittle substance.

mentary conjugate sets which are symmetrical with regard to the strain axes (see Fig. IV–12). On the other hand in a body subjected to rotational strain, one set of such shear-fractures (that more nearly parallel to the applied forces causing the strain) develops more prominently than the other (see Fig. IV–18).

Indeed, notions in the theory of elasticity, and especially that of simple shear, are subject to so many conceptual limitations that they should be used with great restraint in structural geology, where permanent strain is involved. Thus, the gliding of crystal lattices along translation-glide planes (Fig. V–11) has been regarded as simple shear.[2] In this instance, however, the strain has passed beyond the elastic stage, into that of permanent strain. During the elastic stage, it is highly improbable that the displacements in fact took place only in parallel planes. Furthermore, the actual gliding, when it did occur, was limited to certain planes, with slices between undergoing no or little

[1] The word 'shear' is also used for surfaces of rupture along which the parts of a body are displaced by a sliding movement – in popular parlance, 'sheared off'. Indeed, the basic notion of shearing is such a sliding along one plane, and this is incorporated in the concept of shear-stresses.

[2] A. Nadai, *Theory of Flow and Fracture of Solids*, New York, 2nd edn, 1950, p. 110; see Chapter V for glide planes.

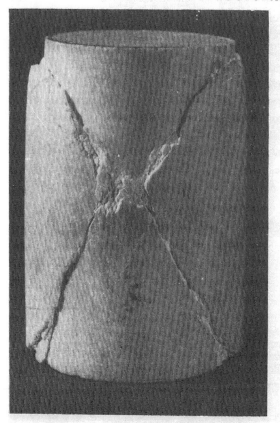

FIG. IV–12. SPECIMEN OF WOMBEYAN MARBLE SUB-JECTED TO 12% SHORTENING BY END-ON COMPRESSION (TRI-AXIAL TEST AT 150 ATMOSPHERES CONFINING PRESSURE) SHOWING COMPLEMENTARY SHEAR FRACTURES

(*Photo: M. S. Paterson*)

permanent strain. The deformation is therefore essentially discontinuous, the crystal itself far from isotropic – in fact the glide planes are preferred directions of ready displacement – and the strain is inhomogeneous. This illustrates the difficulty of relating concepts based on abstract mathematical ideas to real objects and to actual deformations.

Permanent deformation and fracture

It is useful to relate certain concepts about strain and permanent deformation to those materials, especially metals, which have afforded the commonest analogues for rock deformation before adequate studies of rocks at high temperatures and pressures were carried out in the laboratory.

When a ductile metal is subjected to increasing stress in experiments carried out at normal temperatures and pressures and over a period of minutes or hours, there is at first a stage of elastic deformation during which, if the applied forces be removed, the test-piece will recover its original shape and volume and there is thus no permanent strain. During this elastic stage, strain is proportional to stress. The material obeys Hooke's Law and in the general case the strain is said to be 'Hookean'. If then the stress be continually increased, a point is reached – the *elastic limit* – at which the metal begins to deform permanently.[1] At the same time the increase of strain for a given increase in stress also increases greatly and eventually fracture occurs (Fig. IV–11). Between the elastic limit and the point of fracture the metal is said to have yielded by *plastic flow*. Naturally, the stress–strain relationships for different materials vary greatly. Wet clay shows practically no elasticity; a rock like granite shows no plasticity at normal temperatures and pressures.

Solids are classed as *brittle* or *ductile* according to the amount of plastic deformation they can exhibit. With brittle substances the extent of plastic flow is zero or very small (Fig. IV–11); with ductile substances it is large. Many substances are found to be much more brittle in tension than in compression. Furthermore, the strength of brittle substances is much smaller in tension than in compression, and this applies to all the common crystalline rocks. However, the stress–strain curves for many rocks become similar to those of ductile metals at elevated pressures and temperatures, when considerable plastic flow may occur.

As will appear later, permanent deformation may also take place by creep, and the elastic limit is often rather indefinite, especially in tests carried out on rocks at high confining pressure and elevated temperature, or with two-phase systems such as clay. It is therefore undesirable to regard permanent flow as being limited to plastic deformation such as is obtained in metals at room-temperature and -pressure. Macroscopically uniform permanent deformation that occurs without loss of internal cohesion may be described as *flow*.[2]

FIG. IV–13. BRITTLE SUBSTANCE IN COMPRESSION SHOWING TWO TYPES OF FAILURE

S', S'', shear-fractures. C, extension-fractures.

 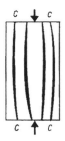

Fracture – When a body tested in compression ultimately breaks, the loss of internal cohesion is shown by the collapse of the specimen. The breaks in compression-tests are of two principal types, the most common being planes inclined at somewhat less than 45° to the direction of compression, along which the pieces slip by a shearing action. These planes are *shear-fractures*.

---

[1] In many instances the limit of elastic strain is ill-defined and it is difficult to determine a *yield-point* with precision.

[2] D. Griggs and J. Handin, 'Observation on Fracture and a Hypothesis of Earthquakes', in *Rock Deformation* (ed. D. Griggs and J. Handin) *Geol. Soc. Amer.* Mem. 79, 1960, pp. 39–104.

With brittle materials, fragments may also flake off parallel to the axis of compression, pulling away along *extension-fractures* (Fig. IV–13). When tested in tension, ductile metals approaching the rupture-point firstly contract to form a 'neck', and then rupture by a combination of shear-fractures and

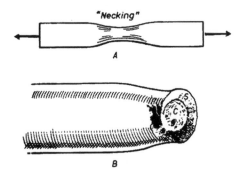

FIG. IV–14. DUCTILE SUBSTANCE IN TENSION

A. Shows necking in ductile metal rod under tension.

B. Enlargement showing details of fractures – S, shear-fracture; C, fibrous-fracture. (*After Nadai, 1950*)

fibrous-fractures, the latter being normal to the axis of extension (Fig. IV–14).[1] Under certain conditions brittle materials in tension break along extension fractures which ideally are normal to the axis of tension.

Relationship between flow and fracture

When plastic deformation begins in ductile substances, the first visible sign is the formation of fine *slip-lines* or broad *slip-bands*, best seen on the surface of polished metals or marble. The slip-bands represent layers which show up in reflected light as being brighter or darker than the rest of the specimen, indicating that their surfaces have been slightly deviated during the test. In mild steel they are known as Lüders' Lines, which on closer examination are seen to represent zones in which strong grain-deformation has occurred, and which have rotated at sharp kinks in the metal surface, due to shearing motions parallel to the bands. Under the microscope it may be seen that the deformed crystals in the bands have developed twinning, and that the twin planes show a strong statistical parallelism or preferred orientation (Fig. IV–15). Slip-lines seen on polished surfaces represent either individual planes of slip (or shear), or narrow slip-bands. In general the term 'slip' is preferred to 'shear' during the stage of plastic deformation while cohesion is maintained.

In experiments carried out by H. Cloos on the plastic flow of clay,[2] a cake

---

[1] In engineering practice these are also called *cleavage fractures*, but this usage is liable to be misunderstood in geology through confusion with slaty cleavage. See J. Gramberg, 'Axial Cleavage Fracturing, A Significant Process in Mining and Geology', *Engineering Geol.*, Vol. 1, 1965, pp. 31–72.

[2] H. Cloos, 'Zur Experimentellen Tekonik', *Die Naturwissenschaften*, Jhg. 18, Hft. 34, 1930, pp. 741–7; ibid., Jhg. 19, Hft. 11, 1931, pp. 242–7; idem, 'Fliessen und Brechen in der Erdkruste und im geologischen Experiment', *Plastische Massen*, Hft. 1, 1931. See also G. Oertel, 'The Mechanism of Faulting in Clay Experiments', *Tectonophysics*, Vol. 2, 1965, pp. 343–91.

of wet clay of suitable consistency, the upper surface of which has been stamped with small circles, is placed on a sheet of wire-netting resting on a table (Fig. IV−16). The netting is then pulled out in a diagonal direction, causing it to contract at right angles to the direction of extension, so that the clay is subjected to a pure shear. The circles are found to be drawn out into ellipses with their long axes parallel to the direction of elongation, and their short axes parallel to that of shortening. Careful examination of the surface of

FIG. IV–15. GRAIN-DEFORMATION IN A LÜDERS'
LINE

Microphotograph of the border of a Lüders' Line developed on a polished specimen of mild steel just after the yield-point has been reached. (*After Nadai, Plasticity*)

Note the tendency for the slip-bands, which mark the intersection of glide planes in the individual crystals with the surface of the specimen, to be oriented parallel to a certain direction. Note also that grain rotation has occurred, some grains now appearing dark and others light under incident light.

the clay under oblique illumination reveals that the ellipses have been formed from the circles by differential sliding or shearing movements along vertical conjugate slip planes.

The slip planes in clay have a relationship to the axes of the deformation ellipsoid broadly similar to shear-fractures, making an angle of approximately 45° with the maximum and minimum strain axes, and intersecting in the B strain axis. Actually with the great majority of substances the dihedral angle between shear fractures or slip planes is not 90°, but somewhat less for the dihedral angle enclosing the minimum strain axis, i.e. the axis of compression in a compression test. For a particular material at a given temperature and confining pressure this shearing angle is a constant and is a characteristic of the material.

In plastic flow, the first strong deformation to occur takes place in slip-bands or on slip planes, and these latter gradually generate throughout the mass as the flow proceeds, but without destroying internal cohesion. As the rupture-point is approached cohesion is locally destroyed, and finally the mass breaks on macroscopic shear planes broadly similar to the slip planes formed during plastic flow. This is seen in both compression and tensile tests of ductile materials.

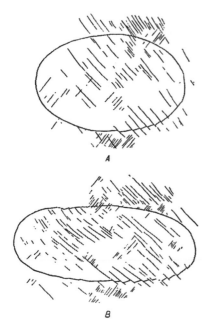

FIG. IV-16. CLOOS' EXPERIMENT WITH CLAY

A. The ellipse was originally a circle impressed on the clay. Vertical slip planes have developed in two complementary or conjugate sets.
B. A later stage of deformation showing larger shearing movements on the slip planes.

With brittle materials the shear fractures along which failure occurs are likewise comparable with slip-bands in ductile materials, but their formation involves cracking and cataclasis rather than slip. Microscopic cracks develop from initial small flaws under stresses considerably less than the ultimate failure strength,[1] and crack-propagation has been shown to be spasmodic and to depend on the history of stress application.[2] The number of shear fractures is limited, whereas ductile deformation involves a large number of small displacements affecting even the lattices of component crystal grains. A range of conditions from extreme brittleness to high ductility is exhibited by different materials, or by the same material under different confining pressures and temperatures, and so the relative importance of the mechanisms of yielding —

[1] W. F. Brace et al., 'Dilatancy in the Fracture of Crystalline Rocks', Journ. Geophys. Res., Vol. 71, 1966, pp. 3939–53.
[2] E. G. Bombolakis, 'Photoelastic Study of Initial Stages of Brittle Fracture in Compression', Tectonophysics, Vol. 6, 1968, pp. 461–73.

cracking with loss of cohesion and local release of elastic strain, cataclasis, slip, crystal deformation, and recrystallization – varies accordingly in actual examples.

Theories of failure

Many theories have been propounded to account for the failure of materials under stress, and in particular for the formation of shear-fractures and the angle at which these are formed. It would obviously be very useful if, from the conditions of stress or strain, it could be predicted at what stage of deformation and at what angle, shear-fractures would appear. A convenient summary of several theories is given by Jaeger.[1] Of late years, engineering practice in soil mechanics and in testing rock, concrete, and many other materials indicates that the Mohr criterion of failure predicts well the shear-plane angles in brittle materials under compression.

If a substance is subjected to compression, any plane within it other than the three orthogonal planes containing the principal stress axes, is subject both to normal stress across the plane, and to shear-stress tangentially along it. It is fundamental that any type of strain other than pure dilatation is accompanied by internal shear-stresses in this way. Further, these shear-stresses are a maximum in planes that make an angle of 45° with the major and minor stress axes, and they intersect in the mean stress axis. This arises from the geometry of stress-distribution. Although it might appear that shear-fractures would occur along these planes simply because they are subject to the maximum shear-stress, this is not found in practice, for in fact the attitude found for most substances including two-phase systems such as clay is that the two complementary shear-fractures enclose an angle of less than 90° about the axis of compression. Mohr suggested that this is so because the normal stresses that act across a shear plane are also concerned in shear-fracturing. It may for instance be conceived that the strength of the material is increased by the normal stresses, so that shearing would actually occur on a plane with a lower normal stress than that which acts on the 45° planes, thus effecting a compromise between the increase of strength and the value of the shear-stresses. Another way of considering the proposition is that the internal friction is increased by the normal stresses, which again suggests that actual shearing should occur on planes inclined at somewhat less than 45° to the maximum compressive stress.

According to the Mohr theory, the maximum or limiting tangential stress which a material can withstand is a function of the normal stress across the plane of shear, but the nature of the dependence is not specified. The shearing angle is a constant depending only on the ratio of the ultimate compressional

[1] J. C. Jaeger, *Elasticity, Fracture and Flow*, London, 1962.

to the ultimate tensile strength.[1] The angle is thus a characteristic of the material at constant temperature and pressure.

Most materials have a greater strength in compression than in tension. According to the stress-relationships established by Mohr, this implies that the angle between the shearing planes must be somewhat less than 90° about the axis of maximum compressive stress. For brittle rocks, the ratio of the compressive strength to the tensile is as high as 20–25 and the dihedral angle is correspondingly acute. Materials that have greater tensile than compressive strength are highly ductile and the dihedral should be obtuse about the axis of compression. The few materials (wood, soap) are, however, special cases of material which have a strong internal structure of fibres or long molecules.

The theory of brittle failure propounded by Griffith[2] also accounts for many observed effects. Measured tensile strengths of most brittle materials are much below those which would be indicated by the values for the molecular cohesive forces, and this reduction in strength is accounted for by the influence of a large number of minute cracks in the material. Very strong stress-concentrations develop at the tips of the cracks, these being responsible for overcoming the cohesive forces, so that the cracks propagate themselves and lead to fracture. The theory can be extended to cover biaxial and triaxial states of stress. In uniaxial compression the most dangerous cracks would be those of 45° to the stress, and these would propagate with very great speed in their own planes, giving the commonly observed paired fractures resembling shear fractures. This may account for conjugate jointing or clean-cut faulting in certain geological examples where brittle rocks are involved.

On the other hand, many geological materials are essentially plastic and yield without significant preliminary elastic strain. Criteria for fracture that assume high stored elastic strain-energy cannot then apply, and Odé[3] has discussed plastic shearing under such conditions, in which, as mentioned above, differential movements occur without loss of cohesion and shearing may be regarded as a result of differences in the rate of strain (flow) across certain surfaces. The theory incorporates the criteria of plasticity of von Mises, and it accounts very well for the geometry of curved and straight shear planes produced in certain earlier experiments, such as those of Hubbert made with sand, to simulate faulting.

Other factors that may affect fracturing are fatigue and the influence of a fluid phase. Fatigue cracks develop in metals under cyclic stress even below

[1] L. Brand, in W. H. Bucher, 'The Mechanical Interpretation of Joints', Pt II, *Journ. Geol.*, Vol. 29, 1921, pp. 1–28 (see p. 9); J. Handin and R. V. Hager, Jr, 'Experimental Deformation of Sedimentary Rocks Under Confining Pressure: Tests at Room Temperature on Dry Samples', *Amer. Assoc. Petrol. Geol.*, Vol. 41, 1957, pp. 1–50.

[2] A. A. Griffith, 'The Phenomena of Rupture and Flow in Solids', *Phil. Trans. Roy. Soc.*, A, Vol. 221, 1920, pp. 163–98; 'The Theory of Rupture', *Proc. 1st Internat. Congr. Appl. Mechanics*, Delft, 1924, p. 55.

[3] H. Odé, 'Faulting as a Velocity Discontinuity in Plastic Deformation', in *Rock Deformation*, Geol. Soc. Amer., Mem. No. 79, 1960, pp. 293–321.

FIG. IV–17A. TENSION GASHES IN CLAY
EXTENDED IN THE DIRECTION OF THE
ARROWS

FIG. IV–17B. GASH VEINS IN SANDSTONE
Tension gashes arranged *en echelon*, are filled with quartz. $\times \frac{1}{4}$.

the elastic limit, and have been suggested as causing the cleat of coal (p. 157). Fluids may act in two ways, either as a separate fluid body in a magma chamber, having an interface with surrounding solid rocks, or as a dispersed fluid phase in the voids of rocks. Fracturing connected with dyke-formation under the first set of conditions is discussed in Chapter XII; the influence of the water in a two-phase system in fracturing is little known but may be critical.

For the special cases of biaxial or triaxial tension such as would appear to be involved in the shrinking and cracking of sand and clay and in cooling igneous rocks, current theories of fracture offer little assistance.[1]

Tension gashes

By flooding with water the surface of the clay cake used in Cloos's experiment, the plasticity of the clay is reduced and it acts rather like a brittle material. Numerous vertical gashes, somewhat irregular and tapering at the edges, open at right angles to the axis of elongation of the cake (Fig. IV–17).

[1] A. Nadai, *Theory of Fracture of Solids*, New York, 1950, discusses many topics of interest.

Likewise in compression tests on brittle materials, extension fractures parallel to the axis of compression form especially if the plates are lubricated, permitting lateral spread of the specimen (Fig. IV–13).

In geological examples comparable gashes and cracks are called *tension gashes*, but these are commonly filled with vein material, forming *gash veins* (Fig. IV–17B). Tension gashes lie in the BC plane of the strain ellipsoid and thus afford valuable information on the strain of rocks (see, e.g., p. 176). Other examples of apparent tensional fractures in rocks include tension joints and certain dyke channels, which are treated in more detail in later chapters.

Shear-fractures and tension gashes in rotational strain

The fact that in many geological examples one set of shear planes is preferentially developed while the other, equally probable according to any theory of failure based on stress-relationships, is subordinated, has long been a puzzle. Experiments show that the result is obtained with rotational strains, in which case it is that shear direction most closely approximating the direction of the tangential applied force that is preferred. The clearest demonstration is that of Riedel,[1] working with a cake of plastic clay placed over two boards that touch each other, and are then pushed sideways to imitate a fault movement with the results shown in Figure IV–18. Shear planes form at an acute angle to the line of differential movement, crossing this in such a way that the acute angles point in the direction of relative displacement of the blocks. Shear planes originating by a similar mechanism are of common occurrence in nature, and have been referred to as 'Riedel shears' in recent works.[2] If the clay be flooded with water, tension gashes open along the line of movement, and these too make an acute angle, indicating the sense of relative movement of the blocks.

Where planar features such as the tension gashes in this experiment are arranged parallel with each other, but oblique to the boundaries of the zone in which they occur, they are said to be arranged *en echelon*. This is common in geology, and is perhaps most familiar in the calcite-filled tension gashes to be seen in marble, but the principle is also very useful in connection with movements on faults (see Chapter VII) and in the localization of certain orebodies.[3]

Combinations of forces

In the laboratory, forces may be applied in an almost infinite number of ways to experimental objects, and likewise in the complex environment of the

---

[1] W. Riedel, 'Zur Mechanik geologischer Brucherscheinungen', *Cbl. f. Min.*, etc., Abt. B, 1929, pp. 354–68.

[2] Further references will be found on pp. 339–40.

[3] H. E. McKinstry, *Mining Geology*, New York, 1948, gives a good discussion.

earth's crust a wide variety of force combinations is possible. An important consideration in nature, the effects of which are negligible in many standard tests on small specimens, is gravity.

Load, geostatic, or confining pressure – The pressure, equal from all sides, to which a particle immersed in a liquid at rest is subjected, is hydrostatic pressure. In geology, the pressure exerted on a particle of rock buried in the

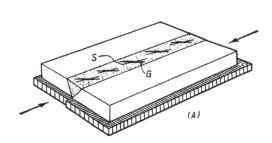

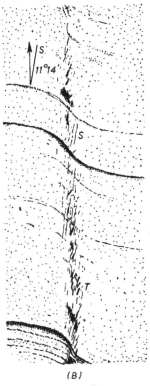

FIG. IV–18. SHEAR PLANES AND TENSION GASHES IN ROTATIONAL STRAIN

A. Set-up with clay cake on boards. S = shear planes; G = tension gashes. (*After Riedel*)

B. Details of the deformed zone. T = tension gashes. (*After E. Cloos*)

crust tends to cause horizontal expansion, but this is resisted by the surrounding rocks which thus exert a horizontal pressure on it. The particle is subjected to equal normal stresses in all directions, simulating hydrostatic conditions, and in general this pressure is equal to the load pressure of the superincumbent rock. This is the *standard state* for rocks, and the pressure is variously called the load, geostatic pressure, or confining pressure.[1] Highly plastic rocks such as clay, shale, or salt may flow into a bore-hole unless the hole is filled with heavy drilling mud to counteract the lateral 'formation pressure'. In porous water-saturated beds the total pressure at a point in depth

[1] E. M. Anderson, *Dynamics of Faulting and Dyke Formation*, Edinburgh and London, 1942, pp. 12, 137. W. Hafner, 'Stress Distributions and Faulting', *Bull. Geol. Soc. Amer.*, Vol. 62, 1951, pp. 373–98.

is due both to the hydrostatic pressure of water filling the voids, and to the load of superincumbent rock.

The effects of unloading a formerly deeply buried rock, as a result of prolonged erosion, are firstly to permit an immediate elastic rebound and secondly to initiate a slow expansion allied to creep. Such movements have been considered to lead to jointing and they may also be of considerable significance in engineering geology and geomorphology.[1]

In discussing faulting, it will be shown later that at depths such that the load-pressure exceeds the crushing strength of rocks, they are unable to sustain tensile stresses and all the principal stresses, are, correspondingly, compressions. In such a case the minimum stress axis is an axis of extension and is often parallel with the direction of applied tensional forces, so that it is effectively an axis of tension, although as to stress it is minimal compression. For reasons such as this it is often convenient in structural geology to think rather in terms of the applied forces than of the stresses, although in the ultimate analysis stress-relationships are of the greatest significance, especially in the more precise aspects of engineering geology.

Applied forces

Where applied forces are balanced so that their vector sum is zero, a rigid body is not translated or rotated by the forces. However, in elastic bodies, even balanced forces cause some translation or rotation of the parts of the body by elastic strain, and in plastic deformation there is considerable movement of the parts, that is, mass translation or deformation. In geology we have to deal with limited strains and translations such as are involved in the deformation of a relatively rigid basement complex of crystalline rocks, also with the mass displacements of highly deformable plastic rocks such as young sediments or the 'cover' over a basement complex, and again with the ruptures of both types of rock-mass. Some simple standard types of force combinations and their general effects are treated below. In each case the deformation discussed is that to be found in homogeneous materials isotropic as to stress and strain, deformed within the elastic limit. These are idealized conditions, and in fact the strain ellipsoid resulting when forces are applied as indicated below to anisotropic materials strained beyond the elastic limit is not predictable except in special circumstances where, as in a single crystal, the nature of the anistropy and the method of yielding by slip or twinning is known.

Uniaxial compression is the squeezing of a body along one axis by opposed forces directed towards each other. This is broadly the condition in a standard test of a rock-specimen to find its *compressive strength*. The strain

---

[1] R. Peterson, 'Rebound in the Bearpaw Shale', *Bull. Geol. Soc. Amer.*, Vol. 69, 1958, pp. 1113–24.

ellipsoid is biaxial, with C parallel to σ_1; $A = B$, with *flattening* in a plane normal to σ_1 and C.

In laboratory compression-testing, special devices may in some cases be used to enclose test-samples and to apply confining pressures by means of oil inside a casing while the specimen is subject to pressure from a plunger at its end. This is called *triaxial compression testing*, although in fact the strain ellipsoid remains biaxial. The major stress is the compression transmitted through the ends of the specimen, but the lateral stresses may also be raised to quite high values, remaining, however, equal since they are applied by a fluid.[1] Tests such as these simulate the conditions to which rocks are subjected at some considerable depth, where the load-pressure is high.

While the deformations described take place under uniform conditions as to constraints and freedom to deform, conditions in the crust are variable, so that regional compression may result in a variety of mass-movements. Often

FIG. IV–19. UNIAXIAL COMPRESSION IN AN INHOMOGENEOUS ENVIRONMENT, WITH ELONGATION ALONG B AND GREATEST RELIEF UPWARDS

(*After Harland and Bayly*, Geol. Mag., Vol. 95, 1958)

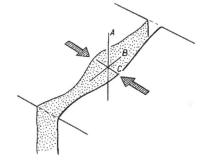

the direction of most ready movement is upwards, while there is a lateral constraint due to the presence of rocks adjacent to the part under consideration. In such a case the deformation may be triaxial, with A vertical, B normal to the pressure and C parallel to it (Fig. IV–19). Other deformations still are possible in regions where variously deformable rocks are present.

Uniaxial tension is the stretching of a body along one axis by opposed forces directed away from each other. Standard laboratory tests give the *tensile strength*. In the elastic stage the specimen contracts in a plane normal to the tension while extending according to Hooke's Law in the direction of the pull. In the plastic stage ductile materials such as metals or clay contract at a restricted part near the middle, producing a 'neck' before rupture occurs. The strain ellipsoid is biaxial with A parallel with the tension; $B = C$ in a plane normal to the tension.

Bending is the bowing of a bar or plate by two opposed force couples acting in the same plane. This is an important operation in the formation of certain types of folds (Fig. IV–7, and Chapter VIII) in which the applied forces act vertically upwards while gravity acts downwards, thus producing a bending moment (Fig. IV–20A).

Cross-bending – When a plate is bent by two opposed couples acting in one plane, and is also bent by two other couples acting in a plane at right angles to

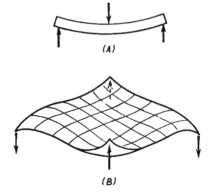

(A)

(B)

FIG. IV–20. BENDING
A. Bending of a slab.
B. Cross-bending of a plate.

the first, a saddle-like figure is produced (Fig. IV–20B). This type of deformation has been studied by Kendall in relation to the cleat of coal (see p. 157).

Buckling – A long, relatively thin bar or plate subject to compression on the ends will generally buckle upwards or downwards due to the development of

FIG. IV–21. TORSION OF A PLATE
Circles drawn on the face before deformation are changed to ellipses. Tension gashes (*g*) tend to develop on *ABCD*, but on the opposite face they form at right angles to *g*.

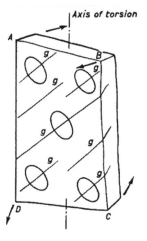

vertical resolutes of the applied forces at places where the plate deviates slightly from the line of application of the compressive forces. This is important in certain types of folding.

Torsion – This is a twisting action by two opposed force couples acting in parallel planes, generally at right angles to the long axis of a rod or bar (Fig. IV–21). Following the experiments of Daubrée, torsion was commonly conceived to be the cause of much jointing and fracturing in rocks, but its importance is not now believed to be great.[1]

Notions in the foregoing discussion on applied forces should be clearly distinguished from those in earlier discussion of stress, recognizing that the stress-distribution within a body which is subjected to the simple combinations of applied forces mentioned may, in fact, be quite complex (see, e.g., Fig. XII – 32B). Again, the same state of stress may be generated at any one point by different combinations of applied forces. It is, however, useful at times to think in terms of applied forces, especially in dealing with the relative movements of various rocks, or rock-masses, any one of which may be considered as applying forces to another contiguous mass, as in the penetration of sediments by rising salt-plugs and the like.

---

[1] It is worth noting that a single force couple does not constitute torsion unless applied to a body held firmly at the opposite end, when the opposed couple is generated as a resistance.

CHAPTER V

Environment, time, and material

In the preceding chapter the general principles of deformation have been treated. We may now turn to the discussion of the effects of variations in the physical and chemical environment of materials, of the duration of time over which forces act, and of differences in the fine structure of the materials themselves, as, for example, between a crystalline solid and a clay–water, or sand–water mixture. This chapter, then, deals with special aspects of the subject of deformation.

Effects of confining pressure

In the classic experiments of Adams and Coker,[1] the effects of high confining pressure were combined with temperatures of the order of 1000°C, and marble was found to yield plastically. Von Kármán carried out tests on many different materials,[2] and of recent years much additional testing has been done on rocks.[3]

All tests show that at high confining pressure the rupture-strength of the great majority of rocks is very greatly raised, and especially for calcareous rocks, the stress–strain curves become similar in type to those for ductile metals, the rocks exhibiting considerable plastic deformation (Figs. V–1, 2).

[1] F. D. Adams and E. G. Coker, 'The Flow of Marble', *Amer. Journ. Sci.*, Vol. 29, Ser. 4, 1910, pp. 465–85; F. D. Adams and J. T. Nicholson, 'An Experimental Investigation into the Flow of Marble', *Phil. Trans.*, Ser. A, Vol. 195, 1901, pp. 363–461.

[2] T. von Kármán, 'Festigkeitsversuche unter allseitigen Druck', *Zeitschr. Vereins deutscher Ingenieure*, Vol. 55, 1911, pp. 1749–57.

[3] D. Griggs, 'Deformation of Rocks under High Confining Pressures', *Journ. Geol.*, Vol. 44, 1936, pp. 541–72. F. J. Turner *et al.*, 'Plastic Deformation of Dolomite Rock at 380°C', *Amer. Journ. Sci.*, Vol. 252, 1954, pp. 477–88. J. Handin and H. W. Fairbairn, 'Experimental Deformation of Hasmark Dolomite', *Bull. Geol. Soc. Amer.*, Vol. 66, 1955, pp. 1257–74. J. Handin and R. V. Hager, Jr, 'Experimental Deformation of Sedimentary Rocks under Confining Pressures; Tests at Room Temperature on Dry Samples', *Bull. Amer. Assoc. Petrol. Geol.*, Vol. 41, 1957, pp. 1–50 (contains a brief review of earlier work). Idem, 'Experimental Deformation of Sedimentary Rocks under Confining Pressures; Tests at High Temperature', ibid., Vol. 42, 1958, pp. 2892–935. *Rock Reformation: Geol. Soc. Amer.*, Mem. No. 79, 1960 (A Symposium: D. Griggs and J. Handin, ed.). I. Borg and J. Handin, 'Experimental Deformation of Crystalline Rocks', *Tectonophysics*, Vol. 3, 1966, pp. 249–368.

The effects of high confining pressure may be visualized as the continual healing of incipient fractures by close apposition of the parts, and Bridgman has pointed out that the total flow of materials without rupture may probably be indefinitely increased with increasing confining pressures.[1] Test-specimens

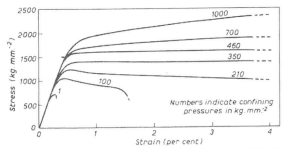

FIG. V–1. STRESS–STRAIN CURVES FOR WOMBEYAN
MARBLE AT VARIOUS CONFINING PRESSURES
(*After Paterson*, Bull. Geol. Soc. Amer., *Vol. 69, 1958, Fig. 3*)

FIG. V–2. PLASTIC DEFORMATION OF MARBLE
From left to right: Undeformed specimen; 20 per cent strain at 280 atmospheres confining pressure; 20 per cent strain at 460 atmospheres; note the increasing closeness of spacing of shear planes with higher confining pressure. (*After Paterson, 1958, Fig. 2*)

that have yielded without actual fracture show shear-surfaces that would, under normal pressures, have allowed the parts to separate from each other. Flow on the one hand and shear-fracture on the other are thus, as we have already noted, related aspects of the same general phenomenon.

[1] P. W. Bridgman, *Large Plastic Flow and Fracture*, New York, 1952.

In geology, however, it is possible to distinguish more or less clearly between at least two types of flow or yielding. The first is the intimate intra-crystalline flow which takes place with the accompaniment of plastic yielding of individual crystals by the mechanisms discussed later, and in many instances also with some recrystallization. This flow leaves its mark in a preferred orientation of crystal elements. The second is movement on many individual shear-surfaces which extend over distances greater than the length of several crystal grains, and between which the rock may preserve its crystal fabric largely undisturbed, or with some shattering (cataclasis). On a micro-scopic scale, the fracturing of grains is an element of the flow of solids and is termed *cataclastic flow*. Equally, megascopic cataclasis with major fractures resulting in a discrete assemblage of blocks ('particles') can also be regarded as 'flow', the extreme case being superficial rock-flows. In order to give any precision to the word 'flow' it is accordingly necessary to describe not only the physical processes involved but also the conditions under which the

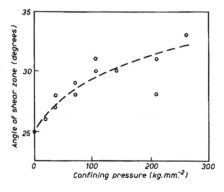

FIG. V–3. CURVE SHOWING THE INCREASE IN ANGLE BETWEEN PLANES OF SHEAR FAILURE IN WOMBEYAN MARBLE AND THE AXIS OF MAXIMUM COMPRESSIVE STRESS, WITH IN-CREASING CONFINING PRESSURE

(After Paterson, 1958, Fig. 5)

processes operated. The greater the depth at which cataclastic flow occurred, the tighter will the fragments be appressed. Open gashes and irregular shatter-ing are evidence of low confining pressure, and again, a brittle rock enclosed within ductile beds may 'flow' by fracturing and cataclasis as the surrounding ductile beds yield plastically.[1]

Changes in angle of shear – As a rock becomes more ductile under the influence of high confining pressure and increased temperature, the dihedral angle between complementary shear-fractures should change, the angle widening about the axis of maximum compressive stress. Such a change has been demonstrated to occur (under the influence of high confining pressure alone) in limestone and marble (Fig. V–3) and in some, but not in all, sandstones and shales. A theoretical angle of shear may be computed from the Mohr criterion of failure, and this shows reasonable agreement in many cases with the measured angle found in experiments on rocks. With our present

[1] D. W. Stearns, 'Fracture as a Mechanism of Flow in Naturally Deformed Layered Rocks', *Geol. Soc. Canada* Paper 68–52, 1968, pp. 79–95.

knowledge the assumption is warranted that the shearing angle about the axis of maximum compressive stress in rocks which were deformed in nature by flow involving crystal plasticity, such as deformed marble and schist, is in general larger than in rocks that were deformed chiefly by cataclastic flow.[1] It seems unlikely, however, that the shear planes could pass beyond the 45° position (dihedral angle 90°) under any conditions while the rocks remain essentially solid. Moreover, with granites and some sandstones, there is virtually no change in the angle of shear up to some few thousand kilograms per square centimetre of confining pressure.

It is essential to distinguish clearly in this connection between the attitude of shear-fractures at failure, and the arrangement of twin and glide planes in the individual crystals of strongly deformed, plastic rocks such as marble at elevated temperature and confining pressure. In these rocks prolonged plastic deformation causes mechanical rearrangement of the crystal planes, which develop a strong preferred orientation along with the stretching and orientation of the crystal grains themselves. The grains are elongated normal to the compression axis, and certain twin lamellae also come to lie statistically normal to this axis, these effects being the more pronounced the greater the compression (or, in extension tests, the elongation)[2] (Fig. V−4). At the same time, it may be judged from the results of other experiments,[3] that the dihedral shear angle at fracture, even after strong plastic deformation, would approximate to 60° or less about the axis of maximum compressive stress.

Indeed, Byerlee[4] has shown that at the confining pressure required to transform brittle rocks to the ductile condition the angle made by fault surfaces with the axis of compression is in fact close to 30° in most common rocks. Internal friction also seems to be almost independent of rock-type, and the brittle–ductile transition pressure at room temperature is the pressure at which the stress required to form a fault is equal to the stress needed to cause slip on the fault surface.

Internal rotation of shear planes, leading to a widening of the acute angle between them, occurs with continuing deformation in clay. The initial angle is 55°–60° in the clay used by E. Cloos[5] and this increases to 90° as

[1] An excellent critical review is given by J. Handin and R. V. Hager, Jr, 'Experimental Deformation of Sedimentary Rocks under Confining Pressures; Tests at Room Temperature on Dry Samples', *Bull. Amer. Assoc. Petrol. Geol.*, Vol. 41, 1957, pp. 1–50.

[2] F. J. Turner et al., 'Deformation of Yule Marble. Part VII: Development of Oriented Fabrics at 300°–500°C', *Bull. Geol. Soc. Amer.*, Vol. 67, 1956, pp. 1259–94; D. T. Griggs, F. J. Turner, and H. C. Heard, 'Deformation of Rocks at 500° to 800°C', *Geol. Soc. Amer.*, Mem. 79, 1960, pp. 39–104.

Note that in a triaxial test with extension along one axis, the grains are elongated along this axis, that is, normal to the compression acting along the other two axes.

[3] J. Handin and R. V. Hager, Jr, op. cit., 1957.

[4] J. D. Byerlee, 'Brittle–Ductile Transition in Rocks', *Journ. Geophys. Res.*, Vol. 73, 1968, pp. 4741–50.

[5] E. Cloos, 'Experimental Analysis of Fracture Patterns', *Bull. Geol. Soc. Amer.*, Vol. 66, 1955, pp. 241–56.

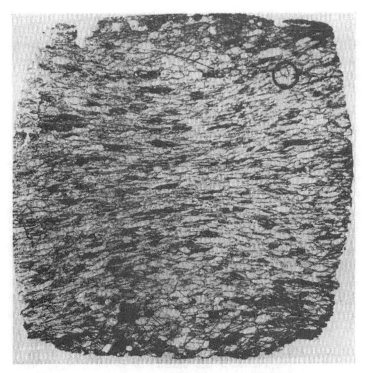

FIG. V–4A. YULE MARBLE SHOWING ELONGATION OF CALCITE
GRAINS NORMAL TO COMPRESSION AXIS
(Confining pressure 3000 Atm.; temp. 400°C.).

FIG. V–4B. CIRCLED AREA IN A, SHOWING STATISTICAL
PARALLELISM OF TWIN LAMELLAE NORMAL TO THE AXIS OF
COMPRESSION
(*After Turner* et al., *1956*)

deformation by extension on a wire mesh is continued. The mechanism by which rotation occurs is that many small displacements on the conjugate shear planes produce a new overall trend in the larger and more prominent initial shears (Fig. V–5).

Rotation of structure or fabric elements such as twin (composition) planes, glide planes, or shear planes is in fact possible only when the enclosing material can yield on another set of such planes. This implies that while lattice

FIG. V-5. CLAY DEFORMATION SHOWING SHEARING OF EARLY-FORMED SHEAR PLANES AND APPARENT WIDENING OF THE DIHEDRAL ANGLE FROM 60° TO 85°
(*After E. Cloos, 1955*)

planes may readily rotate, faults and joints in hard rocks are limited in such movement by the necessity for the compensating movements in the rock, between the planes concerned.

Influence of temperature

Contact metamorphism to form rocks of the hornfels groups, the prismatization of rocks at igneous contacts, the formation of porcellanite by baking, and the coking of coal by dykes afford examples of the influence of high temperatures without accompanying high stress.

Typical unstressed hornfelses and marbles are particularly interesting for, while recrystallization has taken place, plastic flow is in general negligible in them. Contact-altered rocks in which a gneissic or schistose structure is developed are associated chiefly with intrusions at deeper crustal levels, and it appears that the confining pressure must in general be high in order to induce plastic flow at plutonic contacts.

A structure seen at some igneous contacts is *prismatic jointing* of the invaded sediments, especially sandstones. The prisms are, ideally, hexagonal like basalt columns, but many are five-, or even four-sided, and they form at right angles to the contact (Fig. V – 6). Coal that is turned to coke along dyke intrusions may also show prismatization, and in all these instances there is no sign of plastic yielding of the substance of the rock. The development of prismatic jointing in furnace-linings has been interpreted by French[1] as

---

[1] J. W. French, 'The Fracture of Homogeneous Media', *Trans. Geol. Soc. Edinburgh*, Vol. 17, Pt 1, 1922, pp. 50–68.

FIG. V–6. PRISMATIC JOINTING DEVELOPMENT IN SANDSTONE AT THE CONTACT OF
A BASALT SILL, BONDI, NEAR SYDNEY, N.S.W.
(*Photo: Geol. Survey N.S.W.*)

implying that the structure formed on cooling and is thus an effect of shrinkage as with columnar jointing in lava-flows. Indeed, clay and shale may be changed from the plastic condition to intensely brittle porcellanite by near-surface baking, during which water escapes, iron is oxidized, and new aluminium silicate minerals are formed. Baking is commonly found beneath lava-flows, and at coal outcrops where spontaneous combustion has occurred.

Recent experiments with cold-strained single crystals of calcite and Yule marble, and unstressed Solenhofen Limestone (which is very fine-grained), have demonstrated that annealing occurs between 500° and 800°C, at 5 kilobars. Equigranular grains grow, showing random orientation, or slight relict orientation inherited from the deformed marble. Water and CO_2 do not affect the process, which goes on with great rapidity, being in some cases complete in fifteen minutes. Dolomite, on the other hand, does not exhibit annealing effects.[1]

Laboratory tests on rocks, combining the effects of high confining pressure with increased temperature, have yielded different results with different rocks.[2] All the rocks tested except quartzite, but including granite, peridotite,

[1] D. T. Griggs *et al.*, 'Annealing Recrystallization in Calcite Crystals and Aggregates', in *Rock Deformation, Geol. Soc. Amer.*, Mem. No. 79, 1960, pp. 21–38.

[2] D. T. Griggs, F. J. Turner and H. C. Heard, 'Deformation of Rocks at 500° to 800°C': in *Rock Deformation, Geol. Soc. Amer.*, Mem. No. 79, 1960, pp. 39–104.

pyroxenite and basalt, become ductile between 300° and 500°C, at 5 kilobars confining pressure. Likewise all these rocks except quartzite show a decrease in compressive strength at higher temperatures. The reduction is not great (20 kilobars to 7 kilobars) for peridotite, pyroxenite, and granite. Basalt shows a sudden decrease above 600°C. Calcite is not only weak at lower temperatures, but shows a more rapid decrease in strength at higher temperatures, whereas dolomite[1] is stronger and shows less reduction in strength. Heard has shown that Solenhofen Limestone changes from brittle to ductile at lower confining pressures, the higher the temperature, i.e. from 7300 atmospheres at 25°C to 700 atmospheres at 700°C in tension, and from 1000 atmospheres at 25°C to 1 atmosphere at 480°C in compression.[2]

The time-factor

Creep – In short-term tests carried out on polycrystalline metals at normal temperatures and pressures stress–strain curves such as are shown in Fig. IV–11 are obtained. But when stresses are applied for periods of days or months rather than minutes, slow persistent deformation is found to occur, even below the elastic limit. Indeed, with metals approaching the melting-point, it may not be possible to recognize any definite yield-point. This 'slow persistent strain that takes place under stress' is known as *creep*.[3] Creep has been experimentally demonstrated in rocks and minerals, including Solenhofen Limestone, marble, shale, halite, and igneous rocks.[4]

Creep involves three distinct component processes; the first is an elastic effect known as *transient* or *primary creep*, which decreases logarithmically with time and is recoverable on removal of stress; the second is *quasi-viscous* or *secondary* creep, which is proportional to the time over which the stress is applied; the third is *tertiary* creep, in which the strain-rate accelerates with time. A typical creep curve is shown in Fig. V–7.[5] At low temperatures and pressures, creep deformation is limited. Firstly the rupture-strength is much reduced, and secondly the total strain is normally less than in short-term tests. However, under moderate pressures approaching the rupture-strength itself, creep deformation is much greater and may greatly exceed normal plastic flow, and high temperature also favours an increase in the rate and extent of creep (Fig. V–8). These remarks apply particularly to secondary, time-

[1] See also D. Griggs *et al.*, 'Deformation of Yule Marble: Part IV, Effects at 150°C', *Bull. Geol. Soc. Amer.*, Vol. 66, 1951, pp. 1385–406. J. Handin and H. W. Fairbairn, 'Experimental Deformation of Hasmark Dolomite', *Bull. Geol. Soc. Amer.*, Vol. 66, 1955, pp. 1257–73.

[2] H. C. Heard, 'Transition from Brittle Fracture to Ductile Flow in Solenhofen Limestone, etc.', in *Rock Deformation, Geol. Soc. Amer.*, Mem. No. 79, 1960, pp. 193–226.

[3] E. da C. Andrade, 'Creep', *Journ. Roy. Soc. Arts*, Vol. 103, 1955, pp. 487–502 (Cantor Lecture).

[4] D. T. Griggs, 'Creep of Rocks', *Journ. Geol.*, Vol. 47, 1939, pp. 225–51. C. Lomnitz, 'Creep Measurements in Igneous Rocks', ibid., Vol. 64, 1956, pp. 473–79.

[5] A. H. Sully, *Metallic Creep*, London, 1949.

dependent creep, which inductile materials is known to be due principally to movements on grain-boundaries, whereas transient creep is due to intra-granular atomic and lattice displacements. Tertiary creep is connected with reconstitution of the substance, involving grain-growth, the growth of new solid phases, polymorphic phase-changes, and minute cracking, and is obviously likely to be of great importance in rocks under metamorphic

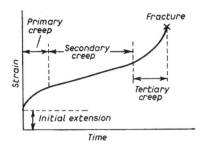

FIG. V–7. GENERALIZED CREEP CURVE
(*After Sully*)

conditions. It has been shown that transient creep in brittle rocks is due to movements on minute fractures, and it is possible that secondary and tertiary creep in such rocks also involves microfracturing.[1]

Cracking and rehealing – The ultimate or rupture-strength of crystalline solids depends to a great extent upon the original soundness of the specimen with regard to the presence of minute cracks, which greatly lower the strength.

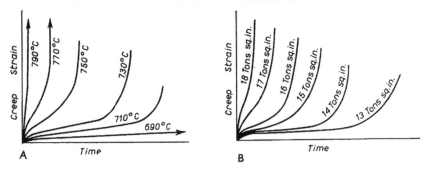

FIG. V–8. CURVES SHOWING INCREASE IN CREEP
A. With increase in temperature at constant stress.
B. With increase in stress at constant temperature. (*After Griffiths from Sully, 1949*)

Rock-salt is a material that normally has a low crushing strength of about 0·25 tons per sq. in., but this appears to be due largely to the presence of many cracks even in single crystals. If a rather long crystal is immersed for a short time in warm water, it may be bent up to 30° simply with the fingers, without cracking. This is partly due to the removal of the outer cracked surface, and partly, perhaps, to rehealing of cracks both old and new, by

[1] C. H. Scholz, 'Mechanics of Creep in Brittle Rock', *Journ. Geophys. Res.*, Vol. 73, 1968, pp. 3295–302.

deposition of salt in the tips of the cracks, which permits plastic deformation. This may account for the well-known deformability of salt in nature (see Salt Domes, p. 275), although it is also probable that halite is intrinsically a readily deformable mineral, especially as to creep under low stress differences. Creep of gypsum and marble is much increased when the specimens are immersed in solvent fluids.[1] With alabaster, a dry specimen shortened 0·03 per cent in a few days and thereafter ceased to yield, whereas in water a specimen shortened 1·75 per cent in 36 days, under compressive stress of 205 kg/cm², which is less than half the normal elastic limit of alabaster. This effect, therefore, is one of creep.

Interstitial fluids in rocks – Rocks of differing granularity, permeability and mineral composition, when strained while the pores contain fluids, react differently, and the same rock will also behave differently, depending on the magnitude of the pore-pressure developed within the fluid in the interstices, relative to the external confining pressure. Ultimate strength and ductility in permeable rocks have been shown to be functions of the *effective stresses*, that is of the difference between the external pressures and the internal pore-pressure.[2] At high effective confining pressures (1·2 kilobars or more) deformation of a quartzo-feldspathic sandstone is, as with the dry rock, largely by cataclasis, but at intermediate effective pressures (400–1200 bars) pore-pressure effectively reduces the frictional resistance across potential shearing planes, and these develop early in the deformation. With low effective pressures (200 bars) there is no cataclasis and shearing occurs at low differential stress. When the pore-pressure and confining pressure are equal the rock is very weak, there being virtually no internal friction, and deformation is purely by intergranular movements. These enlarge the voids as the grains assume open packing, and the rock becomes dilatant (see p. 122).

In metamorphic rocks it is often to be observed that crystals of metallic minerals are shattered whereas others, silicates and carbonates, grew or were plastically deformed during deformation. These effects are due to the different reaction of the two types of mineral to the fluids present (Fig. V–29), the one being mechanically disrupted, the other recrystallizing or rehealing.

Threshold stress for deformation – The vexed question as to whether rocks will continuously deform under extremely small stress-differences over a very long period of time requires further questions before an answer can be given. What kinds of rocks? Under what geological conditions as to stress? And for what order of magnitude of time?

Considering only crystalline rocks, geological evidence strongly indicates

[1] D. T. Griggs, 'Experimental Flow of Rocks under Conditions Favouring Recrystallization', *Bull. Geol. Soc. Amer.*, Vol. 51, 1940, pp. 1001–22. D. Griggs *et al.*, 'Deformation of Yule Marble: Part IV, Effects at 150°C', ibid., Vol. 62, 1951, pp. 1385–405.
[2] J. Handin, R. V. Hager *et al.*, 'Experimental Deformation of Sedimentary Rocks under Confining Pressure: Pore Pressure Tests'; *Bull. Amer. Assoc. Petrol. Geol.*, Vol. 47, 1963, pp. 717–55.

that at depths of the order of 10–15 km or more, deformation may take place continuously under relatively small stresses such as are involved in loading the crust by ice-caps or alluvial deposits, or unloading by erosion or the melting of ice. Such deformation may be broadly explained as creep. At shallower depths, however, strong flowage is apparently limited to certain belts or zones where differential stresses of considerable magnitude have operated, as well as high confining pressures, as in the deeper levels of orogenic belts. The proposition that a crystalline rock at shallow depths will yield continuously by creep under very small stress-differences is negated by the preservation of Pre-Cambrian rock-structures including micro-structures and idiomorphic crystal outlines, and of Palaeozoic fossils in a wide variety of environments where quite strong shear-stress must have affected them at various times, without important flowage. If rocks at shallow depths were subject to creep there would be no structural geology in old rocks, but likewise there would be no continents to support structural geologists, for the land would disappear by lateral flow to the ocean basins.

It is therefore assumed here that if continuous deformation under low stress is theoretically possible, the order of time involved at temperatures and pressures commonly attained in the outer crustal layers is greater than that over which the crust has existed. On the other hand, it is assumed that in the deeper layers of the crust, or elsewhere under conditions favouring creep, slow deformation may proceed to a limited extent in all rocks. But in rocks that have been subject to slow flowage, the results will be preserved in the fabric and especially in the micro-fabric.[1]

Materials

In the physical study of deformation, several types of substances are recognized according to the relationships of stress, strain, and time. For example, the Hookean type is a perfectly elastic material, in which strain is proportional to stress, as previously mentioned. Other types include the perfectly viscous fluid (Newtonian substance), the perfectly plastic solid, and others again with complex behaviour. These are commonly represented diagrammatically by symbols, using for instance a spring for the Hookean effects, sliding weights for friction (implying a threshold value of stress before deformation occurs, as at the yield-point) and so on (Fig. V–9).[2] These various types of 'substances' may be regarded as models for idealized types of 'solids' and 'fluids' in the earth, this permitting a numerical theoretical approach with

---

[1] M. Nishihara, 'Stress–Strain–Time Relation of Rocks', *Doshisha Eng. Rev.*, Vol. 8, No. 2, 1958, pp. 32–55; ibid., No. 3, pp. 85–115, gives an excellent review.
[2] A general account is given by J. C. Jaeger, *Elasticity, Fracture and Flow*, London, 1956.

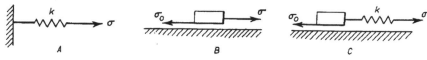

FIG. V–9. STRESS–STRAIN MODELS

A. Hookean substance: extension is proportion to stress $\epsilon = k\sigma$.

B. Model for a yield stress σ_o. If the applied stress is less than σ_o there is no strain. The model represents a frictional contact.

C. Substance that first deforms elastically and with increasing stress, plastically. (St Venant substance, e.g. ductile metal.)

geometrical consequences which may yield fruitful comparisons with actual examples[1] (see also Chapter VIII).

Deformation of crystals

Much fundamental work on single crystals has been done on metals,[2] but the early work of Mügge on mineral deformation provided many basic notions,

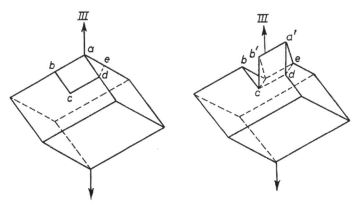

FIG. V–10A. TWIN FORMED IN CALCITE RHOMB BY FORCING THE CRYSTAL APART WITH A KNIFE APPLIED AT *b*

(*After Miers, Mineralogy*)

and more recent studies have afforded precise knowledge of readily deformable minerals, notably calcite and dolomite.[3]

In a crystal, plastic flow takes place by displacement of parts of the lattice beyond the stage of elastic strain, without destroying the bonding between the

[1] J. H. Dieterich and E. T. Onat, 'Slow Finite Deformation of Viscous Solids'; *Journ. Geophys. Res.*, Vol. 74, 1969, pp. 2081–8.

[2] W. Boas, *Physics of Metals and Alloys*, Melbourne, 1947; E. Schmid and W. Boas, *Plasticity of Crystals*, London, 1950.

[3] O. Mügge, 'Über Translationen und verwandte Erscheinungen in Krystallen', *Neues Jahrb. f. Min.*, etc., 1898, Bd. 1, pp. 71–159. F. J. Turner, D. T. Griggs, and H. Heard, 'Experimental Deformation of Calcite Crystals', *Bull. Geol. Soc. Amer.*, Vol. 65, 1954, pp. 883–934. J. Handin and H. W. Fairbairn, 'Experimental Deformation of Hasmark Dolomite', ibid., Vol. 66, 1955, pp. 1257–74.

atoms. The movements are, firstly, *twin gliding*, by which one part of the lattice takes up a twinned position relative to an adjacent part, as in the classic experiment of inducing twinning in a calcite rhomb by a suitably directed

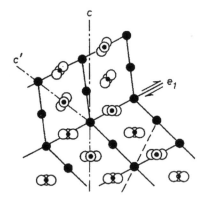

FIG. V–10B. ROTATION OF LATTICE LINES AND IONS IN TWINNING OF CALCITE

Solid black circles, Ca ions; CO_2 groups shown by open circles, reduced in size, e_1, composition-plane of twin, c, vertical crystal axis. (*After Turner* et al., *1954*)

blow (Fig. V–10A). This involves translation parallel to the composition plane of the twinned parts, together with rotation of the ions and ionic groups (Fig. V–10B). Multiple twinning due to deformation is beautifully shown in brass, and also in marble (Fig. V–12A). *Translation gliding* is a sliding of one portion of a lattice relative to an adjacent sector along a lattice plane known

FIG. V–11. TRANSLATION GLIDING
(*After W. Boas, 1947*)

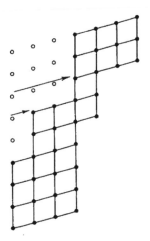

as the *glide plane*, in such a way that after each step the lattice structure is reconstituted (Fig. V–11). With tin, the so-called 'cry' during bending is a summation of clicks, each representing a jump on a glide plane, of one interatomic distance. In each glide plane, there is a definite direction of gliding, the *glide direction*, and there is likewise a corresponding direction of movement for each twin-law in twin gliding, which is also known as the glide direction.

Kink bands, which are similar to Lüders' lines, are megascopic deformation zones that form with the accompaniment of translation gliding, for the most part restricted to the kink band itself. In some instances kink bands are bounded by zones of *bend-gliding* (Fig. V–12B). Kink bands and bend-gliding are well shown in mica, kyanite, and calcite[1] (Fig. V–13).

In polycrystalline materials such as rocks there are, in addition to these intracrystalline displacements, effects due to the sliding and rotation of grains relative to each other. In part these involve simple movements of the whole grain, permitted by movements on the grain-boundaries, as in a pile of roundshot which starts to collapse. However, any rotation of a lattice element, relative to fixed axes in space, is known as an *external rotation*. In Fig. V–14 it is seen that the glide planes have rotated in this way, although

FIG. V–12. KINK BANDS AND
        BEND-GLIDING

A. Shows kink bands of two series (I, II), with appropriate glide planes. (*After Friedel*)

B. Shows kink planes *k*, bounded by areas of bend-gliding. The thin lines are glide planes, as in mica (see Fig. V–13). (*After Orowan*)

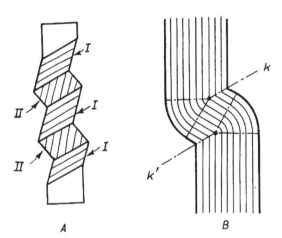

the crystal as a whole, having been held firmly at its ends, was not rotated. On the other hand, rotational displacements of lattice elements relative to each other within a crystal are termed *internal rotations*. The grain-boundaries affect the properties of the substance according to the adherence across them, and the facility with which ionic migration may occur so as to re-establish bonding after a displacement has occurred.

The relatively high ductility of metals is due to the non-directional nature of the metallic bond between the atoms, which permits ready re-establishment of cohesion across glide planes and crystal boundaries. In many minerals, the interatomic bonds are directed in space. The bonds are thus less readily re-established after dislocation, and accordingly most minerals are brittle rather than ductile at normal temperatures and pressures. However, as we have seen, plasticity may be induced under appropriate physical and chemical

[1] E. Orowan, 'A Type of Plastic Deformation New in Metals', *Nature*, Vol. 149, 1942, pp. 643–4; I. Borg and J. Handin, 'Experimental Deformation of Crystalline Rocks', *Tectonophysics*, Vol. 3, 1966, pp. 249–368.

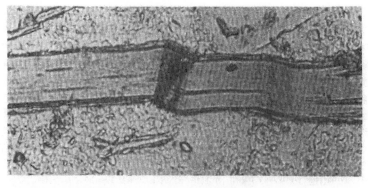

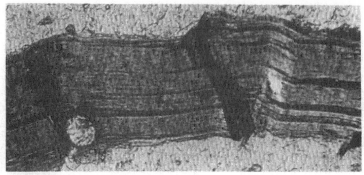

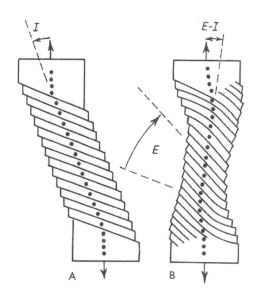

FIG. V-13. KINK BANDS AND BEND-GLIDING IN BIOTITE CRYSTALS
DUE TO VISCOUS FLOW IN DACITE, MARYSVILLE, VICTORIA

FIG. V-14. EXTERNAL AND
INTERNAL ROTATION

Diagonal lines represent glide planes.
E = angle of external rotation;
I = angle of internal rotation.

A. Gliding in specimen free to move at
the ends.
B. Gliding in specimen with ends fixed
in position, resulting in external
rotation due to the constraint.
(*After Turner* et al., *1954*)

conditions, and in rocks that have undergone plastic deformation, the re-arrangement of lattice planes by twinning, gliding, and rotation results in a preferred orientation of crystal elements (*c* axes, cleavage-planes, twin planes, etc.) which, using experimental results as a guide, may permit the determination of principal stress directions[1] (see also Chapter XIII).

Two-phase systems

In a two-phase system of solid grains in a liquid medium, deformation occurs by the translation of grains relative to each other and by grain-rotation. Cohesion is provided by inter-granular forces, which are relatively strong in clay, but very low in quartz sand. These forces are analogous with the forces bonding the grains of a crystalline aggregate, although much smaller. Especially in clay, the interstitial water is involved in the bonding of particles through the effects of unsaturated chemical bonds at the edges of the clay particles, which hold water molecules in a halo around each grain. In dry clay there is no means of re-establishing a mechanical bond once the grains have been separated, and the substance is brittle. Increasing water-content tends to increased bonding, until at a certain point the clay becomes plastic. With moderate water-content the effects of the bonds may extend throughout the interstices between grains so that there is a certain shear-strength, but as the water-content is increased the particles are further removed from each other, reducing the effective bond-strength until the *liquid-limit* is reached, when the clay behaves as a viscous fluid. Wet sand and complex solid aggregates also show a marked reduction of shear-strength at a certain water-content. With low water-content the stresses are transmitted from grain to grain, but at the point when the voids are completely filled, stress-transmission through the pore-water itself begins to play an important role. This stage marks the development of high 'pore-pressures' and a sudden reduction in strength. For maximum shear-strength there is an optimum water-content for most two-phase systems.

It is obvious that the influence of interstitial water in clays and sands is likely to be very significant in relation to the slumping and sliding of newly deposited sediments, in compaction, and in compaction-folding. But even when considerable compaction has been effected, without, however, actual cementation of the grains having occurred, sand and clay (or mudstone, greywacke and similar sand–silt–clay mixtures) cannot be regarded as solids in the accepted sense. Close folding of orogenic type may occur while rocks are in such a condition. Because of the weakness of inter-granular bonds, permanent deformation in such rocks may take place without significant crystal deformation, but rather by rotation of the grains (external rotation).

[1] N. L. Carter, 'Principal Stress Directions from Plastic Flow in Crystals', *Bull. Geol. Soc. Amer.*, Vol. 80, 1969, pp. 1231–64.

Likewise in a complex of sandstones and clays or greywackes and siltstones folded while in the soft-rock stage, the argillaceous beds are so mobile that they may squeeze into any potential space left in folding or faulting of the harder beds (Fig. V–15).

Clearly in such instances, which are numerous, high pressure and temperature to induce flowage are not only unnecessary, but would, at least for the clay, actually reduce plasticity. The effects of stress in such rocks may eventually lead to cataclasis of the quartz-grains which, inhibited from external rotation by the pressure, are sheared into slices as in many folded sandstones and greywackes interbedded with slates (Fig. V–16).[1] In such rocks the argillaceous paste is mainly concerned in maintaining internal cohesion.

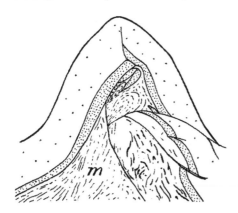

FIG. V–15. FOLD IN SILURIAN MUDSTONES (*m*) AND SANDSTONES (DOTTED)

Studley Park, Victoria. Height of section, 8 in. Shows flowage of mudstone under soft-rock conditions of folding and faulting.

It will be seen that it is useful to distinguish two types of plasticity – *hydroplasticity* for two-phase systems containing water, and *crystal plasticity* for solid polycrystalline aggregates such as hard rocks and metals.

Dilatancy – The curious property of dilatancy of granular aggregates was first studied by Reynolds.[2] It is the expansion, shown typically by close-packed aggregates, that accompanies deformation. The mechanism has been shown by Andrade and others[3] to involve the pushing apart of grains as they override each other along shear planes, whereby any grain must lift itself out of the niche it occupies, to pass over surrounding grains among which its journey lies (Fig. V–17). The result is a change, along the shear planes, from close- to open-packing, with a consequent increase in porosity and total volume. Thus if wet sand is deformed it becomes capable of holding more water, and its surface will appear suddenly dry, as is seen in walking over a

[1] The successive stages of intra-granular strain and eventual slicing of quartz are well described by S. L. Agron, 'Structure and Petrology of the Peach Bottom Slate, Pennsylvania and Maryland, and its Environment', *Bull. Geol. Soc. Amer.*, Vol. 61, 1950, pp. 1265–306.

[2] Osborn Reynolds, *Scientific Papers*, Vol. II, 1901, p. 217.

[3] E. da C. Andrade and J. W. Fox, 'The mechanism of dilatancy', *Proc. Phys. Soc. Lond.*, Vol. 62, Sect. B, 1949, pp. 483–500. R. L. Brown and P. G. W. Hawksley, 'The internal flow of granular masses', *Fuel*, Vol. 26, 1947, pp. 159–73.

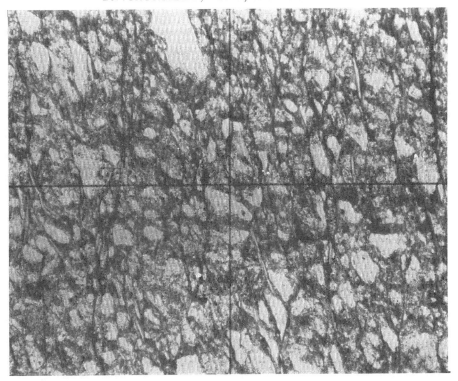

FIG. V–16. SLICING OF QUARTZ GRAINS IN CLEAVED GREYWACKE
Ordovician, Victoria. ×30.

wet beach, due to the absorption of excess water in the increased voids. If sand just saturated with water be placed in a thin rubber balloon and the end tied off, it is found that the mass becomes very rigid, since any deformation then tends to produce a vacuum and movements are resisted by the external air-pressure on the rubber.

In rock-deformation such effects may perhaps account for certain features seen particularly in sandstones interbedded with slates, where shear planes in

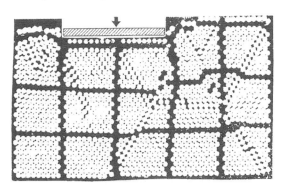

FIG. V–17. MECHANISM INVOLVED IN DILATANCY

Shown by pressing on a pile of ball-bearings. Note the change to open packing along shear directions. (*After Andrade & Fox, 1949*)

the sandstone are darkened for several inches, apparently by carbonaceous particles derived from the slate, although the slate itself need not actually inject the sandstone. This may indicate the flow of water expelled from the argillaceous bed into the sandstone, due to reduction of pore-pressure accompanying dilatancy[1] on shearing. More important, however, may be the relationship of dilatancy to the spacing of shear planes developed in a deformed rock. Since a considerable amount of work is done in developing a single shear plane under dilatant conditions, the principle of least work suggests that the number of planes will be small in coarse-grained materials, whereas with a fine-grained aggregate each shear plane involves less work and the effort may be expended with equal facility on a greater number. Indeed with materials like plastic clay the flaky shape of the particles aids shearing by permitting slip along innumerable planes without dilatancy. Again, clay particles align themselves parallel to a shear-surface in the clay, or normal to the pressure exerted by an object pressing against the clay.[2]

The effects of volume relationships in shearing have already been mentioned and are further discussed under slaty cleavage (p. 307).

Deformation of geological objects

Later in this book we shall be concerned with the deformation of large rock-masses, of regional extent, which have been folded, faulted, or otherwise

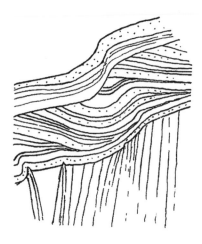

FIG. V–18. DEFORMED CROSS-BEDDING IN
SANDSTONE LAYER IN SLATE
Ordovician, Chewton, Victoria. ×$\frac{1}{2}$.

disturbed and which in general exhibit inhomogeneous strain. The approximately homogeneous strain of small elemental parts of such a rock-mass may, however, be gauged from the distortion of objects including individual

[1] K. S. Lane, 'Pore Pressure Effects on Berea Sandstone Subjected to Experimental Deformation: Discussion', *Bull. Geol. Soc. Amer.*, Vol. 80, 1969, pp. 1587–90.

[2] I. V. Popov, 'The Cryptostructure of Clays during their Deformation', *C. R. Akad. Sci. U.S.S.R. (Doklady)*, Vol. 45, 1944, p. 162.

crystals, oöid grains, pebbles, and fossils. Depositional structures of sediments and the primary structures of igneous rocks afford the most readily available space-data in structural geology, but many such primary structures, for example bedding, present a picture of the deformation of a two-dimensional reference plane, whereas most of the objects to be discussed now give data for three dimensions. Cross-bedding may permit inferences as to deformation normal to the major bedding planes, but its use is limited in quantitative study (Fig. V–18).

FIG. V–19. DEFORMED GRAPTOLITES (*Isograptus caduceus* var. *maxima* Harris) IN BLACK SLATE

Ordovician, Bacchus Marsh, Victoria. Cleavage R–L, bedding in plane of photograph. Note that the spacing of thecae normal to the cleavage is much closer than that parallel (or nearly so) to the cleavage. Original spacing of thecae 9–11/cm. Cleavage approximately at right angles to plane of paper. ×4. (*From Harris and Thomas, 1944*)

The distortion of fossils is readily observable when, for example, normal bilateral symmetry is destroyed (Fig. V–19), but careful study may be needed to distinguish biological variation from *post-mortem* deformation. Fanck [1] reduced the number of bivalve mollusc species from 426 to 62 in the Upper Marine Molasse of St Gallen, Switzerland, after recognizing the effects of deformation, and Breddin,[2] in studying the same beds, demonstrated nearly 28 per cent shortening almost parallel to the stratification although the beds are uniformly dipping rather than strongly folded. Indeed, fossils may be deformed syndiagenetically by pressure of superincumbent sediments and

[1] A. Fanck, Dissert. Zürich, 1929 (quoted by Breddin, 1964).
[2] H. Breddin, 'Die tektonische Deformation der Fossilien und Gesteine in der Molasse von St Gallen (Schweiz), *Geol. Mitt.*, Vol. 4, 1964, pp. 1–68 (with good bibliography).

also by early tectonism[1] as well as by the more radical processes operative in slates and strongly folded limestones.[2]

The methods used to study fossil-deformation must be based on known geometrical relations of the parts of the fossil species, which include the logarithmic spiral growth of ammonities,[3] the spacing and angle of growth of graptolite thecae (Fig. V–19), the bilateral symmetry and proportions of brachiopods[4] and trilobites, and other data.

Distortion may also be estimated for objects that may be assumed to have been approximately spherical before deformation, such as oöid grains and pebbles or boulders, and which have the form of triaxial ellipsoids after deformation. An equivalent sphere of radius r to a triaxial ellipsoid with major axes (semi-diameters) $X > Y > Z$ whose lengths are respectively x, y, and z, may be determined on the assumption that no change in volume has occurred, from the equation $x . y . z = r^3$. Comparison of the major axes with r then gives an indication of the elongation or shortening along these axial directions. The results may be influenced by knowledge of the undeformed shape of pebbles if this can be seen at some nearby locality (Fig. V–20), but this is rarely available. There are, moreover, many considerations that affect the interpretation of the results in terms of stress and applied force.

Results that relate significantly with regional tectonics have been obtained by assuming that deformed pebbles were originally spherical, and calculating the strain from pebble shapes.[5] Statistical analysis using a triangular diagram of sides representing log x/r, log y/r, and log z/r, on which each pebble is represented by a point, permits the recognition of inherited shape-patterns from the undeformed conglomerate and also progressive strain at different stages of deformation.[6] Gay has used the differing deformation of pebbles of different rock-types to estimate the finite strain of the total rock,[7] and fractured pebbles may also yield useful data by analogy with the shearing and

---

[1] G. Langheinrich, 'Syndiagenetische Fossildeformation im untersten Lias (Hettangium) von Göttingen', *N. Jb. Geol. Paläont., Monatshefte*, 1966, pp. 666–80; W. Plessmann, 'Laterale Gesteinsverformung vor Faltungsbeginn im Unterkarbon des Edersees (Rheinisches Schiefergebirge)', *Geol. Mitt.*, Vol. 5, 1965, pp. 271–84.

[2] E. S. Hills and D. E. Thomas, 'Deformation of Graptolites and Sandstones in Slates from Victoria, Australia', *Geol. Mag.*, Vol. 81, 1944, pp. 216–22.

[3] G. Langheinrich, 'Die Bestimmung der tektonischen Gesteinsdeformation mit Hilfe deformierter Ammoniten', *N. Jb. Geol. Paläont., Abh.*, Vol. 128, 1967, pp. 275–93; idem, 'Die tektonische Deformation von *Schellivienella umbraculum* im Givet von Niedersalwey (Westfalen)', *Geol. Mitt.*, Vol. 7, 1967, pp. 159–72.

[4] H. W. Wellman, 'A Graphical Method for Analysing Fossil Distortion', *Geol. Mag.*, Vol. 99, 1962, pp. 348–52.

[5] J. R. Hossack 'Pebble Deformation and Thrusting in the Bygdin Area (Southern Norway)', *Tectonophysics*, Vol. 5, 1968, pp. 315–39.

[6] K. L. Burns and A. H. Spry, 'Analysis of the Shape of Deformed Pebbles,' ibid., Vol. 7, 1969, pp. 177–96.

[7] N. C. Gay, 'Pure Shear and Simple Shear Deformation of Inhomogeneous Viscous Fluids. 2. The Determination of the Total Finite Strain in a Rock from Objects such as Deformed Pebbles', ibid., Vol. 5, 1968, pp. 295–302.

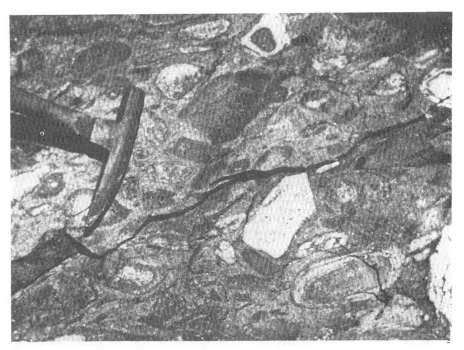

FIG. V–20A. UNDEFORMED CONGLOMERATE
Proterozoic, Yampi Sound district, Western Australia.

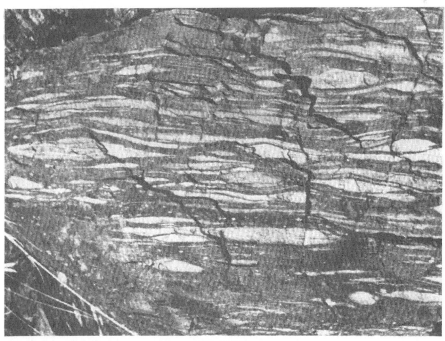

FIG. V–20B

The same bed deformed at the nose of a fold. Note especially the pebbles with concentric banding which demonstrate that the fragments are true pebbles. (*Photos by courtesy Broken Hill Pty. Ltd. and Mr I. Reid*)

fracturing of test specimens.[1] Where movements on one set of closely spaced shear planes are involved, deformation such as is shown in Fig. V−21 for a spherical object cannot be interpreted to imply compression normal to the shear planes, still less, parallel to the short axis of a pebble. This is admirably demonstrated by the deformation of a Pleistocene till in Idaho, containing

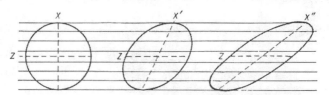

FIG. V–21. DEFORMATION OF A SPHERE BY MOVEMENTS ON ONE SET OF SHEAR PLANES

FIG. V–22. DEFORMATION OF DECOMPOSED BOULDERS IN WEATHERED PLEISTOCENE TILL, IDAHO, BY HILL-CREEP
(*After Capps, 1941*)

soft, weathered granite boulders. These are drawn out into ellipsoids by hill-creep, the amount of which decreases downwards in the soil until the un-deformed till is reached (Fig. V−22). It will be noted that where the boulders are but little sheared, in the lower layers, their long axes are markedly oblique to the shear-direction which is parallel with the hill-slope. In the upper layers

[1] D. M. Ramsay, 'Deformation of Pebbles in Lower Old Red Sandstone Conglomerates Adjacent to the Highland Boundary Fault', *Geol. Mag.*, Vol. 101, 1964, pp. 228–48. (See, however, pseudo-stretched pebbles, p. 132.)

they may be drawn out into bands only an inch or less in thickness, but even these are only sensibly parallel with the surface, and make, in fact, a very acute angle with the shear-direction.[1]

Any inferences that are made concerning the stress-distribution giving stretched or flattened pebbles or oöid grains must be based not only on the form of the pebbles but also on their geometrical relationship to associated structures including especially cleavage or schistosity. Petrofabric data are significant, but in many cases are difficult to interpret in deformed pebbles.[2] Similar considerations apply to planar objects such as trilobites, flat thin-shelled brachiopods, and graptolites. These normally lie in the plane of bedding, and shearing may result in apparent shortening normal to the cleavage, or, with the opposite sense of shear, in apparent elongation (Fig. V–23), This

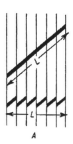

FIG. V–23A. APPARENT SHORTENING OF PLANAR OBJECT BY UNILATERAL SHEAR
Note also apparent rotation of the disjointed parts.

B. APPARENT ELONGATION OF PLANAR OBJECT BY UNILATERAL SHEAR

latter is, however, more obvious where cleavage and bedding make a very acute angle with each other in fold limbs, so that the inference from such 'smeared out' graptolites is normally taken correctly to imply shearing along the cleavage direction. Stronger fossils such as horny brachiopods are not sheared through by such movements, but crumpled into small folds as described by Sharpe.[3] Again, the shearing of a linear or planar object may result in an *en echelon* arrangement of the pieces, often taken to indicate their rotation, but in fact equally likely to be simply an effect of shearing along parallel planes. Despite such difficulties, however, associated features may, in fact, permit more definite inferences to be made.

The deformation of oöids in oölitic limestones has afforded a very intimate picture of the internal readjustments in folded beds,[4] especially in the work by E. Cloos at South Mountain, Maryland. Here, a soft muddy coating surrounds many of the carbonate grains, and this coating is more strongly

---

[1] Some minor effects of vertical compression due to gravity may also be represented in this example, which has not been studied in detail. See S. R. Capps, 'Observations of the Rate of Creep of Soil in Idaho', *Amer. Journ. Sci.*, Vol. 239, 1941, pp. 25–32.

[2] D. Flinn, 'On the Deformation of the Funzie Conglomerate, Fetlar, Shetland', *Journ. Geol.*, Vol. 64, 1956, pp. 480–505.

[3] D. Sharpe, 'On Slaty Cleavage', *Quart. Journ. Geol. Soc. Lond.*, Vol. 3, 1847, pp. 74–105.

[4] E. Cloos, 'Oölite Deformation in the South Mountain Fold, Maryland', *Bull. Geol. Soc. Amer.*, Vol. 58, 1947, pp. 843–917. See also Chapter X.

deformed. Where the elongation is considerable it is pulled out into points, which clearly indicates compression normal to the short Z axis of the deformed grain, rather than unilateral shear (Fig. V–24). Also the long axes of the grains are precisely parallel with the cleavage rather than at an acute angle to it as would arise with simple shear.

In deformed conglomerates, flowage in the matrix results in moulding of bedding or foliation about pebbles, the reorientation of pebbles in the direction of flow, and the opening of small 'eyes' at the ends of pebbles, where

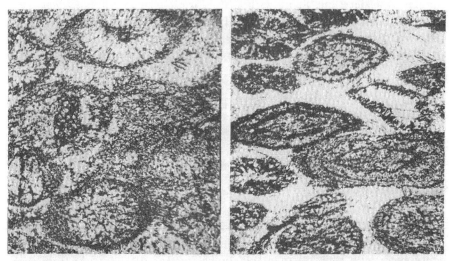

FIG. V–24. DEFORMATION OF OÖLITE GRAINS IN THE SOUTH MOUNTAIN FOLD, MARYLAND

Note the pointed ends, and the greater deformation of the muddy coatings of grains. (*After E. Cloos, 1947*)

secondary minerals, chiefly quartz, tend to grow (Fig. V–25). The symmetry of these features points to the shortening and elongation of the rock-mass along two axes of deformation, but likewise it is clear that actual pebble deformation, may in many instances be quite minor (see also under till-fabric, p. 8). Indeed the interpretation of so-called *stretched pebbles* is beset with difficulties, and many reported results are entirely at variance with the geometrical relationships established for oöid grains and fossils to the geometry of folds and to rock-cleavage.[1]

[1] Many recent statements are in fact also contrary to the clear statements of earlier authors – see especially A. Heim, *Geologie der Schweiz*, Bd. II, 1921, Leipzig, pp. 81–7. For more recent studies: H. W. Fairbairn, 'Elongation in Deformed Rocks', *Journ. Geol.*, Vol. 44, 1936, pp. 670–80; J. J. Runner, 'Pre-Cambrian Geology of the Nemo District, Black Hills, South Dakota', *Amer. Journ. Sci.*, Ser. 5, Vol. 28, 1934, pp. 353–72; A. Kvale, 'Petrologic and Structural Studies in the Bergsdalen Quadrangle. Part II: Structural Geology', *Bergens Mus.*, Arbok, 1946 and 1947; W. F. Brace. 'Quartzite Deformation in Central Vermont', *Amer. Journ. Sci.*, Vol. 253, 1955, pp. 129–45; D. Flinn, 'On the Deformation of the Funzie Conglomerate, Fetlar, Shetland', *Journ. Geol.*, Vol. 64, 1956, pp. 480–505.

FIG. V–25. FLOW OF MATRIX ABOUT A GRANITE PEBBLE IN DEFORMED
PROTEROZOIC TILLITE

Sturt R. gorge, South Australia. Note the 'eyes' at either end of the pebble.
Natural size.

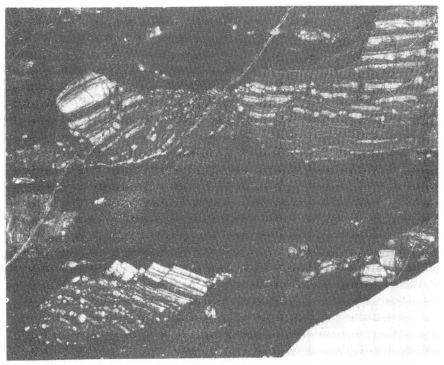

FIG. V–26. SHEARING PRODUCING NECKING AND DETACHED FRAGMENTS

Matrix, submarine tuff; banded rock, cherty mudstone. Cambrian, Pyalong, Victoria. Natural
size.

The possibility of error in failing to distinguish true stretched pebbles from the pseudo-pebbles of crush conglomerates, which are in reality fragments of the original rock torn apart by shearing (Fig. V–26), seems often to have been overlooked, and it is desirable in describing stretched conglomerates to afford adequate proof that a true conglomerate is, in fact, the subject of discussion. The presence of a variety of rock-types is generally an adequate test if these are haphazardly distributed through the conglomerate, but where a series of strata, perhaps including dykes or quartz-veins is crushed, an appearance of admixture of various rock-types results. The 'pebbles' of each type may then be seen to be aligned in the field-exposure, indicating their former continuity (Fig. V–27A). Moreover, the shearing of pebbles in a true

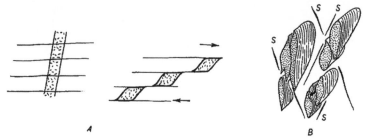

FIG. V–27A. SHEARING AND SEPARATION OF DYKES AND BEDS
Note the alignment of the fragments.

B. SHEARING OF PEBBLES GIVING PSEUDO-STRETCHED PEBBLES

conglomerate may produce fragments apparently elongated either in the shear-direction (with one set of shear planes), or along the intermediate strain axis, if two complementary shears operate (Fig. V–27B). The matrix of crush conglomerates is formed by mashing of the rock along shear planes.

Pressure fringes – Very similar effects to the deformation of conglomerates are to be seen on a microscopic scale. At opposite ends of crystals of pyrites, magnetite, or other hard minerals in slates or schists, a sharply bounded area occupied by minerals such as quartz, calcite, and chlorite may extend parallel to the cleavage or schistosity (Fig. V–28A). These pale-coloured areas, which are called *pressure fringes* (Fr. *halos d'étirement*; Ger. *Streckungshöfe*), clearly represent secondary fillings of potential spaces rather than replacements of the slate or schist. Their boundaries are very sharp, and irregularities in the surfaces of the large crystals may be matched in the boundaries of the pressure fringes, indicating that the two surfaces were once in contact and were pulled apart (Fig. V–28B). From a study of the orientation of the secondary minerals in fringes in schist, Mügge concluded that the spaces were formed as a result of a rolling action affecting hard magnetite crystals between the planes of schistosity, as these sheared over

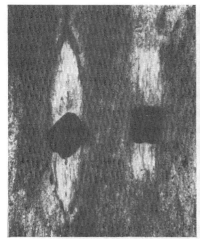

FIG. V–28A. PRESSURE FRINGES AT ENDS OF MAGNETITE CRYSTALS IN SCHIST,
COOMA, NEW SOUTH WALES
Note the correspondence in shape of the fringes with the crystals.

B. PRESSURE FRINGES AT ENDS OF PYRITE CRYSTALS IN SLATE, BENDIGO, VICTORIA
Note the matching shapes of the fringe with the crystal outlines.

each other;[1] but, while some shearing action such as this may be admitted, leading for example to rotation of the magnetite or pyrite porphyroblasts, the fringes owe their formation essentially to extension of the rock parallel to the cleavage or schistosity. This is shown by the matching surfaces mentioned above, by one-sided fringes, by the presence of fringes so closely spaced as to interpenetrate each other, and by the formation of relatively small fringes at the ends of elongate groups of pyrites crystals. The pulling apart of pyrites in pressure fringes is often thought of as an effect of 'shearing' or 'crushing' of the hard sulphide (Fig. V–29),[2] but since the surrounding rock was clearly plastic, a pressure sufficient to crush the pyrites would tend to close or fill the potential spaces, and in fact the effect is one of pulling apart of brittle crystals, in which the cracks have been penetrated by fluids. Some compression normal to the cleavage or schistosity is naturally involved, and its effects are seen in the 'drag' of the cleavage at the edges of the crystals, but this is small compared with the amount of extension involved. The formation of pressure fringes may be simply imitated by stretching a block of plasticine in which wooden match-sticks have been inserted.

In pressure fringes, as in many other examples of 'fissure fillings' in

[1] O. Mügge, 'Bewegungen von Porphyroblasten in Phylliten und ihre Messung', *Neues Jahrb. f. Min.*, etc., Vol. 61, B.B., Abt. A., 1930, pp. 469–510.

[2] I. de Magnée, 'Observations sur l'origine des gisements de pyrite du Sud de l'Espagne et du Portugal', *Congr. Internat. Mines, Metall., et Géol. appliquée*, 7th Session, Paris, 1935, pp. 95–104.

geology, the presence of a fluid component which pervades the whole rock and permits simultaneous growth of new minerals with the opening of the spaces, is an essential factor in the process. Pressure fringes appear to represent a considerable increase in volume of the whole rock, which it is difficult to account for on purely mechanical grounds. However, since fringes are found only in mineralized slates and in schists and gneisses, the possibility of removal of rock-substance by diffusion or transfusion must be envisaged, or a reduction in porosity by recrystallization, which permits the growth of the pressure fringes as a compensating mechanism.

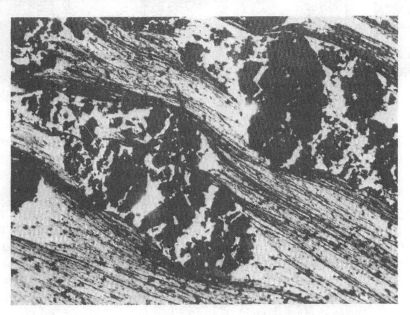

FIG. V–29. FRACTURING OF PYRITES IN PRESSURE FRINGES
(*After Magnée, 1935*)

The pulling apart of belemnite guards in cleaved limestones[1] and the filling of the spaces so created with secondary quartz or calcite is clearly analogous with the formation of pressure fringes, and in these too the evidence indicates both the presence of a fluid phase, and the elongation of the rock by stretching under high confining pressure which induces plastic flow of the rock as a whole.

Boudinage – Differences in the physical properties of rocks deformed in the same environment are responsible for many small-scale structures from which the kind and amount of deformation may be approximately gauged. Boudinage is the term given where a bed, or more rarely a dyke or vein, has been so deformed that it resembles sausages (Fr. *boudins*) lying side by side

---

[1] A. Heim, *Geologie der Schweiz*, Bd. 2, Lief 1, 1921, pp. 86–7, Leipzig.

(Fig. V−30B) or, in cross-section, a string of sausages.[1] Several forms of boudinage are illustrated in Fig. V−30, the essential features being the stretching of a more brittle bed parallel to the plane of bedding, accompanied either by the formation of tension gashes at fairly regular intervals, or by thinning due to necking, while the more mobile host-rock on either side squeezes into the spaces so formed.

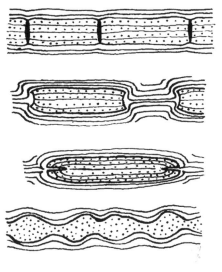

FIG. V−30A. SOME TYPES OF BOUDINAGE IN CROSS SECTION
Boudins are often about 1–2 ft long.

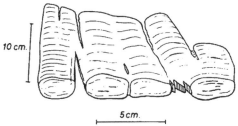

10 cm.

5 cm.

FIG. V−30B. BOUDINAGE SEEN IN THREE DIMENSIONS, ALLESJAURE AREA, LAPLAND
Quartzite with carbonate-filled cracks. (*After M. Lindström, Lunds Univ. Arssk., N.F., Avd. 2, Bd. 54, Nr. 3, 1958, Fig. 12, p. 53*)

Where tension gashes form, the brittle bed is subdivided into separate boudins, but with necking, shearing in the 45° position may result either in thinning, or finally in complete separation of the segments. Striking examples of boudinage are to be seen in banded gneisses, in which it seems clear from the sharp flexing of bands in the enclosing rock where this enters tension gashes, that this rock was plastic while the rock of the boudins was brittle.

Experiments with layered cakes of plasticine, putty, and cheese (which latter represents the brittle bed) reproduce boudinage by compression normal to the bedding, with consequent lateral extension of the mass.[2]

[1] G. Wilson, 'The Tectonic Significance of Small Scale Structures, and their Importance to the Geologist in the Field', *Ann. Soc. Géol. de Belg.*, Vol. 84, 1960–61, pp. 423–548.
[2] H. Ramberg, 'Natural and experimental boudinage and pinch and swell structures', *Journ. Geol.*, Vol. 63, 1955, pp. 512–26.

Boudinage in gneisses may with some confidence be related to the processes of metamorphism and synchronous tectonic deformation. However, it also manifests itself in a variety of ways in little-metamorphosed rocks, ranging from 'pull-aparts' formed at the soft-rock stage where hydroplasticity is involved, to structures formed by deformation during folding, especially in fold limbs. The principles underlying the formation of such structures in soft rocks are similar to those applicable to boudinage in crystalline crocks, but the soft-rock structures are better termed 'pull-apart' rather than boudinage, which latter term carries the implication of tectonic deformation.[1] Examples in soft rocks include stiff clay embedded in more mobile, water-saturated sands (Fig. II–22) or compact sandstone embedded in more plastic and usually argillaceous rocks.

Structures resembling boundinage in strongly folded rocks that are unmetamorphosed or in which slaty cleavage is developed require careful examination to decide whether they are pull-aparts which were present in the unfolded, soft-rock stage, or whether they formed from squeezing and stretching in the fold limbs, in which case the boudins may be elongated in the direction of the longitudinal fold axis, and may contain veinlets of quartz or calcite in the tension gashes.[2] This appears to be the origin of certain *ladder-veins*, including those in lamprophyre dykes in the Woods Point dyke swarm, Victoria. Both tension gashes and shear planes (cross-faults) were formed in these dykes by lateral compression and vertical extension of the surrounding soft slates (Fig. XII–30).[3] The formation of boudinage in altered banded gabbros to which similar principles apply has been described by Scheumann.[4] On the other hand, boudinage is seen to be essentially different from the formation of lenses due to the overall shearing of rocks along complementary shear planes.

Experimental tectonics

Experimental tectonic geology has a long history, but has always developed along two rather distinct lines.[5]

In the first, an attempt is made to imitate the conditions of high confining

[1] For full report, see M. L. Natland and Ph. H. Kuenen, 'Sedimentary History of the Ventura Basin, California, and the Effects of Turbidity Currents', in 'Turbidity Currents', *Soc. Econ. Palaeo. and Mineral.*, Spec. Pub. No. 2, 1951, pp. 76–107.

[2] This is, however, not always found because the planes separating boudins may also intersect along the strike, giving lozenge-shaped masses. However, a geometrical relationship with the fold-form is to be expected.

[3] E. S. Hills, 'The Woods Point Dyke Swarm, Victoria', *Sir Douglas Mawson Anniv. Vol.*, Uni. of Adelaide, 1952, pp. 87–100.

[4] K. H. Scheumann, 'Boudinagen und Mikroboudinagen im Metagabbroischen Plagioklas-amphibolit von Rosswein', *Abh. Sachs. Akad. Wissensch. Leipzig, Math.-naturwiss. Kl.*, Bd. 45, Hft. 1, 1956, pp. 1–18.

[5] W. Paulcke, *Das Experiment in der Geologie*, Karlsruhe, 1912, gives an account of prior investigations.

pressure and elevated temperature that obtain in the deeper levels of the crust. This has led to a greater understanding of the minutiae of crystal and rock-deformation under such conditions, as has previously been mentioned.

The second line of work aims at reproducing major crustal structures such as folds, faults, salt-domes, and the like in model experiments. Although earlier research yielded many informative results, as for instance in the classic experiments of Hall, Daubrée, and others,[1] the principles of the use of scale models have only of recent years been more fully enunciated. Hans Cloos, who used rather wet clay of the consistency of thick cream, argued that normal rocks at a depth of 10–20 km in the crust would yield under their own weight, and thus in an experiment in which the scale is reduced to one or two feet, the strength of the experimental material should correspondingly be reduced.[2] Greater precision has been given to such notions relating to scale models by the use of dimensional analysis, developed for geological models by Hubbert.[3]

The *dimensions* of a physical quantity are expressed by the number of times certain primary quantities such as mass, force, length, and time, are involved in the expression for the chosen quantity. Thus if M, L, and T represent mass, length, and time,[4] then area $= [M^0 L^2 T^0] = [L^2]$. Force (mass × acceleration) is MLT^{-2}. Following a relationship established by Bridgman, it can be shown that the ratio any physical property of the material of a model should hold to the corresponding property of the original, may be calculated from the ratio of easily measurable primary properties and quantities of the model to the corresponding properties and quantities of the original, which may be used in dimensional formulae. These ratios are the *model ratios*, and following Hubbert may be represented by Greek letters such as μ (mass), λ (length), τ (time), δ (density), and so on.

Bridgman's relationship establishes that where an expression for a model ratio in terms of μ, λ, etc., has the form $\mu^a \lambda^b \tau^c$, etc., the dimensional expres-

[1] J. Hall, 'On the Vertical Position and Convolutions of certain Strata and their Relation with Granite', *Trans. Roy. Soc. Edinburgh*, Vol. 7, 1815, p. 79. A. Daubrée, *Études synthétiques de géologie expérimentale*, Paris, 1879. B. Willis, 'The Mechanics of Appalachian Structure', *U.S. Geol. Surv.*, 13th Ann. Rept., Pt 2, 1891–2, pp. 211–89.

[2] H. Cloos, 'Künstliche Gebirge', *Natur und Museum (Senck. Naturf. Gesellsch.)*, Vol. 59, 1929, pp. 225–72; Vol. 60, 1930, pp. 258–69. 'Zur Experimentellen Tektonik', *Die Naturwissenschaften*, Jhg. 18, 1930, pp. 714–47; Jhg. 19, 1931, pp. 242–7. See also the important works of G. Koenigsberger and O. Morath, 'Theoretische grundlagen der experimentellen tektonik', *Zeitschr. Deutsch. Geol. Gesellsch.*, Monatshefte, Vol. 65, 1913, pp. 65–86; R. Maillet and F. Blondel, 'Sur la similitude en tectonique', *Bull. Soc. géol. France*, Ser. 5, t. 4, 1934, pp. 599–602.

[3] M. K. Hubbert, 'Theory of Scale Models as applied to the Study of Geologic Structures', *Bull. Geol. Soc. Amer.*, Vol. 48, 1937, pp. 1459–521. P. W. Bridgman, *Dimensional Analysis*, Yale University Press, 1931, develops the general theory. M. B. Dobrin, 'Re-creating Geological History with Models', *Journ. App. Phys.*, Vol. 10, 1939, pp. 360–71, gives a most useful account of tectonic models.

[4] Dimensional formulae are normally written in square brackets, unless the meaning is otherwise obvious.

sion has the form $M^a L^b T^c$. Thus the model ratio for velocity $[LT^{-1}]$ will be $\lambda \tau^{-1}$, and so on for other quantities involved. According to the principles of dynamic similitude of models and prototypes, not only must the model be geometrically similar, but its physical characteristics must be such that the forces, motions, and deformations are similar also. Since dimensional analysis permits physical properties to be expressed in terms of measurable quantities, the possibility exists, having chosen a model ratio, say, for length, to determine what the magnitude of some other physical property such as stress should be to ensure similitude. The interest may be, for instance, in the yield-point, or elastic limit, which has the same dimensions as stress. The dimensions of stress are $\dfrac{MLT^{-2}\ (\text{force})}{L^2\ (\text{area})} = ML^{-1}T^{-2}$, and the model ratio for stress, $\sigma = \mu \lambda^{-1} \tau^{-2}$.

Suppose the model is to be a static one representing simply a mountain system 100 km long, and composed of rocks as strong as granite, with a strength, say, 2×10^9 dynes/cm². As the model is static, accelerations due to motion are zero, and the dimensions for acceleration LT^{-2} may be factored out. Thus σ becomes $\mu \lambda^{-1} \tau^{-2}$ divided by $\lambda \tau^{-2} = \mu \lambda^{-2}$; but if $\delta = $ density $\delta = \mu \lambda^{-3}$, and $\mu = \delta \lambda^3$. Then $\sigma = \delta \lambda$.

The materials that are commonly used in such models, e.g. wet clay, have density but little different from granite, thus $\delta = 1$ and $\sigma = \lambda$. The model ratio for strength must thus be the same as for length. In the example chosen, $\lambda = 10^{-5}$, hence strength must be $2 \times 10^9 \times 10^{-5}$ dynes/cm², which means that a substance such as vaseline, soft wax, or wet clay should be used, rather than cloth or paper, which were used in some older experiments on mountain structure. It is seen that Cloos was essentially correct in using such clay for his experiments, where these represent structures of the order of 100 km in length, but it has been pointed out that the minutiae of his models do not properly represent the features of a graben only a few miles in length and a few hundred feet in depth. The shear-strength of the clay is then too great, and such features are better represented by loose sand, which has a lower shear-strength than clay.[1]

Moreover, when the length of the prototype mountain-chain is even greater, and the movements involved in its formation are also introduced into the model, the model ratios become different again. Griggs[2] chose model ratios for length, strength, density, and viscosity, in representing the formation of geosynclines and mountain-chains by the drag of convection currents in

---

[1] H. G. Wunderlich, 'Bruche und Graben im tektonischen Experiment', *Neues Jahrb. f. Geol. u. Pal.*, Monatschefte, Jhg. 1957, Hft. 11, pp. 477–98. M. K. Hubbert also demonstrated the suitability of dry sand for representing faulting in models. 'Mechanical Basis for Certain Familiar Geologic Structures', *Bull. Geol. Soc. Amer.*, Vol. 62, 1951, pp. 355–72.

[2] D. Griggs, 'A Theory of Mountain-Building', *Amer. Journ. Sci.*, Vol. 237, 1939, pp. 611–50.

the substratum, on the base of the crust. The time and velocity ratios were fixed when the other ratios mentioned had been determined. Griggs used a heavy oil and sand mixture for the crust, and viscous water-glass for the fluid substratum in one experiment, currents simulating convection cells being set up by rotating drums. In a model 24 in. long an event taking place over one million years was reproduced in one minute: in a model 3 in. long, in two seconds.

Experiments on salt-domes have also been based on dimensional analysis,[1] and E. Cloos has used clay in simulating the pattern of faults in the Kettleman Hills and the San Andreas–Garlock fault system.[2]

References to recent Russian work in this field are now becoming available in English.[3]

Sanford has more recently combined in a most informative way an analytic study of stress trajectories with dimensionally controlled scale-model experiments for simple structures, including broad up-warps and again local up- or down-warps with a step in the displacement at the lower boundary. The results closely simulate many actual structural associations found in nature.[4]

APPENDIX TO CHAPTER V

MOHR DIAGRAMS

1. The Mohr stress-circle

Graphical representation of the values of the normal and tangential stress-components on any plane containing the axis of mean stress (values of which are not considered) is conveniently done by means of the *stress-circle*, devised by Mohr and now known as a Mohr circle. Let the angle between the chosen plane and the minimum stress axis be α. It may then be shown by reasoning similar to that previously used for uniaxial compression (p. 82) that the following equations hold:

$$\sigma = \frac{\sigma_1 + \sigma_3}{2} + \frac{\sigma_1 - \sigma_3}{2} \cos 2\alpha \tag{1}$$

$$\tau = \frac{\sigma_1 - \sigma_3}{2} \sin 2\alpha \tag{2}$$

[1] L. L. Nettleton, 'Fluid Mechanics of Salt Domes', *Bull. Amer. Assoc. Petrol. Geol.*, Vol. 18, 1934, pp. 1175–204. T. J. Parker and A. N. McDowell, 'Model Studies of Salt Dome Tectonics', ibid., Vol. 39, 1955, pp. 2384–470.

[2] E. Cloos, 'Experimental Analysis of Fracture Patterns', *Bull. Geol. Soc. Amer.*, Vol. 66, 1955, pp. 241–56.

[3] See M. V. Gzovsky, 'The Use of Scale Models in Tectonophysics', *International Geology Review*, Vol. 1, 1959, pp. 31–47.

[4] A. R. Sanford, 'Analytical and experimental study of simple geologic structures', *Bull. Geol. Soc. Amer.*, Vol. 70, 1959, pp. 19–52.

where σ is the normal stress and τ the shear-stress on the chosen plane, and σ_1 is the maximum, and σ_3 the minimum principal stress.

If values of τ are represented on the ordinal and of σ on the abscissa of a graph, then σ_1 and σ_3 are points on the abscissa. If a circle be drawn containing them at opposite ends of its diameter, this circle is the locus of points that satisfy equations (1) and (2), as may be seen by inspection of Figure V–31.

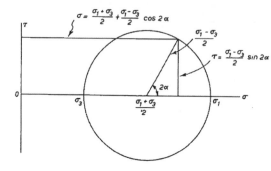

FIG. V–31. GEOMETRICAL REP-
RESENTATION OF STRESS BY
THE MOHR STRESS CIRCLE
(*After Hubbert*)

This possibility arises because if σ and τ are regarded as rectangular co-ordinates, equations (1) and (2) above are the parametric equations of a circle of centre $\left[\left(\dfrac{\sigma_1 - \sigma_3}{2} \right), \ 0 \right]$ and radius $\dfrac{\sigma_1 - \sigma_3}{2}$. The consequential geometry enables magnitudes of the normal and of the shear-stresses on a plane orientated at an angle α to the principal stress to be determined by inspection, as was first realized by Otto Mohr.[1] It may be noted that the shear-stress is a maximum when $2\alpha = 90°$ ($\alpha = 45°$), and that the value of the maximum shear-stress is $\dfrac{\sigma_1 - \sigma_3}{2}$, that is the arithmetic mean of the maximum and minimum principal stresses.

2. The Mohr diagram

Mohr's criterion for failure assumes that shear-fracturing is influenced not only by the magnitude of the shear-stress on a potential shear plane, but also by the normal stress across the plane, which may be conceived to press the two parts together, thus involving the internal friction, which is a character-istic of the material. Accordingly the ratio $\dfrac{\tau}{\sigma}$ at failure is approximately a constant for a given material. The ratio is best determined for loose materials such as soil by varying both σ and τ in a test-box, the upper half of which can be pushed over the lower, while pressure is applied vertically at the top. The value of the horizontal force required to shear the sand may thus be obtained

[1] For more complete explanations see M. K. Hubbert, 'Mechanical Basis for Certain Familiar Geologic Structures', *Bull. Geol. Soc. Amer.*, Vol. 62, 1951, pp. 355–72.

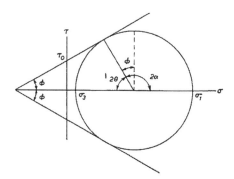

FIG. V-32. STRESS CONDITIONS FOR SHEAR-FRACTURING IN COHESIVE MATERIALS

Faults occur at an angle $= 45 + \dfrac{\phi}{2}$ to the axis of least stress. (*After Hubbert*)

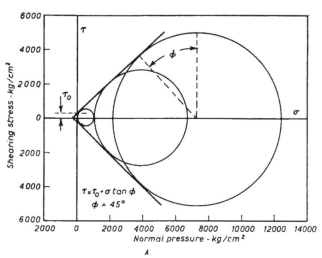

A

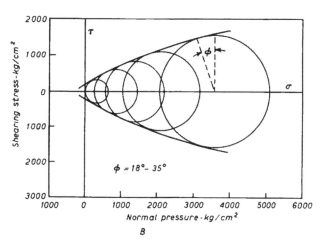

B

FIG. V-33A. MOHR DIAGRAM FOR A SANDSTONE, SHOWING STRAIGHT ENVELOPE CURVES

B. MOHR DIAGRAM FOR A SHALE, SHOWING ENVELOPE CURVES CONCAVE TO THE STRESS AXIS

at different pressures. A graph of values of T (the shearing force), against N (the pressure) for material such as soil generally approximates to a straight line whose angle of slope is ϕ degrees, and which is known as the *fracture-line*. It is seen that $\dfrac{T}{N} = \tan \phi$, and the stresses τ and σ at the moment the material shears through may be regarded as the components of a force acting on a plane inclined at $\phi°$ to the compression, which is by analogy with surface friction, called the *angle of internal friction*. In loose, running sand, which lacks cohesion, the graph $\dfrac{\tau}{\sigma}$ passes through the origin, but cohesive material such as rocks have a *transverse* or *shear-strength* which must be overcome even with zero normal stress before the material will yield. Thus for such materials, the graph intersects the ordinate representing $\tau(\tau_0)$ (Fig. V–32). Since shear-failure will occur, at various values of σ, only for a point that lies on the fracture-line, it is clear that Mohr circles for various values of σ_1 and σ_3 in a triaxial test, must be tangent to the fracture-line. It then follows geometrically that the angle of shear α measured from the axis of minimum stress σ_3 is also represented on the diagram, or that it may be calculated from the primary equations.

For various values of σ_1 and σ_3 a succession of Mohr circles at failure may be constructed, and the envelope to these then represents the line of fracture. For substances that do not change their internal friction under high confining pressure, the envelope is a straight line. This is approximately true for granites and many sandstones (Fig. V–33A). However, if the material becomes more ductile at high pressures, the angle ϕ should decrease, and the envelope will correspondingly flatten, and become convex upwards. This is seen in Fig. V–33B, for a test on shale.

CHAPTER VI

Planar and linear structures and jointing

The geometrical data for structural interpretation in geology become available through plotting the attitude of planar or linear structures and textures, including both those which being visible to the naked eye may be mapped in the field, and those that can be determined only under the microscope, and must accordingly be studied in the laboratory. In this chapter we shall deal with macro-structures, and later (Chapter XIII) with structural petrology, which latter, although it does include reference to macro-structures, is more particularly concerned with those that are seen under the microscope.

REGIONAL GEOMETRY OF MACRO-STRUCTURES

Planar components to be observed in the field include bedding, joints, faults, cleavage, and foliation; *linear components* are less precisely definable, but they include the linear intersections of planes, as well as linear elements *sui generis*, having their own distinctive individuality as structures or textures of linear habit. Slip-striae on slickensided faults or bedding planes, and linear fabrics due to the parallel orientation of elongated crystals, xenoliths, or mineral groups, are examples. All such linear features are commonly termed *lineations*, but if the term is to be applied in this comprehensive way it is obvious that it can, unless more particularly specified, have no precise structural significance, since lineations may form in so many ways having different relationships to the overall form and origin of major structures. Moreover, the intersection of a planar structure on any chance plane observed in the field appears as a lineation, so it is also clear that careful study will be required in such a case to reveal its true geometrical significance.[1]

The structural map — There are two distinct ways in which the geometry of planes and lines is studied. In the first, the attitude of a plane or line at its

[1] A more detailed account of lineation is given in Chapter XIII and lineations of various origins are referred to in their appropriate contexts, especially under faulting, folding, and igneous rocks. See also P. Collomb, 'La linéation dans les roches', *Bull. trimestriel S.I.G. du B.R.G.M.*, 12e Année, 1960, Nr. 48, pp. 1–11. E. Cloos, 'Lineation', *Geol. Soc. Amer.*, Mem. No. 18, 2nd impression, 1949.

place of observation in the field is plotted at the point on the map representing its location. In this way the observations made at the surface are built up into a structural map, which is a geological map on which structural observations are incorporated. In the course of mapping for this purpose, the strike and dip of planes, and the plunge or pitch of lineations must be observed and recorded.

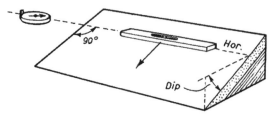

FIG. VI–1. BEDDING PLANE SHOWING MEASUREMENT
OF STRIKE

The spirit-level touches the bed and is horizontal; the prismatic compass is sighted along the strike. The dip is measured at right angles to the strike, using a clinometer.

Strike is the direction of the intersection of an inclined geological plane with an imaginary horizontal plane. It is determined on suitable exposures by laying a spirit-level on the outcrop of the plane, and measuring the bearing of the line along which the level rests (Fig. IV–1).

Dip is the angle of inclination of a plane from the horizontal. It is measured in a vertical plane, at right angles to the strike (Fig. VI–1).

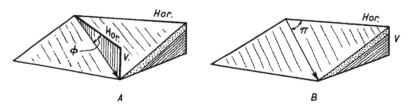

FIG. VI–2. PLUNGE AND PITCH OF LINEATIONS
A. ϕ is the plunge.
B. π is the pitch or rake.

Plunge is the inclination of a linear structure-element, measured from the horizontal in a vertical plane containing the line. It is also necessary to record the *trend* of the plunge, which is the strike of the vertical plane containing the line, and the sense in which the downward inclination occurs (Fig. VI–2A). *Pitch* was formerly used for plunge but this is no longer recommended, because of possible confusion with its use, still practised in mining geology, where the pitch of an ore-shoot is measured from the horizontal in the plane of an ore-body, which itself may not be vertical (Fig. VI–2B). Accordingly it is considered preferable to use *rake* for the inclination of a linear structural feature, measured in an inclined plane, as with lineation (q.v., p. 436) on a cleavage plane dipping at any angle. Rake cannot be fully described without recording also the dip and strike of the plane in which the rake is recorded.[1] Pitch is still widely used for rake.

The integration of data on dip and strike with the other standard observations that are made on rocks in mapping permits the interpolation and extrapolation of outcrop data to give the regional pattern of formations, and

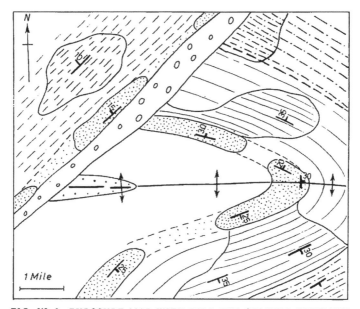

FIG. VI–3. EXPOSURE MAP WITH DIPS AND STRIKES, WITH THE OUTCROP OF THE MAPPED FORMATIONS COMPLETED

The areas enclosed by lines, and marked by close-spaced symbols, are areas where rock is exposed and dips can be measured.

affords the primary data on which the location and geometry of the folds is based, as illustrated in Fig. VI–3. It is in regional mapping and the related exercises of section-drawing, constructing isometric blocks, and the like, that most of the practical results of structural geology accrue, not only in general geology but also in petroleum and engineering geology. The region concerned is often quite small, as in a mine or at a dam site, and many standard methods for the geometrical study of rocks, as for instance the drawing of precise sections, location of the intersection of faults and lodes, and so on, apply to problems that arise in the course of such work.

[1] Standard textbooks on field methods give information on the techniques that may be used in a variety of circumstances. See especially F. H. Lahee, *Field Geology*, 1961, McGraw-Hill, New York; J. W. Low, *Geologic Field Methods*, 1957, Harper, New York.

Statistical study

A second type of structural interpretation is the statistical examination of the attitude of planes and lines, whereby all or most of the attitude-data for a selected region are gathered together, irrespective of their actual location within the area, and imagined to be concentrated at one point. This permits ready determination of the range in attitude of similar planes, the extent to which they approximate to parallelism or are grouped systematically about preferred directions or orientations. In general, where the geology is complex, such statistical study must be made on a succession of small areas within which there is a certain homogeneity of major structures, and a statistical diagram may be drawn on each sector, thus enabling the regional changes in trend or intensity to be visualized.

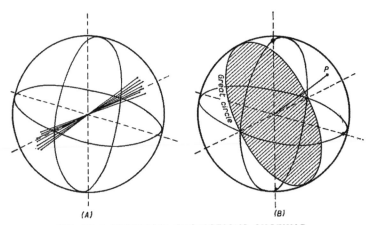

(A)   (B)

FIG. VI-4. SPHERICAL PROJECTIONS, SHOWING –

A. Lineations grouped as at the centre of the sphere and intersecting it in a group of penetration points.
B. A plane shown as a diametral plane, intersecting the sphere in a great circle. The pole to the plane intersects the sphere at *P*.

A much used method of examining the data is to construct a plane projection, either stereographic or equal-area, of the spherical projection of the planes or lines.

In the spherical projection, all planes and lines measured in the field in a given area are imagined to be located in their correct orientation at the centre of a sphere (Fig. VI–4). The planes thus become diametral, and their intersections with the sphere are accordingly great circles of the sphere. The lines become diameters, intersecting the sphere at points known as *centres of penetration*. An alternative method of representing planes is to draw, from the centre of the sphere, a *pole* normal to each plane, which like the structural line intersects the sphere at a point. The position of a plane in space is fully represented by the position of the centre of penetration of its pole. A diagram showing planes as great-circular intersections is known as a cyclographic or β-diagram, and one using poles a *pole* or π-diagram.[1] Lineations are analogous to poles and penetrate the sphere at *point intersections*.

The spherical projection being three dimensional, it cannot be used in routine studies, for which purpose projections onto one plane are made. The projection most susceptible to ready geometrical handling is the stereographic, the use of which for statistical analysis is, however, limited because equal areas on the projection do not represent equal areas on the sphere. The stereographic projection is, nevertheless, of great value in determining the orientation of intersections of individual planes such as lodes and faults, and for many such purposes it replaces the more cumbersome trigonometric methods.[2]

Plotting on the stereographic projection is facilitated by the use of a plat or net, on which is shown the projection of the parallels and meridians of the sphere, at suitable intervals, generally of two degrees. This enables the location of a point whose spherical co-ordinates (latitude and longitude) are known, within limits of accuracy that are appropriate for many geological examples.

A stereographic plat such as the above is referred to as a Wulff Net. On the other hand, where a study of the distribution of points or planes in space is to be made on a statistical basis, Lambert's equal-area plat, now known in structural geology as the Schmidt Net, is used (Fig. VI−5). The geometrical basis for these two nets is shown in Fig. VI−6, from which, if desired, the location of points may be calculated. If the angular relationship to the N.-S. and E.-W. diameters is known, the point may be located using the net lines. The strike of a plane is a diameter on the bearing of the strike, but to insert the great circle corresponding to a dipping plane, a great circular arc must be drawn as shown in Fig. IX−9. On the stereographic net oblique great circles are represented by truly circular arcs, but on the Schmidt Net they are not. Because of the characteristics of the projection, oblique great circles may be drawn on the Schmidt Net by using on a transparency the meridional great circles, rotated to the desired position.

Since, in structural geology, most field data are obtained while the observer is above the rocks and looking down, it is convenient to represent structural data by the intersections on the lower hemisphere of a spherical projection, but the convention used must nevertheless be stated when work is presented in reports or publications. The distribution or spread of point intersections is a *scatter* or *point diagram*, but generally the data are more graphically shown by means of contours which surround areas in which the dots have a certain

[1] B. Sander, 'Geologie des Tauern-Westendes I: Ueber Flächen und Achsengefüge', *Mitt. Reichsamt f. Bodenforsch.*, *Wien*, Vol. 4, 1942, pp. 1−94.
[2] F. C. Phillips, *The Use of Stereographic Projection in Structural Geology*, London, 1954. D. V. Higgs and G. Tunell, *Angular Relations of Lines and Planes*, Dubuque, Iowa, 1959.

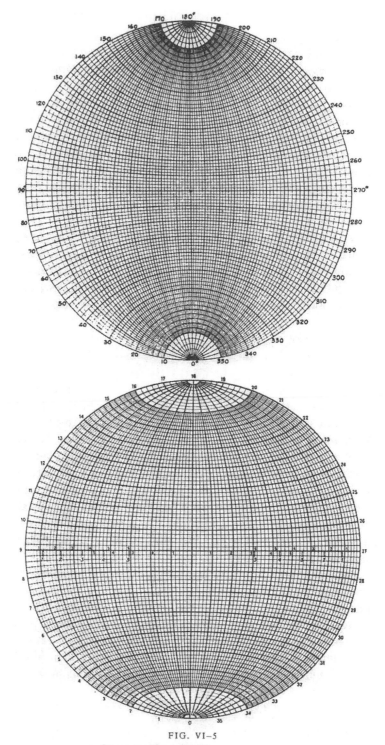

FIG. VI–5

A. Stereographic or Wulff net.
B. Lambert's equal area plat, or Schmidt net.

density of distribution. This is calculated on the basis of a fixed percentage of the total number of dots which lie in one per cent of the area of the net.

In practice the point diagram is made by inserting the dots on a transparent overlay placed on the Schmidt Net, so as to preserve the net for further use. The dots are counted by moving a square of material, in which a small circle is cut of area equal to one per cent of the Schmidt Net (Fig. VI–7) in regular steps over the point diagram. Suppose 300 points are plotted. Then if 15 points appear within the small circle when it is situated at some particular place on the net (shown by a superimposed rectangular grid) this represents 5 per cent of the dots, and the figure 5 is pencilled in at the intersection of the

FIG. VI–6. STEREOGRAPHIC AND SCHMIDT NETS

Left half of diagram shows stereographic projection of points $P'P''$ to X and Y respectively. Right half shows Schmidt projection. For P', $OV = \dfrac{P'N}{\sqrt{2}}$; for P'',

$OZ = \dfrac{P''N}{\sqrt{2}}$. *(After Fairbairn)*

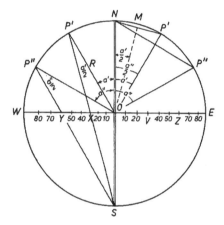

grid axes, centred on the small circle. The centre counter is moved across the whole diagram, and appropriate pencilled numbers inserted, for each unit move made. Generally a centimetre grid is used, with a Schmidt Net 20 cm in diameter and a centre counter of 2-cm diameter.

For points lying close to the circumference of the net, a peripheral counter is used as shown in Fig. VI–7.[1] Point intersections lying just within the net represent poles of lines that deviate but little from each other, except that some lie above and some below the diametral plane of the spherical projection. Those above or below this plane on one side have their congeners in those that lie below or above on the other and accordingly the point intersections lying above the opposite ends of a diameter of the net circle should be grouped together. When the centre counter (Fig. VI–7), in being moved from intersection to intersection of the grid axes, intersects the periphery of the net circle, the peripheral counter B must be substituted for it. By using the slot in this counter, one counter circle can be centred on an intersection such that the circle intersects the periphery. The opposite counter circle likewise intersects the periphery, and the number of points within each counter circle is added

[1] See H. W. Fairbairn, *Structural Petrology of Deformed Rocks*, Cambridge, Mass., 1949, for an excellent account of these procedures.

together. The corresponding percentage is entered on the appropriate grid intersection for later contouring. Finally, the points on the actual periphery are counted by centring the peripheral counter, so that the centre of each of its two circles lies on the periphery, and, after adding the points in each counter circle, the appropriate percentage figure is pencilled in *outside* the net. The counter is rotated at approximately 1-cm intervals around the periphery to complete the so-called counting-out diagram.

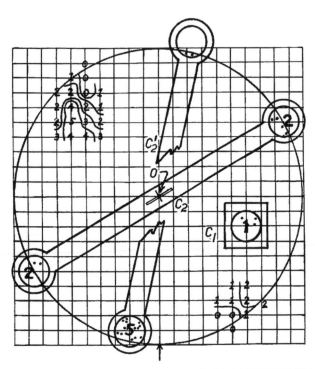

FIG. VI-7. USE OF COUNTING DEVICES IN CONTOURING
SCATTER-DIAGRAM
(*After Fairbairn*)

Contouring is done by lines drawn according to selected percentage intervals, generally with not less than four and not more than ten contours. Although the notion is simple, there are various refinements for more precisely locating the contour lines.[1]

In order to make the diagram more readily understandable, the contour areas are appropriately shaded or stippled, with the optical density of shading matching the density of the dots.

[1] See, e.g., E. B. Knopf and E. Ingerson, 'Structural Petrology', *Geol. Soc. Amer.*, Mem. No. 6, 1938, and H. W. Fairbairn, *Structural Petrology of Deformed Rocks*, Cambridge, Mass., 1949.

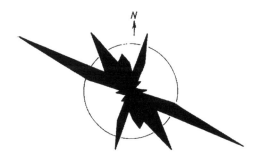

FIG. VI–8. CONSTRUCTION
OF JOINT ROSE

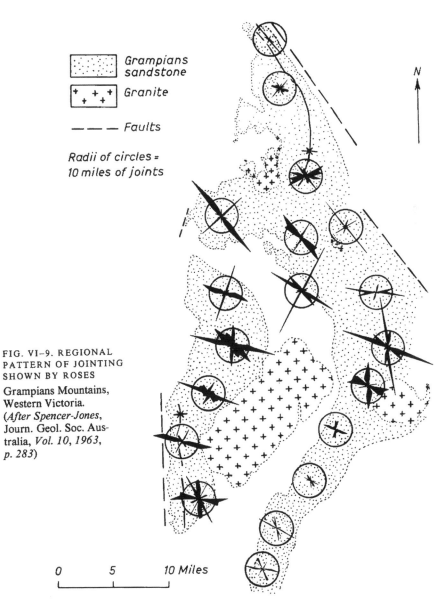

Grampians
sandstone

Granite

— — — Faults

Radii of circles =
10 miles of joints

N

FIG. VI–9. REGIONAL
PATTERN OF JOINTING
SHOWN BY ROSES

Grampians Mountains,
Western Victoria.
(*After Spencer-Jones*,
Journ. Geol. Soc. Aus-
tralia, *Vol. 10, 1963,
p. 283*)

0   5   10 Miles

A method of representing the regular variation in orientation of lineations throughout a region has been described by Elliott.[1] Points of equal dip are joined on the map by lines called *isogons*.

The direction-rose – Where the attitude of planes is not fully determinable, say for instance the strike is readily seen as in aerial photographs of strongly jointed rocks but the dip cannot be determined, it is convenient to use a type of statistical diagram known as a *rose*,[2] which has been used chiefly for representing joints in structural work. All joints occurring in a given sector of the compass circle, say within 10° or 5°, are counted, and a radial line is drawn representing them, in the median bearing of each sector. The length of the radial line is determined by a scale of concentric circles, and often is

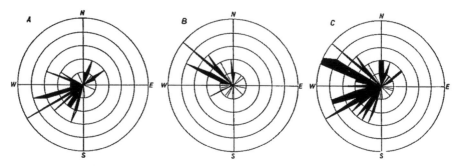

FIG. VI–10. FAULT-DIP ROSES IN STRONGLY FOLDED SILURIAN ROCKS, STUDLEY PARK, VICTORIA

Frequency-diagrams show directions of dip of 100 reverse faults, in two areas *A* and *B*. *C* is a composite diagram combining *A* and *B*. Interval for bearings 5°; circles at unit distance apart, one fault per circle.

shown as the *number* of joints occurring in each sector.[3] This, however, may profitably be modified, as for example to represent the *total length* of joints in the sector. Difficulties due to identification of joints across areas covered with soil or vegetation will be experienced using either method, but are inescapable. The rose is completed by joining the ends of the lines representing the joints, and may be made more obvious by shading (Fig. VI–8). Prominent joint directions may then be clearly revealed, and a series of roses, each for a given sector of a region, will show changes in the trend and number of joints from place to place (Fig. VI–9). Similar roses may be drawn for faults, representing their direction of dip, where the exposures are mainly in road or

[1] D. Elliot, 'The Quantitative Mapping of Directional Minor Structures', *Journ. Geol.*, Vol. 73, 1965, pp. 865–80. A more complete account of the principles of statistical treatment is given on pp. 398–404. See also A. B. Spencer and P. S. Clabaugh, 'Computer Program for Fabric Diagrams', *Amer. Jour. Sci.*, Vol. 265, 1967, pp. 166–72; W. H. Lam, 'Computer Method for Plotting Beta-diagrams' (Discussion), *Amer. Journ. Sci.*, Vol. 267, 1969, pp. 1114–17.

[2] This is perhaps best known in the wind-rose used by meteorologists.

[3] In fact this generally refers to the *number of observations* made on joints rather than the *number of joints* having a certain bearing.

railway cuttings. The fault diagram shown in Fig. VI–10 demonstrates the predominance of fault dips in the N.W.–S.E. sector. In this diagram, only one end of each 'petal' was shown, since in dealing with dip, the sense as well as the direction or bearing is significant.

JOINTS

Joints are the commonest of secondary structures in rocks, and although apparently the simplest, being merely cracks, they include a variety of types of diverse origin. The word *joint* was first used by coal-miners for smooth,

FIG. VI–11. CLOSELY 'JOINTED' GRANITE, MT MARTHA, VICTORIA
Many of the 'joints' can be shown to be small faults. (*Photo: E.S.H.*)

straight cracks separating blocks of rock that appeared to them to have been 'joined' as in brickwork, the 'joints' being at right angles to the bedding planes. This type of joint, that is to say, a fracture which is normal or nearly normal to the bedding, and along which there has been no visible displacement, is the most typical, but also the least understood, of all structures currently classified as 'joints'. 'The joint in the rock, thin as a hair, and straight as a measuring-rod, that piece of petrified geometry, promises much, yet discloses little.'[1]

[1] Quoted from C. M. Nevin, *Principles of Structural Geology*, New York, 3rd edn, 1942, p. 131.

The lack of visible displacement in a direction parallel to the plane of a joint distinguishes it from a fault, but planes that appear to be incipient faults with no visible displacement are also classified as joints.

In massive igneous rocks such as granite, fractures are often called joints because there are no structures that permit fault displacements to be observed, although the planes may be slickensided and are in reality faults (Fig. VI–11). It is preferable to record such fractures as 'shears' or 'shear-fractures' rather than as joints, where the evidence permits such interpretation. A further complication arises because in a rock that was jointed before it was subjected to later deformation, sliding and readjustment will take place on the existing joints, giving rise to slickensides.

FIG. VI–12. AERIAL PHOTOGRAPH OF JOINT SET IN SANDSTONE,
WONDERLAND RANGE, GRAMPIANS, VICTORIA
(After Spencer-Jones, 1963)

Regional mapping of joints will generally reveal a geometrical relationship to faults, folds, warps, intrusions, and other tectonic elements, that will suggest genetic connections with the stress-distribution and strain-pattern, but it is rare that an entirely satisfactory synthesis can be made.[1]

Joint sets and systems – Joints occurring in a parallel or nearly parallel array constitute a *set*, but in most regions there are at least two, and perhaps more, prominent sets which together form the *joint system* of the area. Joints that conform with an overall pattern within a region, as opposed to irregular or

[1] Additional information on joints will be found in the chapters dealing with folding, faulting, and igneous rocks, to which accounts the present chapter afford a broad introduction.

purely local joints, are termed *systematic*. Two sets that are believed to represent complementary shear sets, whether regionally or locally as within a fold, are referred to as *complementary* or *conjugate* sets.[1] Master joints, which penetrate several strata and persist for long distances, even for miles (Figs. VI–12, 14), are important in regional tectonic analysis while the minor joints related to local structures and to particular strata are of interest in detailed structural interpretation.

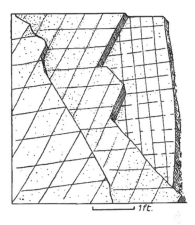

FIG. VI–13. JOINTING OF DIFFERENT PATTERNS IN DIFFERENT FORMATIONS
Three beds of sandstone in strongly folded Silurian rocks, Studley Park, Victoria.

Jointing in superficial rocks – At and near the surface, joints may either be weathered to produce open channels, or cemented by secondary deposits of minerals such as limonite, calcite, quartz, or opaline silica (Fig. VI–15). Major jointing is often strongly reflected in the drainage-pattern as the prominent joints are followed by streams and gullies (see Morphotectonics, Chapter XIV), and again the jointing of particular strata may be sufficiently characteristic of them to permit identification of the strata in aerial photographs.

The *intensity* or closeness of spacing of jointing in any bed is generally greater in the vicinity of structures that influence or determine the joints, a relationship that may assist in the location of buried limestone reefs, salt domes, and other structures.

Some fractures present at and near the surface do not persist in depth, and thus appear to have been formed by superficial processes such as expansion and contraction on weathering, ice action, relief of load on erosion, and so on. Where fractures can be clearly demonstrated to be of superficial origin, it is preferable to distinguish them by specific terms, such as, for example, *sheeting* in granites.[2] *Mud cracks* formed by shrinkage in drying argillaceous

---

[1] P. L. Hancock ('Joints and Faults: The Morphological Aspects of their Origins', *Proc. Geol. Assoc.*, Vol. 79, 1968, pp. 141–51), redefines certain terms, but his distinction between conjugate and complementary systems (preferably, sets) seems too subtle to be generally acceptable.

[2] See Igneous Rocks, Chapter XII.

FIG. VI–14. MASTER JOINTS IN GENTLY FOLDED QUARTZITES
Arnhem Land Plateau, North Australia. The joints are weathered and eroded by streams.
(*Vertical aerial photo by R.A.A.F.: reproduced by permission*)

sediments are likewise not classified as joints. Shrinkage is also involved in
the closely spaced cracking of cherts, which, where not of tectonic origin,
may be connected with the slow dehydration of gelatinous hydrated silica;
and again, curving cracks that originate in blocks of building-stone some
weeks after they have been quarried likewise appear to be caused by shrinkage, due to drying out of the pore-water or 'quarry-sap'. Such curving and
haphazardly arranged cracks are better referred to as 'fractures' than as
'joints', but the regular cracks formed in igneous rock due to shrinkage on
cooling are, on the other hand, usually classed as joints.[1]

The majority of joints are of tectonic origin and represent either tension
cracks or incipient shear-fractures, but jointing is often prominent in little-

---

[1] The desire to reserve the term 'joint' for straight, smooth fractures occurring in parallel
sets is seen in the common tendency to refer to columnar 'jointing' in lavas as columnar
'structure', and it may eventually be possible to restrict the use of 'joint' to tectonic fractures.

disturbed rocks including flat-lying sediments as well as in folded and faulted beds. It has been suggested that such jointing may represent fatigue cracks formed by small alternating strains due, for instance, to earth tides.

Cleat in coal – Coal is noted for the regularity of its jointing. Usually there are two directions approximately at right angles to each other and to the bedding, one of which is better developed than the other, exhibits lustrous surfaces, and is specifically termed the *cleat* or *bord*, the other being the *end* or *head*. Because of the ease of hewing coal across the cleat, its direction

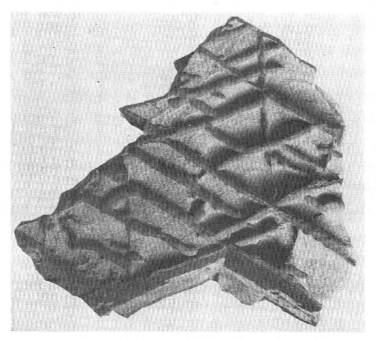

FIG. VI–15. JOINTS IN SANDSTONE, CEMENTED BY LIMONITE
Silurian, Victoria. ×⅓.

influences the layout and working of coal-mines.[1] Cleat is sharper in the bright than in the dull layers of coals, and is to be seen in even the thinnest films of coalified vegetation embedded in shale and sandstone, in which it has the same direction as in the coal-seams themselves. The direction of cleat is also remarkably constant over large areas, not only in the one coal-basin, but in others in the same country. According to Kendall, the trend of cleat in the Northern Hemisphere is predominantly northwest–southeast from China to Spitzbergen, Europe, and North America.[2]

[1] A. Raistrick and C. E. Marshall, *The Nature and Origin of Coal and Coal Seams*, London, 1939.
[2] P. F. Kendall and H. Briggs, 'The Formation of Rock Joints and the Cleat of Coal', *Proc. Roy. Soc. Edinburgh*, Vol. 53, 1933, pp. 167–87.

FIG. VI–16. SHEAR JOINTS PARALLEL TO SMALL FAULTS IN SILURIAN ROCKS
Studley Park, Victoria. (*Photo: E.S.H.*)

Detailed study, however, demonstrates that cleat and head include, in the one coal-field, joints of different origin, both tension- and shear-fractures which are geometrically related to the regional fold and fault pattern.[1] Thus any regional persistence of trend may reflect only the tendency for many persistent geological trends to occur in the N.W.–S.E. or N.E.–S.W. quadrants. Kendall suggested, partly because of the parallelism of cleat over large areas, that it might be caused by fatigue-cracking under the influence of the body tide of the earth.

Shear and tension joints

Although the chief distinguishing feature of a joint is that there has been no detectable movement on it, it is usual to recognize *shear joints* as well as *tension joints*, shear joints being incipient shear fractures. Some fractures on which there is no visible displacement are, however, clearly of the same origin

---

[1] J. M. Deenen, 'Breuken in Kool en Gesteente', Uitkomsten van nieuwe Geol.-Pal. Ondersuchungen van den Ondergrond van Nederland. *Med. Geol. Stichting, Maastricht*, Ser. C, 1–2, No. 1, 1942, p. 104.

as nearby faults, and where this is obvious as in Fig. VI–16 it would be preferable to classify them as incipient faults rather than as 'joints'.[1]

Most field-workers would agree that joints have certain features which distinguish them from faults, and in Fig. VI–16 the passage of the same minor fracture through several beds of different lithology is one such feature. Furthermore, local joints are typically confined to one bed, in which they are normal or nearly normal to the bedding even in folds, and the spacing of such joints in any one locality is virtually proportional to the bed-thickness, in beds of the same lithology. The surfaces of many joints where freshly exposed

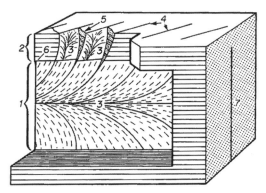

FIG. VI–17. SCHEMATIC BLOCK DIAGRAM SHOWING PRIMARY SURFACE FEATURES OF A JOINT PLANE

1. Main joint plane. 2. Fringe. 3. Plumose structure. 4. F-joints. 5. Cross-fractures. 6. Shoulder. 7. Trace of main joint on a transverse surface. (*After Hodgson, 1961*)

Note: the angle between f-joints and the main joint plane is often much more acute than is shown in the diagram. (From R. A. Hodgson, *Amer. Journ. Sci.*, Vol. 259, 1961, Fig. 1.)

exhibit plumose or herringbone markings which are unknown on faults. The details are shown in Fig. VI–17, in which it should be noted that the curving marks of the plume result from the intersection of *en echelon* minor joints (f-joints) with the main joint face.[2] The V formed by the plumose marks points towards the end from which the joint was propagated.

[1] See P. L. Hancock, 'Joints and Faults: The Morphological Aspects of their Origins', *Proc. Geol. Assoc.*, Vol. 79, 1968, pp. 141–52.

[2] J. B. Woodworth, 'On the Fracture System of Joints, with Remarks on Certain Great Fractures', *Proc. Boston Soc. Nat. Hist.*, Vol. 27, 1896, pp. 163–84 (quoted by R. A. Hodgson, 'Classification of Structures on Joint Surfaces', *Amer. Journ. Sci.*, Vol. 259, 1961, pp. 493–502); J. C. Roberts, 'Feather-fracture and the mechanics of rock jointing', ibid., pp. 480–92. P. Bankwitz, 'Über Klüfte', I, *Geologie*, Jhrg. 14, 1965, pp. 241–53; and II, ibid., Jhrg. 15, 1966, pp. 896–941.

The distinction between shear and tension joints is not easy and is often made on somewhat arbitrary grounds.

Joints that form two complementary conjugate sets are often regarded as shear joints purely because of the geometrical similarity with conjugate shear fractures in deformed solids, but such an assumption is certainly not valid.

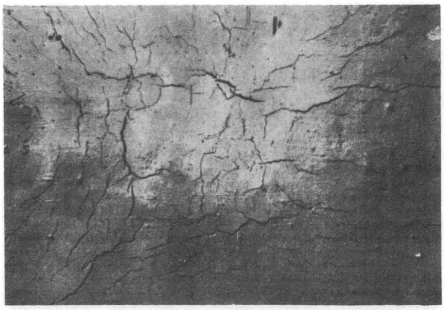

FIG. VI–18. TENSION JOINTS IN THREE DIRECTIONS PRODUCED BY THE UP-DOMING OF A THIN SHEET
(*After Seidl and H. Cloos*)

Analogues with the joint pattern developed on elongated domes have been made experimentally by Seidl and by H. Cloos,[1] and in these two sets of complementary tension cracks were produced with a third set also in certain parts (Fig. VI–18). The closest experimental analogue is, however, that made by Mead using a thin coating of paraffin wax over a rubber sheet (Fig. VI–19), and this suggests that the commonly found close polygonal jointing of sedimentary strata may be connected with their sheet-like form.[2] Tension joints would be expected to form normal to an axis of tension, or, as with extension fractures in test specimens, parallel to an axis of compression. It is generally thought that they would exhibit rougher surfaces than shear joints and would tend to become open and so filled with vein material such as quartz or calcite. However, any simple general explanation of jointing is not to be

[1] See Chapter IX, pp. 273–77.
[2] W. J. Mead, 'Notes on the Mechanics of Geologic Structures', *Journ. Geol.*, Vol. 28, 1920, pp. 505–23.

expected since the term 'joint' is applied to fractures of different origin formed in different environments.

Jointing may be developed in superficial deposits such as alluvium or calcareous sand dunes (aeolianite); it may result from regional stress and also from local strains related to individual folds, faults, and warps; in metamorphic rocks it must form chiefly as a result of the relaxation of stored strain energy. As the rocks are elevated to higher crustal levels and erosion removes part of the load, it is probable that many joints in folded but unmetamorphosed rocks also result from similar stress-relaxation. Near the surface jointing may also be induced by differential volume changes on weathering.

Despite these and other conceptual difficulties, jointing constitutes an important element of structural patterns, both regionally and locally. Not only the trends and relationships of systematic joints but also the intensity of

FIG.   VI–19.   ARTIFICIALLY   INDUCED FRACTURES IN THIN PARAFFIN WAX SHEET ADHERING TO RUBBER, AND COMPARABLE WITH JOINTS

(*After Mead*)

development of jointing can be related to folds and faults, and may yield important information where cover-rocks have been affected by movements in a concealed basement. Fracture analysis is an exploration tool with important practical applications.[1]

[1] P. H. Blanchet, 'Development of Fracture Analysis as Exploration Method', *Bull. Amer. Assoc. Petrol. Geol.*, Vol. 41, 1957, pp. 1748–59; R. A. Hodgson, 'Genetic and Geometric Relations between Structures in Basement and Overlying Sedimentary Rocks . . .', ibid., Vol. 49, 1965, pp. 935–49; C. W. Brown, 'Comparison of Joints, Faults and Airphoto Linears', ibid., Vol. 45, 1961, pp. 1888–92. G. Gol'braykh *et al.*, 'Tectonic Analysis of Megajointing: A Promising Method of Investigating Covered Territories', *Internat. Geol. Rev.*, Vol. 8, 1966, pp. 1009–16.

FIG. VI–20. JOINT SETS IN GRANITE, GUTHEGA TUNNEL, SNOWY
MOUNTAINS AREA, NEW SOUTH WALES
(*Photo by courtesy D. G. Moye and the Hydroelectric Commission*)

Lensen[1] introduced the concept of *isallo-stress lines* in relation to the study
of conjugate faults and Scheidegger[2] has extended the method to shear-
fracture analysis in general. The angle in plan between the two conjugate
systems, which contains the greatest horizontal pressure, is called the compressional angle γ and from lineament study it is possible to determine γ and
plot it on a map. Lines of equal γ are called *isallo-stress lines*, and their values

[1] G. J. Lensen, 'Measurement of Compression and Tension: Some Applications', *N.Z.
Journ. Geol. Geophys.*, Vol. 1, 1958, pp. 565–70.
[2] P. H. Lu and A. E. Scheidegger, 'An Intensive Local Application of Lensen's Isallo Stress
Theory to the Sturgeon Lake South Area of Alberta', *Bull. Canadian Petrol. Geol.*, Vol. 13,
1965, pp. 389–96.

and pattern are indicative of regions of strike-slip, normal, or reverse faulting, also of the direction of principal stress.[1]

Long study by Nickelsen and Hough of jointing in the Appalachian Plateau shows that the joint pattern is cumulative, being a record of all events that have produced stress differences sufficient to cause fracture; systematic joints formed early, independent of folds and faults. Older joints may be tilted, displaced by bedding-slip during folding, or dragged out of alignment by faulting, while younger joints develop, some by stress-relaxation.[2]

Practical aspects of jointing – Jointing is of great practical importance in applied geology, especially in relation to the strength of rocks in engineering works, in quarrying and mining, and also in connection with the flow of fluids such as water leaking from reservoirs or oil flowing from a fractured coral reef. Many unexpected changes in the joint pattern both laterally and in depth are encountered, sometimes beneficial but perhaps more often detrimental in engineering geology (Fig. VI–20).

[1] R. E. Ranken and A. E. Scheidegger, 'An Application of Isallo Stress Analysis to Areas in Tanzania, Texas and Alaska', *Pure and Applied Geophys.*, Vol. 75, 1969, pp. 102–16.

[2] R. P. Nickelsen and V. D. Hough, 'Jointing in the Appalachian Plateau of Pennsylvania', *Bull. Geol. Soc. Amer.*, Vol. 78, 1967, pp. 609–29; see also F. Moseley and S. M. Ahmed, 'Carboniferous Joints in the North of England and their Relation to Earlier and Later Structures', *Proc. Yorkshire Geol. Soc.*, Vol. 36, Pt 1, 1967, pp. 61–90.

CHAPTER VII

Faults

Where rocks have undergone observable displacement along a macroscopic shear or fracture plane in the earth they are said to have been *faulted*. The word fault refers both to the plane itself, which is called the *fault plane*, and to the displacements that have gone on along it. In conformity with the ideas on plastic slip and shear-fracturing expressed in Chapter III, fault planes may be shear-fractures along which the internal cohesion of the rocks has been destroyed, or slip surfaces formed during plastic yielding and across which cohesion was maintained during the movements. Again, under certain conditions of rupturing, the fault plane may be a tension crack, as is suggested for some large fault troughs and perhaps for ring-fractures associated with igneous activity. The majority of fault planes are, however, either shear-fractures or slip planes, and may be said to be shear planes, the geometry of which conforms with the principles of stress–strain relationships already discussed. The stress-conditions that induce faulting in relatively homogeneous rocks do, however, vary widely, and faulting is also strongly influenced by inhomogeneities, particularly where strata of markedly different physical properties are interbedded.

Nomenclature of faults

The word 'fault' was originally used in coal-mining where, at a certain point in the workings, a coal-seam was found no longer to continue in its expected position. It was soon realized that a fault in a coal-seam is due to displacement along a fracture and that the continuation of the seam can generally be found at a higher or lower level after the fault is crossed.

Because of the need for precision in mining geology regarding the direction and amount of displacement of faulted beds or lodes a somewhat involved nomenclature has been developed to describe the movements. Some authorities recommend that faults should be named according to the apparent relative displacement of strata or lodes, but this bears no necessary relationship to the true relative displacement. This is shown in Fig. VII–1 where a fault with only lateral relative displacement gives rise to apparent displacements changing in vertical sections from right-hand-side down to left-hand-side down. In fact the true relative displacement can be confidently decided in most instances, but not, however, the actual movements involved. Thus in Fig. VII−1 it may be clear that the right-hand block has moved northwards relatively to the left-hand block, but whether this occurred by actual north-ward movement, by southward displacement of the left-hand block, or by movements of both blocks is quite uncertain.

Evidence regarding actual movements is best obtained with young faults in which the movements affect the earth's surface, so that an uplifted block appears as an elevation and a subsided block as a depression. Effects on streams may also give some indication, but even so, a full knowledge of the

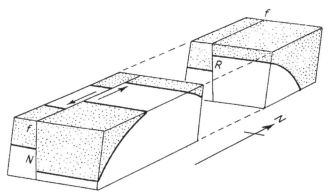

FIG. VII–1. BLOCK-DIAGRAM SHOWING A FAULT WITH ONLY LATERAL RELATIVE DISPLACEMENT (STRIKE-SLIP FAULT), CROSSING AN ANTICLINE

At N the fault appears in cross section to be a normal fault; at R, a reverse fault. (*After Gill, 1935*)

whole river-regime as well as of the regional geology is required, and in most cases we have to be content with a discussion of relative displacements. This affords the basis for the most widely used classification of faults into *normal faults*, *reverse faults*, and *strike-slip faults*.

Faults in which the displacement of the block lying above an inclined fault is downwards relative to the lower block are normal faults, and reverse faults are those in which this displacement is upwards. In strike-slip faults the movement is essentially horizontal, along the strike of the fault.

In all faults the movement may in fact be somewhat oblique on the fault plane, but it is the dominant direction of movement that determines the term to be applied to the fault. Thus a strike-slip fault may show several hundreds of feet of downthrow, but if its lateral shift is some thousands of feet, or even miles, it is essentially a strike-slip fault and is so named. Only where the lateral and vertical displacements are of the same order of magnitude is a fault called an *oblique-slip fault*.

Faults in bedded rocks – A factor that greatly influences the geometry of
faulted beds is the attitude of the fault plane in relation to the stratification.
This is incorporated in a terminology that is independent of the movement on
the fault and relates only to the strike of the fault as compared with the
regional strike and dip of the rocks. A fault is a *strike fault* if it strikes in the
general direction of the beds, a *dip fault* if it strikes approximately parallel to
their dip, and an *oblique fault* if it is markedly oblique to them (Fig. VII–2).

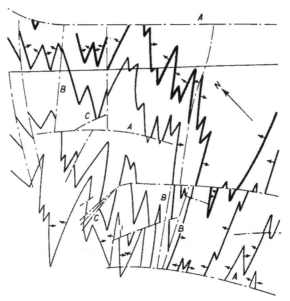

FIG. VII-2. FAULTING IN RELATION TO DIP AND
STRIKE–COAL BASIN, AACHEN
A. Dip fault. B. Strike fault. C. Oblique fault. Ruhr coal basin.
(*After Wagner, from O. Stutzer and A. C. Noé*, Geology of
Coal, *Chicago, 1940, Fig. 96*)

The effects of faulting on the outcrop of stratified rocks depend on the
relative movements of the blocks, on whether the faults are strike, dip, or
oblique faults, and on the extent of erosion after faulting. The topic will not be
treated in detail,[1] but it is important to note that strike faulting, especially in
regions of unfossiliferous rocks where the stratigraphy is uncertain, may be
very difficult to evaluate.

Bedding faults follow the stratification in well-bedded rocks. They are recognizable only with difficulty, but play an important role in the formation of
certain major structures (see Reverse Faults below).

[1] Good accounts are available in many standard works. See K. W. Earle, *Dip and Strike
Problems*, London, 1934; B. C. Brown and F. Debenham, *Structure and Surface*, London,
1929; C. F. Tolman Jr., *Graphical Solution of Fault Problems*, San Francisco, 1911; M. H.
Haddock, *Disrupted Strata*, London, 3rd edn, 1953.

FIG. VII-3. FAULT ZONES ALONG REVERSE FAULTS
Silurian mudstones and sandstones, Studley Park, Victoria. Note the 'drag' affecting strata for
several feet from each fault. (*Photo: E.S.H.*)

The parts of a fault – The fault surface which is often called the *fault plane* even though it may be markedly curved, or in some cases undulating, may also be called simply the *fault*. Instead of one clearly defined shear plane, there may be a number of parallel shears, whereby the fault movement is distributed in a *shear zone*.[1]

Fault zone implies the zone along a particular fault, within which movements associated with it are distributed (Fig. VII–3). These may include shearing, jointing, fracturing leading to brecciation (fault breccia), cleavage formation, or the bending of strata by 'drag', as discussed below.

If the fault is not vertical, the face of rock lying above it is the *hanging wall*, and that below it the *footwall*.

The fault is said to *hade* in the direction of the dip of the fault plane, the angle of hade being the complement of the dip, that is, the angle between the fault and the vertical (Fig. VII–4).[2]

[1] The term shear zone is used in various senses in different regions, and especially in mining geology it means what the local geologists intend, and may thus be applied to a strongly stretched fold limb, or to a zone of schistose rocks with large or small fault displacement, and so on.

[2] Terms such as *hade* of faults and *list* of folds, which refer to deviations from the vertical, arose in mining from the importance attaching to such deviations from the mineshaft, and are little used in structural geology.

Slip (also known as *net-slip*, or *displacement*) and *shift* – Considering two blocks that have been displaced obliquely to each other along a fault, the measure of this relative displacement is the *slip* (Fig. VII–5), which may be considered to be the resultant of two components, that along the strike of the

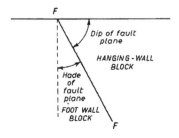

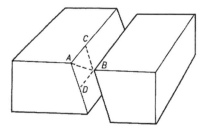

FIG. VII–4. HADE OF A FAULT   FIG. VII–5. SLIP ON A FAULT

The fault is an oblique-slip normal fault
AB, net slip; *AC*, strike-slip; *AD*, dip slip.

fault (*strike-slip*), and that parallel to its dip (*dip-slip*). These terms are used for movements on a single fault at any one place, as in detailed mine geology. Where the movement is distributed over a fault zone it is necessary to consider points lying outside this to arrive at the total displacement (Fig. VII–6), for which the term *shift* has been used in corresponding ways to slip.[1]

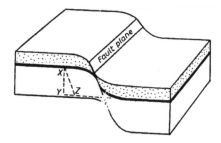

FIG. VII–6. SHIFT AND PHYSIOGRAPHIC
THROW

XZ is the dip shift; for physiographic purposes, *XY* is the throw.

Apparent displacements on faults

It may be seen by reference to Fig. VII–1 that the apparent displacement of planar structures such as beds or lodes as seen in cross-section at any particular place may be completely different from the actual movement on the fault. In the general case this applies with all kinds of faults, although it is true that with each type there is a special case for which the apparent and actual

---

[1] *Shift* has also been used in coal-mining for the apparent horizontal displacement of a bed (i.e. the distance separating the displaced parts of the bed, measured horizontally along the fault as seen in plan) – see Walcot Gibson, *Coal in Great Britain*, London, 1920, p. 81. This usage is, however, now obsolete. M. H. Haddock, *Disrupted Strata*, London, 3rd edn, 1953, gives standard current British usages.

displacements are the same, as for example with a normal dip-slip fault affecting either flat-lying beds or dipping beds that strike parallel with the fault-strike (Fig. VII–6).

For many purposes, particularly in mining, it is of immediate practical importance to recognize the relative position of the parts of a faulted bed or lode in a particular part of the workings, on either side of the fault. In general, the distance separating comparable parts of a bed or lode (e.g. top to top, bottom to bottom), as measured in a desired plane at any locality, is the *separation* in that plane.

Types of separation – The *vertical separation* is the separation measured in a vertical direction (Fig. VII–7C). The *horizontal separation* may be measured in the horizontal plane in any desired direction (e.g. north–south, or east–west) but an important direction is that normal to the strike of the faulted beds, the measure of which is the *horizontal normal separation* or *offset*.[1]

The *stratigraphic separation* is the true thickness of strata normally separating a particular bed from that bed or horizon with which it is brought into contact by the fault. The stratigraphic separation is related to the vertical separation by the formula $S = V . \cos \delta$, where S is the stratigraphic separation, V the vertical separation, and δ the true dip of the beds. The stratigraphic separation is also called the *stratigraphic throw*.

Throw and heave – *Throw* is a term that was formerly used in two senses, the one referring to the displacement of fault blocks relative to each other, the second to the apparent displacement of a bed or lode in cross-section at right angles to the fault, which, as has been noted above (Fig. VII–1), may be entirely different from the actual movement on a fault, and where the geology is complex may change radically at different places along the fault. In mine-workings, however, it is the relative position of the faulted parts of a bed or lode that is of greatest immediate practical value, and the use of throw for the relative displacement of the fault blocks is now superseded except in geomorphology (Fig. VII–6). In mining geology, the *downthrow* or *upthrow* as the case may be is the distance separating the faulted bed or lode, measured on the fault, in a plane at right angles to it. This is also sometimes called the *total throw* – the *throw*, or preferably the *vertical throw*, being the vertical component of the total throw.[2]

Heave is the horizontal distance, measured normal to the fault, separating the parts of the faulted bed or lode. It is used mainly in mining, especially since,

---

[1] The way in which the outcrops of a bed or its intersections with a horizontal plane in depth are displaced relatively to each other on either side of a fault has been used to designate *right-handed* or *left-handed separation* along the strike of a fault (M. L. Hill, 'Classification of Faults', *Bull. Amer. Assoc. Petrol. Geol.*, Vol. 31, 1947, pp. 1669–73). This usage leads to confusion with the naming of strike-slip faults and is not recommended for geological purposes, although it will doubtless be retained by miners.

[2] See O. Stutzer and A. C. Noé, *Geology of Coal*, Chicago, 1940, p. 362. J. Challinor, 'The Throw of a Fault', *Geol. Mag.*, Vol. 70, 1933, pp. 385–93; idem, 'The Primary and Secondary Elements of a Fault', *Proc. Geol. Assoc. Lond.*, Vol. 57, 1946, pp. 153–40.

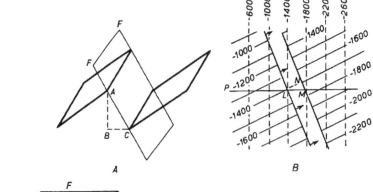

FIG. VII-7. DISPLACEMENTS OF FAULTED BEDS

A. Normal Fault showing Total throw *AC* (dip separation), Vertical throw *AB*, and Heave *BC*.

B. Structure contours showing contours on marker bed (full lines) and on fault plane (broken lines). *LM* = Heave; Vertical throw = difference of level between *L* and *M* in plane of section *P* at right angles to fault (= 360); Vertical separation = difference in level between *L* and *N* = 300. (*By permission Shell Petroleum Co.*)

C. Block diagram showing *horizontal normal offset* (*AB*) and the *vertical separation* (*CD*).

where normal faults are concerned, there is a barren zone or 'want' separating the faulted parts. Heave is also used with reverse faults if the movement is not large, but with low-angle reverse faults of great displacement it is more usual to speak of the amount of *translation* of the upper block, rather than the heave.

These several measurements may be made from structure contour maps as well as from reconstructed sections or actual observations (Fig. VII–7B).[1]

Minor structures associated with faults

Drag – Beds in a fault zone generally show flexures giving an appearance as if the beds had been dragged back by frictional resistance on the fault plane.[2] This so-called drag includes, however, several distinct effects. In the first, a flexure occurs before the fault develops, and is subsequently sheared through as the displacement increases. This is seen for instance in *stretch thrusts*

[1] For further information consult, 'Report of the Committee on the Nomenclature of Faults', *Bull. Geol. Soc. Amer.*, Vol. 24, 1913, pp. 163–86; J. E. Gill, 'Normal and Reverse Faults', *Journ. Geol.*, Vol. 43, 1935, pp. 1071–9; H. W. Straley III, 'Some Notes on the Nomenclature of Faults', ibid., Vol. 42, 1934, pp. 756–63.

[2] The effect is called 'drag', but the resulting flexures are not drag folds in the commonly accepted sense of this term.

found on the 'sheared middle limb' of some asymmetrical folds (Fig. VII–8A), in monoclinal flexures that pass into faults, and in *shear-flexures* (Fig. VII–8B). These latter are flexures confined to a narrow zone, which may be sliced through by a fault if the displacement exceeds a certain amount which is determined by the possible elastic and plastic yielding of the rocks. The continuation of the shear-flexure beyond the fault demonstrates its earlier development than the fault whereas drag due to friction on a fault plane itself must of necessity be a secondary result of the primary fracturing and translation.

At the stage when fracturing occurs the fault movement is suddenly accelerated, and the total displacement may then be considerable. Irregularities due for instance to the fault crossing beds of different physical properties may then be torn off and crushed, forming mylonite, pug, and polished slickensided surfaces, and some true drag-flexures may also form at this stage.

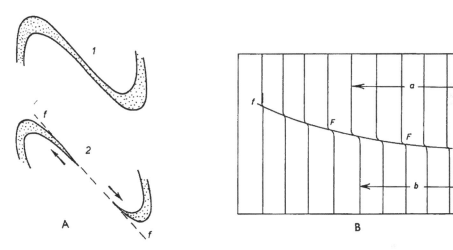

FIG. VII–8A. DEVELOPMENT OF 'STRETCH THRUST' WITH APPARENT STRONG DRAG, BY SHEARING OF THE MIDDLE LIMB OF AN ASYMMETRICAL FOLD
B. RELATIONSHIP OF A 'SHEAR-FLEXURE' TO A REVERSE FAULT, WITH 'DRAG' DUE TO SHEARING OF THE FLEXURE
Flexing at *ff*, faulting between *F* and *F*.

A related effect is seen in faulted gneisses and schists, where a pre-existing foliation is strongly turned into the fault zone and a secondary foliation results (Fig. VII–9). This implies faulting at considerable depths in the crust. *Mylonite* – Strongly sheared, crushed, and cemented *mylonite* on fault planes is found chiefly with great low-angle reverse faults, for which the pressure of the thrust block on the fault must be very great. Mylonite is of entirely cataclastic origin, lacking the crystallographic effects that give rise to schistosity and related fabric orientation. Mylonite is thus likely to be formed in cold

FIG. VII-9. DEVELOPMENT OF SECONDARY FOLIATION IN A NARROW FAULT ZONE
CROSSING GNEISSIC BANDING, NATURAL SIZE

rocks, although it is formed under considerable pressure as is evident from
the development in it of structures resembling flow planes due to the shearing,
as described by Read,[1] and the frictional heat generated must be considerable.
Mylonite may also form in rocks that are in a condition suitable for metamorphic recrystallization if the displacements involved are very rapid, or if the
chemical environment necessary for the recrystallization of particular rocks
is lacking.

In softer rocks a tough leathery layer of ground rock, generally dark
coloured especially where carbonaceous slates are involved, may line the
fault plane. This is *fluccan*, *gouge*, or *pug*. It is frequently polished and
striated by the fault movements.

Crush breccia, crush conglomerate – In referring to the shattered or fractured
rocks in fault zones, a distinction may be drawn between *fault breccias*, in
which there is a jumbled mass of fragments with a high percentage of voids
formed largely by the continued opening of tension gashes, and sheared rocks
in which a regular arrangement of fractures and of the resulting separate
rock-fragments is to be seen. These latter include *crush breccias*, with
angular, generally lozenge-shaped fragments bounded by intersecting shear
surfaces, and *crush conglomerates* in which prolonged movement has rounded
off the solid angles of the fragments, which, however, still lie with their long
axes parallel and resemble stretched conglomerates (see p. 132). The distinction implies that fault breccias are formed under conditions such that the
normal stresses across the fault are low, and accordingly they belong to a

---

[1] H. H. Read, 'Mylonitization and Cataclasis in Acidic Dykes in the Insch (Aberdeenshire)
Gabbro and its Aureole', *Proc. Geol. Assoc.*, Vol. 62, 1951, pp. 237–47. See also C. Lapworth,
'The Highland Controversy in British Geology', *Nature*, Vol. 32, 1885, pp. 558–9.

FIG. VII–10. DISJUNCTIVE FOLD IN HEMATITE-QUARTZITE
Middleback Ranges, S. Australia. ×$\frac{1}{2}$.

superficial crustal zone and are commonly associated with normal faults formed by mechanisms such as permit the considerable volume expansion that this type of brecciation implies (see, for example, Fig. V – 29).

Crush breccias and conglomerates, on the other hand, demonstrate the operation of strong normal stresses across the fault. In general, two complementary shear-directions are present, slicing the rock into lenses between which the shearing produces a fine-grained matrix.[1] The rock is squeezed as well as stretched, and considered as a whole it becomes elongated along one of the planes bisecting the shear surfaces, this being the plane of flattening,[2] which generally appears to lie in the acute bisectrix of the shear planes (AB plane of a local strain ellipsoid). Parts of rock-masses such as beds or dykes may be pulled widely apart, this increasing the resemblance to true conglomerates,[3]

[1] A. Thurmer, 'Entstehung von Linsen in Gesteinen', *Centralbl. f. Min.*, etc., Abt. A, 1928, pp. 147–58.
[2] B. Sander, *Gefügekunde der Gesteine*, Vienna, 1930, pp. 173–5. Also *Einführung in die Gefügekunde der Geologischen Körper*, Vienna, 1948, pp. 155 ff.
[3] The resemblance to tillite is even stronger, because of the 'unsorted' arrangement of the pebbles and matrix.

but the disjointed pieces may be 'lined up' in the field, thus affording an indication as to the real origin.

Crush conglomerates and mylonite may be associated with any type of fault but they are best known on reverse faults. Lensing and flattening may, however, extend broadly throughout a region, being more obvious when

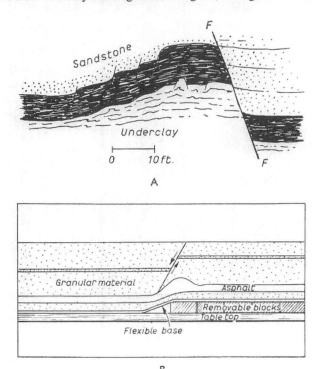

FIG. VII–11. 'BULGING' ADJACENT TO FAULTS

A. Bulging of coal on upthrown side of normal fault, Wonthaggi coalfield, Victoria. Due mainly to the migration of the underclay. (*After Edwards* et al.)

B. Bulging of asphalt against normal fault after erosion of up-thrown block. (*From Parker and McDowell, 1955*)

brittle beds are included among more plastic rocks. This often leads to the lensing of the beds in fold limbs, forming *disjunctive folds* (Fig. VII–10).

Bulging and reverse drag – Many of the coal-seams in the Wonthaggi field, Victoria, are bulged upward for several feet as a fault, downthrown in the opposite direction, is approached (Fig. VII–11A).[1] Experiments imitating salt domes show exactly similar structures (Fig. VII–11B),[2] which are due to

[1] A. B. Edwards, G. Baker, and J. L. Knight, 'The Geology of the Wonthaggi Coalfield, Victoria', *Proc. Aust. Inst. Min. Met.*, N.S., No. 134, 1944, pp. 1–54.

[2] T. J. Parker and A. N. McDowell, 'Model Studies of Salt Dome Tectonics', *Bull. Amer. Assoc. Petrol. Geol.*, Vol. 39, 1955, pp. 2384–470.

the flowage of mobile strata under differential pressure adjacent to the fault. The rise of strata in the upthrown block as a normal fault is approached is in the reverse direction to the normal 'drag', and similar 'reversed drag' in both upthrown and downthrown blocks is seen in the Gulf Coast oilfields (Fig. VII–12).[1] Hamblin[2] proposes as a general explanation the collapse of the unsupported edge of a downthrown block pulled away from a curving fault plane, but this does not account for the 'rise to the downthrow' in an upthrown block, or the bulges developed near many faults which exhibit reverse drag. Experiments by E. Cloos[3] simulating Gulf Coast structures reproduce reverse drag, this being caused by antithetic faulting.

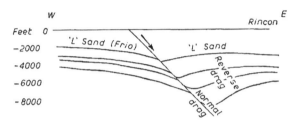

FIG. VII–12. 'REVERSE DRAG', RINCON FIELD, GULF COAST, U.S.A.

(*Based on Bornhauser, 1958*)

Vertical and horizontal scales the same.

Feather joints and tension gashes – Shear and tension joints, secondary faults, and cleavage are commonly developed adjacent to major faults, and analogous structures have been reproduced in experiments with clay, carried out by Riedel and by E. Cloos.[4]

In these experiments, two boards are placed on a table, side by side and touching one another, and covered with a cake of clay. One board is then pushed relative to the other, so that shearing stress is transmitted to the clay above the line of contact of the boards (Fig. IV–18). If the plasticity of the clay is reduced by spraying a film of water onto its upper surface, tension fractures are found to form in the clay above the contact of the two boards. When they first appear, these fractures make an angle of 45°–47° with the direction along which the relative displacement occurred, but if the experiment is continued, they open out into gashes, and finally may make an angle

[1] M. Bornhauser, 'Gulf Coast Tectonics', *Bull. Amer. Assoc. Petrol. Geol.*, Vol. 42, 1958, pp. 339–70.

[2] W. K. Hamblin, 'Origin of "Reverse Drag" on the Downthrown Side of Normal Faults', *Bull. Geol. Soc. Amer.*, Vol. 76, 1965, pp. 1145–64.

[3] E. Cloos, 'Experimental Analysis of Gulf Coast Fracture Patterns', *Bull. Amer. Assoc. Petrol. Geol.*, Vol. 52, 1968, pp. 420–44.

[4] W. Riedel, 'Zur Mechanik geologisher Brucherscheinungen', *Centralbl. f. Min.*, etc., Abt. B, 1929, pp. 354–68. E. Cloos, 'Feather Joints as Indicators of the Direction of Movements on Faults, Thrusts, Joints, and Magmatic Contacts', *Proc. Nat. Acad. Sci. America*, Vol. 18, 1932, pp. 387–95 (with Bibliography).

of as much as 60° with the direction of relative displacement. This is brought about by the rotation of those portions of the clay lying between the first-formed fractures (Fig. VII–14). By reference to patterns impressed on the upper surface of the clay before the experiment, it can be seen that the tension fractures first form at right angles to the direction of greatest elongation in the deformed zone of clay. If the clay is not flooded with water but allowed to remain plastic, then shearing planes form in the deformed zone. One dominant set generally appears, making an angle of 12°–17° with the direction of relative displacement of the boards. The complementary set is, however, sometimes feebly developed.

In these experiments it is found that the acute angles enclosed between the tension gashes or the shearing planes, and the direction of relative displacement of the blocks of clay, point in the direction in which the blocks moved. It should be noted that the clay does not shear through along a single surface parallel with the direction of relative displacement of the boards, but that tension gashes and shearing planes are developed *en echelon* in a zone lying between the blocks.

A similar arrangement of faults or tension gashes arranged *en echelon* is commonly found in the field, and may be of special importance in mineralized areas where ore deposits are localized along the faults or in tension gashes (Fig. VII–13).

Shear and tension joints developed in the zone of deformation between crustal blocks that have moved relatively to each other along a fault are known as *feather joints* (*Federklüfte*). The name is given because of the re-

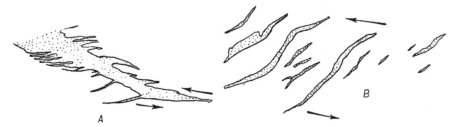

FIG. VII–13. *En echelon* QUARTZ VEINS IN THE 60-FT FLOOR, A1 MINE, WOODS POINT, VICTORIA
(*After Whitelaw*, Mem. Geol. Surv. Vict., *No. 3, 1905*)

semblance of the joints and the fault, as seen in cross-section, to the barbs and shaft of a feather. More specifically we may refer to feather joints as *pinnate shear joints* on the one hand, and *pinnate tension joints* (or gashes) on the other.

Both field and experimental evidence indicate that pinnate joints or tension gashes may be used to determine the direction of relative displacement along faults or fault zones. The rule is that the acute angles enclosed between pinnate structures and the fault plane or the medium plane of the structures point in the direction of movement of the blocks (Fig. VII–15A). However, the continued movement of blocks after tension gashes are first formed causes these to be dragged into an S-shape, in which case it is the parts of the gashes more distant from the fault that indicate the movement according to the rule given above, although the drag also substantiates this (Fig. VII–14). Roering, however, argues that in brittle quartzites some *en echelon* arrays develop

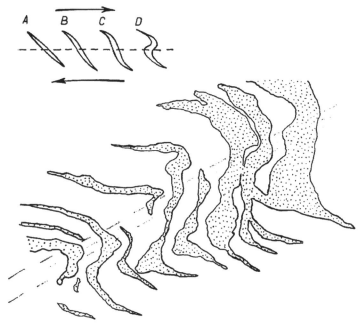

FIG. VII–14. SIGMOIDAL *en echelon* QUARTZ VEINS IN SLATE, BENDIGO, VICTORIA

Inset shows scheme for development of sigmoidal *en echelon* tension gashes in stages from *A* to *D*. (*After Dunn, Special Rept. Nos. 1 and 2, Mines Dept. Vict., 1896*)

parallel with the axis of maximum compressive stress, the cracks within the zones representing shear planes filled with vein quartz.[1]

The marginal thrusts along intrusive contacts (see p. 374) have been interpreted in a similar way, and analogous pinnate structures also occur in the marginal zones of glaciers. In each of these examples we have to deal with masses, of which at least one is plastic and between which there is some degree of cohesion, that have moved relatively to each other like fault blocks.[2]

If pinnate shear planes are very closely spaced they constitute a type of

[1] C. Roering, 'The Geometrical Significance of Natural En-Echelon Crack-Arrays', *Tectonophysics*, Vol. 5, 1968, pp. 107–23.

[2] H. Cloos, 'Zur Mechanik der Randzonen von Gletschern Schollen und Plutonen', *Geol. Rundschau*, Bd. 20 Hft. 1, 1929, p. 66; also idem, *Einführung in die Geologie*, Berlin, 1936, pp. 235–6.

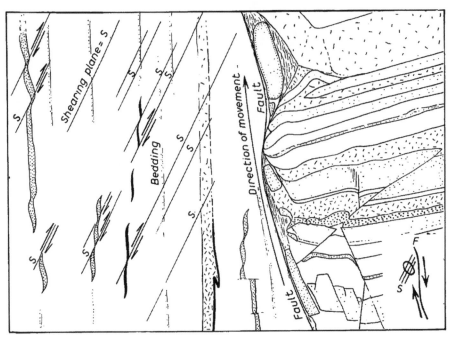

FIG. VII–15A. PINNATE SHEAR PLANES
Pinnate shear planes *S* due to secondary stresses on a fault
in Silurian rocks, Victoria.

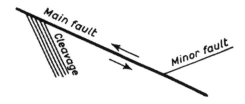

FIG. VII–15B
Pinnate 'cleavage' along a thrust fault,
probably representing the complementary
shear direction. (*From Nevin, after
Sheldon, 1928*)

FIG. VII–15C
Sheeting parallel to a fault. (*From Nevin, after
Dale, 1896*)

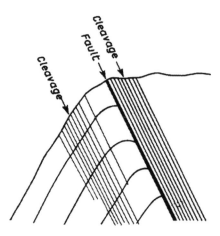

FIG. VII–16. SLICKENSIDED QUARTZ FILM ON FAULT PLANE, STUDLEY PARK,
VICTORIA, SHOWING MOVEMENTS IN TWO DIRECTIONS. ×¼.

fracture cleavage (or close-joints cleavage) but the angular relationships of locally developed cleavage to fault planes commonly differs from that of pinnate shears, as in the example described by Sheldon.[1]

In cases where the fault plane is parallel to the cleavage in the adjacent rocks,[2] it may be that the fault and the cleavage planes are all shearing surfaces induced by the same stress, or that the cleavage planes afforded a predetermined direction of ready shear, which was made use of by later faulting. A *sheeting* structure parallel to the fault plane and confined to the zone of deformation may, however, be a direct result of the faulting, especially in incompetent rocks (Fig. VII–15C).

Slickensides – Slickensides are striations (*slip-striae*) on a fault plane, or on the pug or mineral deposits lining a fault, that are formed in the direction of relative movement of the fault blocks. Two or more sets of slickensides generally intersecting at an acute angle may sometimes be seen, especially in layers of laminated quartz, indicating successive movements in slightly different directions or a sudden deviation in the movement during one displacement (Fig. VII–16). Slickensides may form under quite superficial conditions, as for instance on the clay coatings of blocks of rock involved in soil slides, at the base of landslides, or even in masses of ore dislodged during stoping in

[1] P. Sheldon, 'Note on the Angle of Fracture Cleavage', *Journ. Geol.*, Vol. 36, 1928, pp. 171–5.
[2] T. N. Dale, 'Structural Details in the Green Mountain Region, and in Eastern New York', 16th Ann. Rept., *U.S. Geol. Surv.*, 1896, Pt 1, pp. 543–70.

FIG. VII-17. SECONDARY PINNATE SHEAR PLANES, WITH SLICKENSIDES, DEVELOPED ALONG A FAULT IN DIORITE, MORNING STAR MINE, WOODS POINT, VICTORIA. ×⅘.

The apposed block moved down relative to the photo.

mines. Quite small movements, of the order of an inch or less, produce excellent slickensides, so that they offer no guidance as to the total displacement on faults.

In hard relatively brittle rocks slickensided surfaces are often highly polished and feel smooth to the touch, although in some instances each slickensided groove is deeper at one end and 'tails out' towards the other, in the direction of displacement of the adjoining block. The surface feels correspondingly somewhat rougher when stroked in the opposite way, towards the deeper ends. This rule is, however, not applicable in the more common examples in which secondary pinnate shear planes branch off from the main fault. Such shears are prominent and are themselves slickensided; they conform with the geometrical arrangement of pinnate shear planes in experimental studies, and consequently their tips in one block face in the direction *opposite* to that of the movement of the adjoining block (Fig. VII–17). The tips often break off, forming steps that roughen the fault plane and afford a very reliable guide to the sense of the relative movements of the blocks.[1]

---

[1] Pinnate shears and slickensides may be very simply reproduced by squeezing a block of moderately soft kitchen soap in the jaws of a vice. There has been a minor controversy over

Slickensides are developed on conical shear surfaces in the structure known as *cone-in-cone*, which is best seen in calcareous rocks and in coal. Cone-in-cone consists of two arrays of cones, one set with its bases on a bedding plane and the apices facing up, the other on a parallel bedding plane with the apices facing down and interpenetrating between the first set. The flanks of the cones are slickensided and in calcareous rocks they generally

A   B

FIG. VII–18. CONE-IN-CONE
A. In impure Cretaceous limestone, Queensland. ×⅔.
B. In coal (one set of cones removed). ×⅔.

also show concentric rings around the circumference of the cones (Fig. VII–18). It is believed that cone-in-cone is formed by compression between two parallel bedding planes, and presumably the solid angle of the cones represents the dihedral angle between shear planes in the rock, although this angle varies widely, from 15° to 100°.[1]

these matters (see M. S. Paterson, 'Experimental Deformation and Faulting in Wombeyan Marble', *Bull. Geol. Soc. Amer.*, Vol. 69, 1958, pp. 468–75; R. E. Riecker, 'Fault Plane Features: an Alternative Explanation', *Journ. Sed. Petrol.*, Vol. 35, 1965, pp. 746–70; E. Rod (discussion), ibid., Vol. 36, 1966, pp. 1163–5). Only Rod has understood that it is the delicate stroking of slickensides themselves, not involving the coarse breaks due to pinnate shears and other effects, that has consistently been used especially in mines to give preliminary guidance as to the displacement on faults (see M. P. Billings, *Structural Geology*, New York, 2nd edn, 1954, p. 149).

[1] W. A. Tarr gives the best general account of cone-in-cone, in W. H. Twenhofel, *Treatise on Sedimentation*, 2nd edn, 1932, pp. 716–33. The pinnate shear planes along faults developed as described above often show a tendency to originate from points, giving an effect of strongly asymmetrical cones, and suggesting some unexplained analogy with cone-in-cone.

Stress distribution in faulting

Consideration of the states of stress that cause faulting in homogeneous relatively rigid rocks affords a general basis for the discussion of actual examples, in which various complexities appear. Stress analysis, however, tells us little about the mechanisms that may be concerned in the generation of the stresses, for which we must turn to the regional tectonic setting of the faulted rocks.[1]

Assuming that of the three principal stresses, two are horizontal and the third vertical and due to gravity alone,[2] three sets of conditions in which all the stresses are compressions, may first be recognized. The faults in each case are complementary slip-surfaces or shear-fractures, the acute dihedral angle between them enclosing the maximum principal stress axis, while the mean stress axis is the direction along which the planes intersect.

1. Maximum stress horizontal: mean stress horizontal: minimum stress vertical. This produces reverse faults. The section of faulted rocks is shortened in the direction of maximum compression and the faults dip at less than 45° (Fig. VII–19A).

2. Maximum stress horizontal: minimum stress horizontal: mean stress vertical. This produces complementary strike-slip faults. One set has a clockwise sense of movement (right-handed or dextral), the other an anticlockwise (left-handed or sinistral) sense (Fig. VII–19B).[3]

3. Maximum stress vertical: minimum and mean stress horizontal. This produces normal faults, dipping at more than 45° (Fig. VII–19C). This stress environment represents the conditions at such a depth in the crust that the lithostatic pressure exceeds the ultimate or crushing strength of the rocks. Any mechanism that reduces the horizontal compression along one axis leaves gravity as the active compression, and the rock fractures along shear planes, resulting in horizontal extension of the section.

At shallower depths brittle rocks subjected to extension (as for instance on the extrados of an arch) may rupture along tension cracks at right angles to the axis of extension. While this generally leads to widely distributed rupturing as tension joints, removal of support from below may permit blocks isolated by tension cracks to settle, giving rise to faults. This mechanism may be involved in the formation of ring fractures (see p. 395) and even some great grabens appear to be bounded by tension cracks. The presence of

[1] The account that follows is based largely on E. M. Anderson, *The Dynamics of Faulting and Dyke Formation*, London, 2nd edn, 1951, and M. K. Hubbert, 'Mechanical Basis for Certain Familiar Geologic Structures', *Bull. Geol. Soc. Amer.*, Vol. 62, 1951, pp. 355–72.

[2] See, however, the comment on p. 183 regarding the effect of uplift pressure due to hydraulic pressure in water-saturated rocks.

[3] If one faces the fault, the movement of the distal block is to the right in right-handed and to the left in left-handed faulting.

magma appears, however, to be necessary to permit tension-cracking at considerable depths in the crust.[1]

Reverse faults of the kind known as overthrusts, in which relatively thin wedges, some of which are many miles in length, are translated over great distances on faults dipping at 10° or less, have proved difficult to explain in physical terms. Hubbert and Rubey[2] have shown that high pore pressure developed in water-saturated rocks makes the mechanics feasible because of

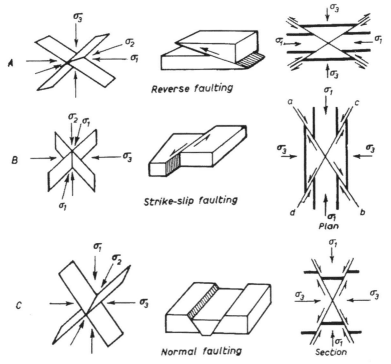

FIG. VII–19. INITIAL STRESS DISTRIBUTION CAUSING FAULTING, ACCORDING TO ANDERSON

σ_1 maximum, σ_2 mean, σ_3 minimum (compressive) stress. In B, ab = dextral (clockwise) and cd = sinistral (anticlockwise) displacement.

the reduction of sliding friction along the thrust plane, especially when the pore pressure approximates to the overburden pressure. Important additional factors are the *toe-effect*[3] whereby the advancing front of a thrust would be retarded by the toe riding upwards as it must do, and perhaps breaking up: also the cohesive strength of rocks, which was regarded as negligible by

[1] See the discussion of dykes in Chapter XII.

[2] M. K. Hubbert and W. W. Rubey, 'Role of Fluid Pressure in Mechanics of Overthrust Faulting', *Bull. Geol. Soc. Amer.*, Vol. 70, 1959, pp. 115–66 and 167–206.

[3] C. B. Raleigh and D. T. Griggs, 'Effects of the Toe in the Mechanics of Overthrust Faulting', ibid., Vol. 74, 1963, pp. 819–30.

Hubbert and Rubey once movement on the fracture has started, is, according to Hsü,[1] in fact significant and would tend to reduce the possible maximum length of overthrust sheets. He distinguishes between slow movements of cohesively bound blocks and catastrophic movements of cohesionless blocks, which latter may be represented in the remarkable Heart Mountain slides described by Pierce,[2] and in many landslides. All the above, and other factors such as the presence of lubricating horizons are relevant both to the formation of overthrust sheets by gravitational sliding and to overthrusts formed by end-on compression in orogenic belts.

Normal faults

Dip of fault plane – The assumption that for much normal faulting, gravity is the maximum compressive force is based in part on the knowledge that the fault planes in most block faulted regions dip at angles greater than 45°, but

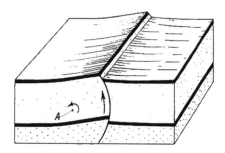

FIG. VII–20. A CYLINDRICAL FAULT SHOWING THE CHANGE FROM NORMAL FAULTING AT THE SURFACE TO HIGH-ANGLE REVERSE FAULTING IN DEPTH
(*After Tolman*)

the apparent sinking of the downthrown blocks also suggests movement under gravitational forces, and accordingly normal faults to which these notions apply have long been called *gravity faults*.

In fact the dip of normal fault planes ranges widely, from vertical to nearly horizontal, but statistical studies, especially those made in coalfields and oilfields where there are numerous faults known to be typical shear fractures, demonstrate that in such regions the dip averages between 45° and 70°.

Again, in *cylindrical faults* the movement of the upthrown block is actually upwards, and the fault passes into a high angle reverse fault in depth (Fig. VII–20).[3] The possibilities of alteration of the dip of early-formed faults by

[1] K. J. Hsü, 'Role of Cohesive Strength in the Mechanics of Overthrust Faulting and of Landsliding', *Bull. Geol. Soc. Amer.*, Vol. 80, 1969, pp. 927–52, with discussion by Hubbert and Rubey, and reply, pp. 953–60; idem, 'A Preliminary Analysis of the Status and Kinetics of the Glarus Overthrust', *Eclogae Geol. Helv.*, Vol. 62, 1969, pp. 143–54.

[2] W. G. Pierce, 'Heart Mountain and South Fork Detachment Thrusts of Wyoming', *Bull. Amer. Assoc. Petrol. Geol.*, Vol. 41, 1957, pp. 591–626; idem, 'Role of Fluid Pressure in the Mechanics of Overthrust Faulting: Discussion', *Bull. Geol. Soc. Amer.*, Vol. 77, 1966, pp. 565–8.

[3] C. F. Tolman, Jr, *Graphical Solution of Fault Problems*, San Francisco, 1911, pp. 36–7. H. Cloos, *Einführung in die Geologie*, Berlin, 1936, pp. 198–200.

compaction of sediments, or of some normal fault planes being tension cracks, afford partial explanations for the wide range of dips.

A normal fault dies away along its length by decrease of the throw towards

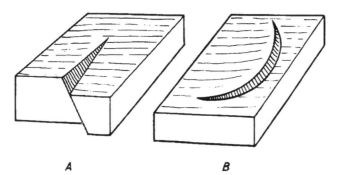

A   B

FIG. VII–21. HINGE FAULTING, AND CURVING OF NORMAL
FAULTS

either end, and where the fault commences the two blocks are said to *hinge* (Fig. VII–21). At the same time, the fault plane often curves towards the upthrown block, and accordingly, if the fault is short, it is curved in plan (Fig. VII–21). Again, the fault may pass laterally into a monoclinal flexure, or it

FIG. VII–22. PASSAGE OF A NORMAL
FAULT IN DEPTH TO A MONOCLINE AT THE
SURFACE IN BEDS OF DIFFERENT PHYSICAL
PROPERTIES

Note the 'splintering' of the fault at various levels.
(*After Wunderlich, 1957*)

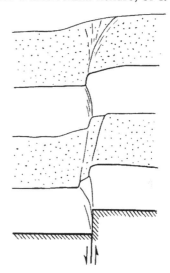

may degenerate into a series of smaller faults, either parallel with the main fault or splitting off from it (Fig. VII–22). A normal fault in more brittle basement rocks may also be replaced by a monocline in softer cover-rocks, often with many associated smaller fault splinters.[1]

---

[1] The association of normal faults with monoclines is further discussed under Monoclines: see Chapter VIII. High-angle reverse faults may, however, also pass into monoclines both laterally and vertically.

Block faulting – The subdivision of the crust into blocks by normal faults is *block faulting*. Where the faults are complementary shears the blocks are long and relatively narrow, with some remaining high as *fault ridges* and others

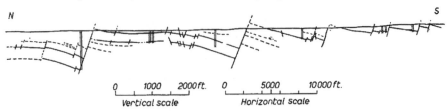

FIG. VII–23. BLOCK FAULTING IN THE WONTHAGGI COAL-BASIN (JURASSIC) VICTORIA, SHOWING TILTING OF THE FAULT BLOCKS
(After Edwards et al., *1944)*

sinking to form *fault troughs.*[1] A series of parallel faults thrown in the same direction gives *step faulting*, and where the blocks are tilted they form *tilt blocks*. All these features may be revealed in the topography of recently faulted regions, but they are also represented in the attitude of originally flat

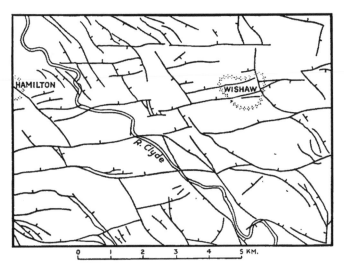

FIG. VII–24. BLOCK FAULTING IN THE MIDLAND VALLEY OF SCOTLAND (COALFIELD), SHOWING TWO CROSSING SETS OF NORMAL FAULTS
(After Anderson)

strata and may thus be recognized after erosion has affected the region (Fig. VII–23).

Two intersecting sets of normal faults give rise to a mosaic of rectangular or lozenge-shaped blocks. This is well exemplified in the central coalfield of

---

[1] The German terms *Horst* and *Graben* have certain special connotations that make it undesirable for them to be used indiscriminately for fault ridges and fault troughs (respectively) of all kinds.

the Midland Valley of Scotland,[1] where one set strikes nearly east–west, and the other northwest–southeast (Fig. VII–24). Anderson infers that the east–west faults are the earlier. Each set exhibits complementary faults, dipping in opposite directions. In other areas, however, the two sets of intersecting faults appear to have developed simultaneously, as in South Gippsland, Victoria (Fig. VII–25), where faults on one trend link with those of the other. The synchronous development of intersecting faults in South Gippsland most

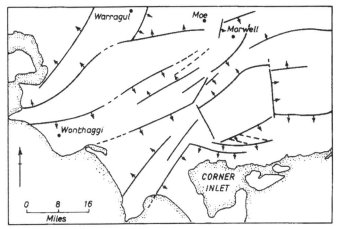

FIG. VII–25. PRINCIPAL FAULTS AND MONOCLINES IN SOUTH GIPPSLAND, SHOWING CHANGE IN FAULT TRENDS, AND DEVELOPMENT OF LOZENGE-SHAPED FAULT BLOCKS

For details at Wonthaggi, see Fig. VII–23; for Morwell, see Fig. VIII–7.

probably results as a secondary effect of more deep-seated movements, perhaps connected with strike-slip faulting in the basement, and almost certainly involving high-angle reverse rather than normal faulting.

Fault troughs

Long and relatively narrow troughs bounded by faults with roughly parallel strike are major features of the topography of continents and are often referred to as *rift valleys* (Ger. *Graben*). The term *rift* (Ger. *Bruch*) has an implication of pulling apart, that is, of tensional origin which is appropriate for many rift valleys. Nevertheless some major structures which are sometimes called 'rifts' (e.g. the Dead Sea rift) involve major strike-slip faulting and others again may perhaps be bounded by reverse faults. Accordingly, certain special types of major fault-troughs have been recognized, among which may be mentioned grabens, ramp valleys, sphenochasms, orthochasms, and aulacogenes.

[1] See especially One-inch Sheets 23, 30, 31 and 39 (Scotland). E. M. Anderson, *The Dynamics of Faulting and Dyke Formation*, London, 2nd edn, 1951.

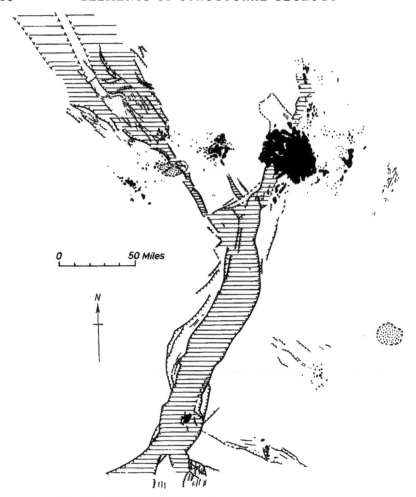

FIG. VII-26. THE RHINE GRABEN IN ITS TECTONIC SETTING
Tertiary volcanic areas shown in black. (*After H. Cloos*, Geol. Rundsch., *Vol. 30,
1959*)

Grabens – The Rhein graben may be taken as the type example and it will be
seen that it is of regional extent (Fig. VII–26), being indeed part of a more
widely developed system of faults with which Cainozoic vulcanism is associated, in North Africa and Europe.[1] The boundary faults and the great
majority of faults affecting the sediments in the trough are normal faults, but
there is evidence of horizontal displacements also. Any great structure is
likely to be affected by a variety of earth movements, but the tensional
origin of the Rhein graben has not been questioned, and its features have been

[1] J. H. Illies, 'An Intercontinental Belt of the World Rift System', *Tectonophysics*, Vol. 8,
1969, pp. 5–9; L. Picard, 'On the Structure of the Rhinegraben', *Israel Acad. Sci. &
Humanities*, Proc. Sectn Sci., No. 9, pp. 1–34.

closely simulated by Hans Cloos in clay models (Fig. VII−28). Many grabens occur in plateaux or are formed on the crests of broad swells (Fig. VII−32) or of elongated mountain belts such as the Andes[1] (Fig. VII−27), in

FIG. VII−27
GRABENS IN THE ANDEAN CHAIN
(*After Gerth*)

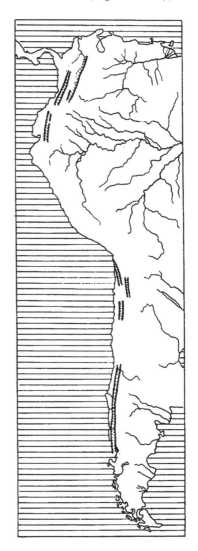

such a way that Gregory called them *keystone faults*.[2] Cloos showed that in the swells the main graben lies along the axis but bifurcates at the ends, and this he matched in experiments in which a clay cake is arched by the inflation

[1] H. Gerth, 'Die Bedeutung der Alten Kerne für die geologische struktur der jungen Kordillere', *Geol. Rundsch.*, Vol. 45, 1956–7, pp. 707–21.
[2] J. W. Gregory, 'The African Rift Valleys', *Geogr. Journ.*, Vol. 56, 1920, pp. 13–47; 327–8; idem. *The Rift Valleys and Geology of East Africa*, London, 1931.

of a rubber balloon below (Fig. VII–28B).[1] The edges of plateaux adjacent to grabens are often tilted away from the trough, as is well exemplified in East Africa (Fig. VII–29).

Taber attributed this tip-tilting to isostatic effects induced by the release of sialic blocks from mutual contact, but Cloos explained it as being due to

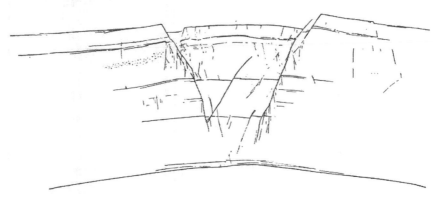

FIG. VII–28. EXPERIMENTAL GRABENS

A. Produced in a cake of clay subjected to tension.
B. Produced in a cake of clay arched over a balloon. (*After H. Cloos*)

so-called antithetic movements. He has designated as *antithetic* those faults associated with major faults of a graben edge, that hade in the opposite direction to the boundary fault, and as *synthetic* those that hade

---

[1] H. Cloos, 'Zur Experimentellen Tektonik', *Die Naturwissenschaften*, Jhg. 18, 1930, pp. 714–47; ibid., Jhg. 19, 1931, pp. 242–7. Idem, 'Zur Mechanik grosser Brüche und Gräben', *Centralbl. f. Min.*, etc., Abt. B., 1932, pp. 273–86; idem, 'Künstliche Gebirge', *Natur und Museum (Senck. Naturf. Gesellsch.)*, Vol. 59, 1929, pp. 225–72; Vol. 60, 1930, pp. 258–69. Idem, 'Hebung-Spaltung-Vulkanismus', *Geol. Rundsch.*, Vol. 30, 1939, Zwischen-Hft. 4A, pp. 405–527.

in the same direction. Originally flat-lying beds are rotated between the minor faults, the result being the production of a series of small tilt-blocks[1] (Fig. VII–30).

The east African rift valleys (Fig. VII–31), impressive as they are, are now regarded as forming but a part of a worldwide system of fractures which are

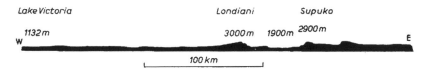

FIG. VII–29. TILTING OF GRABEN EDGES IN EAST AFRICA
(After Krenkel)

best displayed in the oceanic ridges with which crestal grabens are normally associated (see Chapter XI). Their essentially tensional origin resulting in the splitting apart and displacement of crustal plates has been established only of recent years. However, where such splitting affects a continent, trends and other local details may be determined by pre-existing fractures, foliation, and

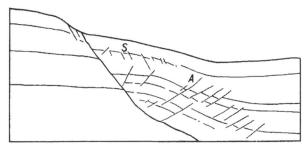

FIG. VII–30. ANTITHETIC AND SYNTHETIC FAULTS
ALONG THE EDGE OF AN ARTIFICIAL GRABEN IN CLAY
(After H. Cloos) (Compare Fig. VII–23)

folding, and this has been demonstrated locally for the east African rifts, movements on which commenced at least as long ago as the Mesozoic, and were in fact influenced by Pre-Cambrian basement structures.[2]

---

[1] The small faults are perhaps better reproduced in experiments using sand, of lower internal friction than clay. See H. G. Wunderlich, 'Brüche und Gräben in tektonischen Experiment', *Neues Jahrb. f. Geol. u. Pal., Monatshefte*, 1957, pp. 477–98. M. K. Hubbert, 'Mechanical Basis for Certain Familiar Geologic Structures', *Bull. Geol. Soc. Amer.*, Vol. 62, 1951, pp. 355–72.
[2] F. Dixey, 'The East African Rift System', *Colon. Geol. and Min. Res.*, Suppt. Bull. No. I, 1956.

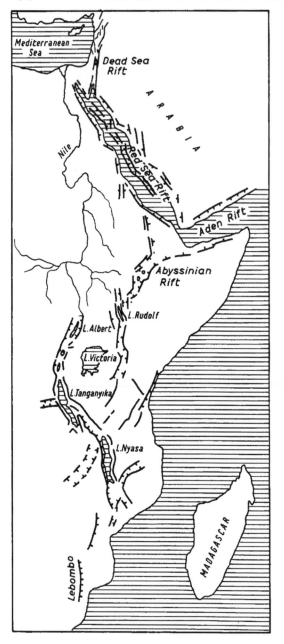

FIG. VII–31. THE EAST
AFRICAN RIFT SYSTEM
AND SOME RELATED
STRUCTURES

Recent work suggests that great fault troughs may originate in various ways. The Red Sea, which, broadly regarded, appears to be a typical graben with Pre-Cambrian rocks sharply faulted in places against thick Cainozoic sediments, shows a strong positive gravity anomaly beneath a central deep trough some 30–40 miles wide. This indicates that basic rocks occupy much of this narrow central zone.[1] Here it seems that after upwarping in the lower Tertiary the Arabian sialic block has moved away from the North African, the basic rocks in the crack representing sub-sialic (or 'oceanic') crust. Major structures of this type have been called by Carey *rhombochasms* if they are parallel-sided, and *sphenochasms* if wedge-shaped.[2] The association of vulcanicity with rift valleys is well known, and may thus be fundamental. The Great Dyke of Southern Rhodesia may represent a similar though smaller and much older infilled tension crack related to the rift valley system.[3]

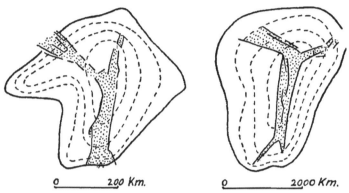

0 _____ 200 Km.   0 _____ 2000 Km.

FIG. VII–32. THE RHINE AND THE RED SEA RIFTS

The Rhenish (left) and Afro-Arabian upwarps, with their associated grabens (dotted), reduced to the same width and with the Red Sea in reverse orientation north–south, to show analogous patterns. (*After H. Cloos, 1932*)

The Dead Sea rift, which lies on the northerly continuation of the Gulf of Aqaba,[4] has been regarded as a ramp valley, a rift valley with normal faulting, and as a structure associated with a great strike-slip fault zone.[5] A feature distinguishing it from the other grabens of the Red Sea and East African systems is its asymmetry, except in the Gulf of Aqaba where Pre-Cambrian rocks rise high on either side. In Palestine the great fault on the east, bounding the Arabian massif, is the dominant structure, and the Dead Sea trough, although 2600 ft below sea-level, appears as a relatively minor feature along

---

[1] R. W. Girdler, 'The Relationship of the Red Sea to the East African Rift System', *Quart. Journ. Geol. Soc.*, Vol. 114, 1958, pp. 79–105; I. G. Gass and I. L. Gibson, 'Structural Evolution of the Rift Zones in the Middle East', *Nature*, Vol. 221, 1969, pp. 926–30; D. G. Roberts, 'Structural Evolution of the Rift Zones in the Middle East', *Nature*, Vol. 223, 1969, pp. 55–7.

[2] S. W. Carey, 'The Tectonic Approach to Continental Drift', *Symposium on Continental Drift*, Hobart, 1958, pp. 177–355.

[3] D. R. Grantham, *Proc. Geol. Soc. Lond.*, Session 1957–8, No. 1555, p. 7.

[4] The Gulf of Aqaba, although a branch of the Red Sea rift, has negative gravity anomalies and may in fact have a different fundamental structure.

[5] A. M. Quennell, 'The Structural and Geomorphic Evolution of the Dead Sea Rift', *Quart. Journ. Geol. Soc. Lond.*, Vol. 114, 1959, pp. 1–24; R. Freund, 'A Model of the Structural Development of Israel and the Neighbouring Countries', *Geol. Mag.*, Vol. 102, 1965, pp. 189–205.

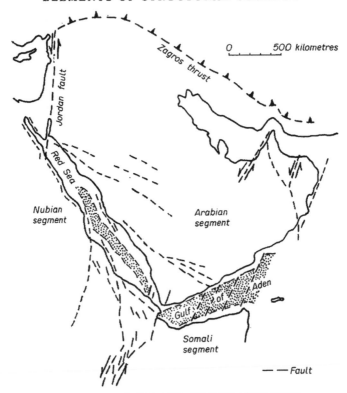

FIG. VII–33. RED SEA AND GULF OF ADEN RIFTS

Formed during the updoming of the Afro-Arabian upwarp, the segments subsequently moved apart, exposing oceanic crust (dotted) in the Red Sea and the Gulf of Aden. The Arabian segment underthrusts to the north along the Zagros Thrust. (*Compilation after Gass and Gibson, 1969, and Roberts, 1969*)

this marginal fault, which may, in fact, be as old as Proterozoic, although it may well have been inactive for long periods and active at others (Fig. VII–35A). Faulting marginal to a continental block, and with a deep trough containing Palaeozoic, Mesozoic, and Tertiary sediments, is also seen in Western Australia along the Darling Scarp, where some 40,000 ft of sediments lie on the downthrown side, and bedrock structures indicate the presence of an asymmetrical trough lying between the continent and the Indian Ocean (Fig. VII–35B and C).

Ramp valleys

At its northern end the Red Sea bifurcates into the Gulf of Aden and the Gulf of Aqaba, the latter forming part of the Dead Sea trough. It was for this latter that Bailey Willis [1] postulated the structure to be a *ramp valley*, a fault trough

[1] B. Willis, 'The Dead Sea Problem, Rift or Ramp Valley', *Bull. Geol. Soc. Amer.*, Vol. 39, 1928, pp. 490–542.

bounded by reverse faults which pass into low-angle thrusts in depth, the central wedge-shaped block being forced down (Fig. VII–34B). Although this concept is now known not to apply to the Dead Sea trough, the Jordanian boundary fault on the east being in fact a strike-slip fault and there being no

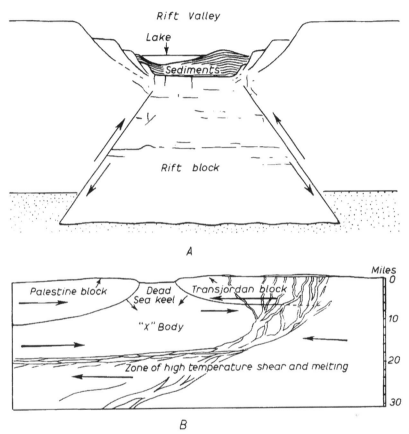

FIG. VII–34. RAMP VALLEY HYPOTHESES

A. An east African rift valley (Lake Albert). *After Wayland.* Showing the up-
  thrust of lateral blocks following the forcing down of the central wedge.
  Marginal gravity faults produce the rift valley as seen.
B. Willis' hypothesis for the Dead Sea trough as a ramp valley.

corresponding fault on the west (see Fig. VII–35A), the ramp valley hypothesis remains a valid general concept. Bullard's study of gravity anomalies in the Lake Albert rift of East Africa revealed an excess of light rocks which was regarded by Wayland as being due to the forcing down of a sialic wedge in compression between reverse faults (Fig. VII–34A). Afanas'yev has ascribed crestal grabens in the Saur Range, Kazakhstan, to compression, and high-angle reverse faulting is probably involved in the formation of grabens and uplifted horsts in eastern Australia (see Fig. VII–37). Reactivation of old

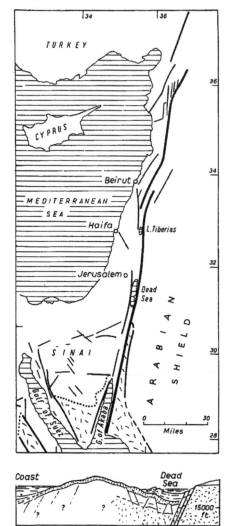

FIG. VII–35A

The Palestinian (Dead Sea) Rift showing the Jordan Fault on the east at the edge of the Arabian Shield. (*After L. Dubertret, 'Géologie des Roches Vertes du Nord-ouest de la Syrie et du Hatay (Turquie)', Mus. Nat. d'Hist. Nat., Paris, Notes et Mémoires sur le Moyen-Orient, T. VI (Extrait), 1953. Section compiled from various sources*)

The Jordan Fault is now generally believed to be a dextral strike-slip fault

faults under new stress systems complicates the issue, especially as it is rare to find precise data as to the dip of boundary faults, even for the well-studied Rhine graben.

Aulacogenes – Ancient grabens of great length (up to 1000 km) that have been filled with young rocks and subsequently deformed were called *aulaco-genes* by Shatsky, and Bogdanoff[1] cites the Donbas trough as a typical

[1] A. A. Bogdanoff, 'Sur Certains Problèmes de Structure et d'Histoire de la Platforme de l'Europe Orientale', *Bull. Soc. Géol. France*, Ser. 7, Vol. IV, 1962, pp. 898–911; A. A. Bogdanoff, M. V. Mouratov and V. E. Khain, 'Éléments Structuraux de la Croûte Terrestre', *Rev. Géog. Phys. et Géol. Dynam.*, Vol. 5, 1962, pp. 263–85; A. V. Chekunov, 'Mechanism Responsible for Structures of the Aulacogen Type (Taking the Dnieper–Donets Basin as an Example)', *Geotectonics*, No. 3, 1967, pp. 137–44.

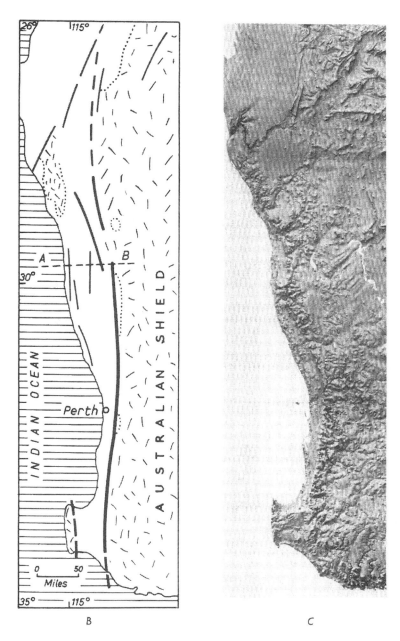

FIG. VII–35B AND C

B. Simplified geological map showing the Darling Fault at the edge of the Australian shield. (*After* Geol. Soc. Aust. Journ., *Vol. 4, Pt 2, 1958*)

C. Photograph of relief model of the same area.

example. The eastern end is more strongly developed and has been more affected by later folding than the western, where the trough dies away in the Russian Platform.

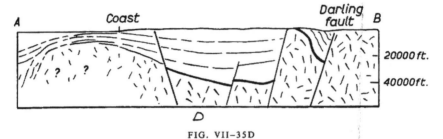

FIG. VII–35D

Section across *A–B* on Fig. VII–32B, showing the rise of the basement complex towards the west. (*By courtesy of J. R. H. McWhae*)

Lag faults – These are low-angle faults with normal fault displacement, that originate from the upward movement of the footwall block in a region of general thrusting. The hanging-wall block appears to have lagged behind in the regional movements (Fig. VII–36). Many such faults are recognized in the South Wales and the Belgian coal-basins.[1]

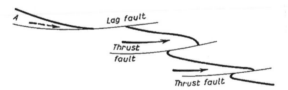

FIG. VII–36. A LAG FAULT, DUE TO RELATIVE RETARDATION OF BLOCK A, IN THE GENERAL FORWARD MOVEMENT SHOWN BY THE THRUST FAULTING

Reverse faults

In reverse faults the hanging-wall block is displaced up the fault plane relative to the footwall block, and the length of the section of faulted rock is short-ened. According to the stress analysis previously outlined, reverse faults may be expected to dip at somewhat less than 45°. In fact, many of the greatest reverse faults dip at very low angles, approaching horizontality rather than a '45°' shear angle, whereas others again dip very steeply, approaching the

---

[1] J. E. Marr, 'The influence of the geological structure of English Lakeland upon its present features', *Pres. Addr. Geol. Soc., Quart. Journ.*, Vol. 62, 1906, see p. lxxvi for definition. F. M. Trotter, 'The structure of the Coal Measures in the Pontardawe–Ammanford Area, South Wales', ibid., Vol. 103, 1947, pp. 89–123.

vertical. Accordingly these types are described as *low-angle* and *high-angle reverse faults* respectively.

Equal development of complementary reverse faults is rare. It is sometimes found in the case where the two sets are symmetrical or nearly so, about a horizontal axis of compression. With both low-angle and high-angle reverse faults, one set of shear planes is much more strongly developed than the other, and therefore it may be recognized that rotational strain is involved in such examples. This also accounts for the dip of low-angle and high-angle reverse faults. With the former, the direction of the force couple producing rotational strain approaches the horizontal, and the set of shears that lies closest to the couple is preferentially developed.

High-angle reverse faults cutting already folded rocks of complex geology also indicate the action of a force couple, but this is directed upwards rather

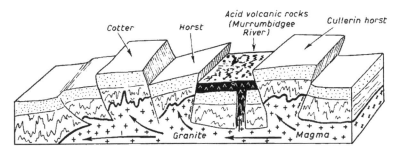

FIG. VII–37. POSSIBLE RELATIONSHIP OF GRANITIC MAGMA TO HIGH-ANGLE REVERSE FAULTING

Canberra district, Australia. (*Based on section by L. Noakes*)

than horizontally. The possibility that this may be associated with the injection of magma is indicated by the close relationship between parallel meridional high-angle reverse faults, and elongated granitic intrusions in southeastern Australia. Lower and middle Palaeozoic igneous action in New South Wales and Victoria suggests the introduction from the east of magmatic wedges moving towards the west and southwest, which may account for the component of uplift involved in the prevalent high-angle reverse faults of the region[1] (Fig. VII–37).

High-angle reverse faults may also be formed as secondary shears (pinnate shears) in a zone affected by monoclinal warping (Fig. VII–38). They are, however, most prominently displayed in regions where they are clearly related to great low-angle faults. Because the geostatic pressure decreases upwards, most reverse faults are concave upwards, the dip increasing at the

[1] L. C. Noakes, 'The Significance of High-Angle Reverse Faults in the Canberra Region', *Aust. and N.Z. Assoc. Adv. Sc.*, Dunedin, 1957 (abstr.). E. S. Hills, 'Cauldron Subsidences, Granitic Rocks and Crustal Fracturing in S.E. Australia', *Geol. Rundsch.*, Vol. 47, 1959, pp. 543–61.

end distal from the applied forces in the more active sector of the diastrophic belt. In such cases a low-angle reverse fault passes into a high-angle fault.[1]

Low-angle reverse faults – With low-angle reverse faults the lateral displacement is so pronounced that they are often termed *thrust faults*. The term *overthrust*, also commonly applied to them, carries the implication that the superior thrust block underwent actual translation rather than merely a displacement relative to the rocks below, but *underthrusting* is in most cases equally probable, and it is therefore preferable to use '*overthrust*' only when it is desired to specifically indicate such a movement.

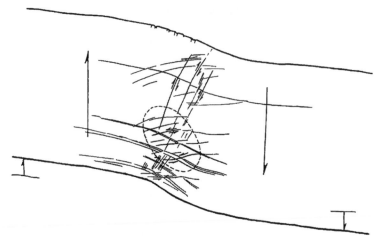

FIG. VII–38. FORMATION OF HIGH-ANGLE REVERSE FAULTS IN A
MONOCLINAL FLEXURE
(*After E. Cloos*)

The geometry of thrust faults is greatly influenced by inhomogeneities in the faulted rocks, especially where a shear plane attempts to form nearly parallel to the bedding, in which case it may be induced to follow the bedding planes rather than to cut acutely across them. The various possibilities may best be discussed in relation to an actual example.

Appalachian Valley and Ridge Province – The outer zone of deformation of the Appalachian chain is well displayed in the Central and Southern Appalachians from Pennsylvania to Alabama, where Cambrian to Pennsylvanian rocks rest on a relatively little-deformed basement complex of Pre-Cambrian age. The cover rocks are folded and faulted more simply than in the core of the Appalachian folds, and the surface exhibits long strike-ridges with intervening valleys, characterizing the Valley and Ridge Province[2] (Fig. VII–39).

[1] See the examples discussed by S. Kienow, 'Mechanische Probleme bei Auffaltung der subvariszischen Vortiefe', *Zeitschr. Deutsch. Geol. Ges.*, Bd. 107, 1956, pp. 140–57; M. K. Hubbert, 'Mechanical Basis for Certain Familiar Geologic Structures', *Bull. Geol. Soc. Amer.*, Vol. 62, 1951, pp. 355–72. W. Hafner, 'Stress Distributions and Faults', ibid., pp. 373–98.

[2] Figure XI–4, shows the whole Appalachian chain in cross-section.

At and near the surface, reverse faults slice the sedimentary cover, and are chiefly high-angle faults or upthrusts, but the interrelationships of flat-lying beds in synclines, with the geometry of faulted anticlines indicates that these high-angle thrusts are frontal prows of bedding thrusts, originating by the mechanism shown in Fig. VII–40. The faults are thought to have developed

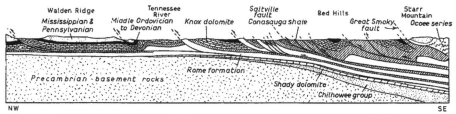

FIG. VII–39. THE APPALACHIAN VALLEY AND RIDGE PROVINCE
(After J. Rodgers)

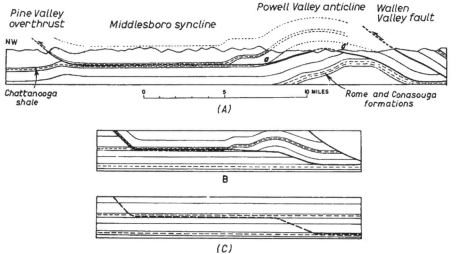

FIG. VII–40. SECTIONS SHOWING THE RELATION OF HIGH-ANGLE REVERSE FAULTS
TO BEDDING FAULTS AND FOLDS IN THE APPALACHIANS
(After P. B. King)

in sequence, those farther east initiating the movements in the more remote western segments. The basement rocks are unaffected by the faulting.[1]

Erosion and surface thrusts – The intersection of low-angle thrusts with the earth's surface over which the thrust mass advances, gives a *surface thrust*. On a small scale these may arise by erosion of the crest of a rising anticline,

[1] P. B. King, *The Tectonics of Middle North America*, Princeton, 1951. J. D. Rogers, 'Evolution of Thought on Structure of Middle and Southern Appalachians', *Bull. Amer. Assoc. Petrol. Geol.*, Vol. 33, 1949, pp. 1643–54; also idem, 'Mechanics of Appalachian Folding', ibid., Vol. 34, 1950, pp. 672–81. J. L. Rich, 'Mechanics of Low-angle overthrust faulting as illustrated by Cumberland Thrust-block, Kentucky and Tennessee', ibid., Vol. 18, 1934, pp. 1584–96. W. H. Bucher, 'Deformation in Orogenic Belts', *Geol. Soc. Amer.*, Special Paper 62 (*Crust of the Earth*), 1955, pp. 343–68.

allowing a competent member in the other limb to slip forward on underlying incompetent beds [1] (Fig. VII–41A) but the phenomenon is also developed on a regional scale. Goldschmidt showed that such conditions existed during the growth of the Caledonian thrust nappes in Norway, and that erosional conglomerates from the front of the nappe were over-ridden by its later advance. [2]

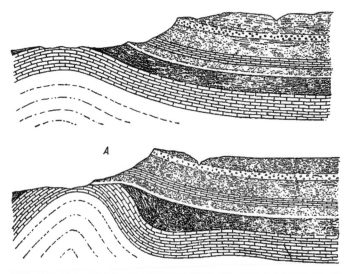

FIG. VII–41A. EROSION THRUST DUE TO THE BREAKING OF
A MAJOR ANTICLINE
(*After Willis*)

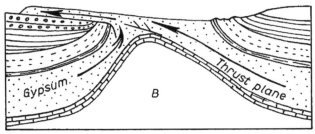

FIG. VII–41B. EROSION THRUST IN PERSIA, THE
MOVING BLOCK BEING ROCK SALT EXTRUDED FROM AN
ANTICLINAL ARCH
(*After Busk*)

In Persia the extrusion of highly mobile gypsum and gypseous shales at anticlinal crests gives rise to masses that advance over the surface, but the forces generating such extrusions are probably local rather than regional, and

---

[1] B. Willis, 'The Mechanics of Appalachian Structure', 13th Ann. Rept. *U.S. Geol. Surv.*, Pt 2, 1893, see pp. 222–3.
[2] See review in *Geol. Mag.*, Vol. 4, 1917, pp. 130–2.

due to the squeezing up of the gypsum by the weight of superincumbent rocks in neighbouring synclines (Fig. VII−41B).[1]

Surface thrusting on a regional scale is perhaps best known in southern Nevada and California in a belt some 100 miles wide in which thick Palaeozoic and Mesozoic sediments are affected by late Cretaceous or early

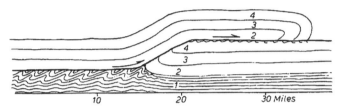

FIG. VII−42. EROSION THRUSTS IN NEVADA AND CALIFORNIA
(After Longwell)

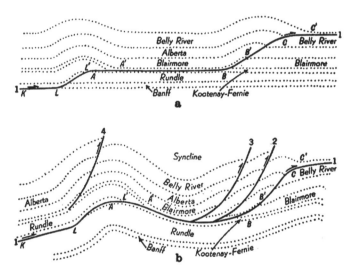

FIG. VII−43. FOLDING AND THRUSTING IN THE EASTERN
ROCKY MOUNTAINS, ALBERTA
(Compare Appalachians above) *(After Haine, 1957)*

Tertiary (Laramide) diastrophism.[2] The low-angle thrust sheets are relatively simple structurally, and much of the movement took place on the bedding, within weak shales over which the stronger limestones were forced, as shown diagrammatically in Fig. VII−42. The thrust sheets took on their great displacement when the limestones folded and the folds sheared through.

[1] H. G. Busk, *Earth Flexures*, Cambridge, 1929, pp. 86−95. For the explanation here given, see C. A. E. O'Brien, 'Salt Diapirism in South Persia', *Geol. en Mijnb.*, N.S. Jhg. 19, 1957, pp. 357−76.
[2] C. K. Longwell, 'Mechanics of Orogeny', *Amer. Journ. Sci.*, Vol. 243A, 1945, pp. 417−47.

Faulting of this type, which begins to develop when the beds are little-disturbed and continues with the accompaniment of folding and further slicing by high-angle upthrusts, must now be recognized as a type of characteristic of rather thick deposits of diversified lithology including thick formations of differing competency. It is very finely exemplified in the eastern Rocky Mountains in Alberta[1] (Fig. VII-43).

Somewhat similar structural associations of reverse faults and folds in the Ruhr and the Belgian coalfields[2] (Fig. VII-44) are less influenced by lithological differences, so that the faults transgress the bedding at low angles and are in places folded sharply as the deformation continues (Fig. VII-45).

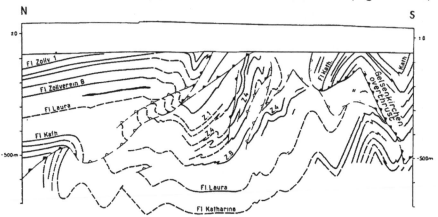

FIG. VII-44. CROSS-SECTION OF PART OF THE RUHR COAL-BASIN, SHOWING STRONGLY FOLDED FAULTS

(*After Honermann, from Kienow, 1956*)

These various effects are very expressive of the high mobility of the faulted and folded rocks and may indicate the overall influence of a field of force, which may perhaps be gravity, rather than of an end-on push to generate the faults. Kaisin[3] in discussing these topics for the Ardennes points out that the faults arise because of the difference in the rate of advance of slices separated by faults, a view that accords with Odé's analysis of faulting in a plastic medium.[4]

The Moine thrust – The Moine-Thrust in the North-West Highlands of Scotland affords a classic example of thrusting associated with regional metamorphic rocks. This great thrust, on which the translation is known to exceed

[1] G. S. Haine, 'Fault Structures in the Foothills and Eastern Rocky Mountains of Southern Alberta', *Bull. Geol. Soc. Amer.*, Vol. 68, 1957, pp. 395–412.

[2] S. Kienow, 'Mechanische Probleme bei der Auffaltung der subvariszischen Vortiefe', *Z. deutsch. geol. Ges.*, Vol. 107, 1956, pp. 140–57. F. Kaisin, 'Le Bassin Houiller de Charleroi', *Mem. Inst. géol. Uni. Louvain*, T.15, 1947.

[3] F. Kaisin, 'Poussées tangentielles ou "Champs tectoniques"?', *Bull. Soc. Belge de Géol.*, Vol. 53, 1944, pp. 228–63.

[4] See Chapter IV, p. 97.

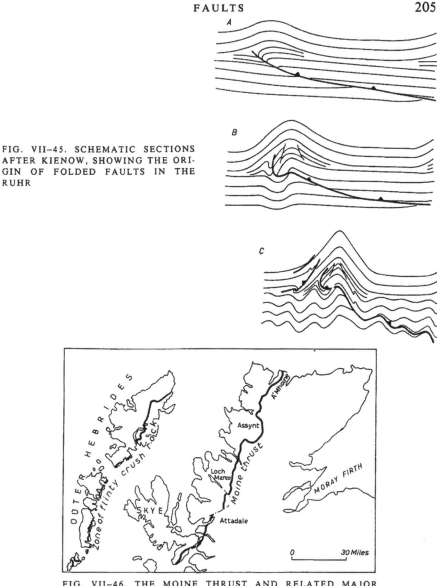

FIG. VII-45. SCHEMATIC SECTIONS AFTER KIENOW, SHOWING THE ORIGIN OF FOLDED FAULTS IN THE RUHR

FIG. VII-46. THE MOINE THRUST AND RELATED MAJOR STRUCTURES
(*After McIntyre*)

ten miles,[1] carried the Moine schists over Pre-Cambrian and little-altered Cambrian rocks. The movement on the Moine Thrust most probably involved a general translation towards the northwest, with local variation due to inhomogeneity of the great mass of rocks involved and to variation in the

[1] The account given is based largely on the review article by D. B. McIntyre, 'The Moine Thrust – its discovery, age, and tectonic significance', *Proc. Geol. Assoc. Lond.*, Vol. 65, 1954, pp. 203–23.

stresses from place to place. It belongs to a broad zone of strong deformation to which a parallel zone of crushed rocks in the Outer Hebrides is also attributed (Fig. VII–46). In places a clear-cut *sole* or single fault plane is observable, but parallel thrusts also exist and in places there is a wide fault-zone. A set of small parallel thrusts dipping more steeply slice the over-ridden

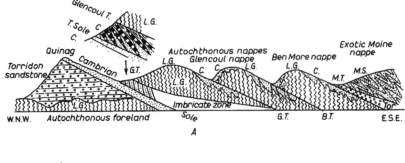

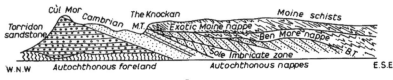

FIG. VII–47. SKETCH SECTIONS OF THE MOINE THRUST ZONE, SCOTLAND, SHOWING THE MAJOR LOW ANGLE THRUST-PLANES, AND IMBRICATE STRUCTURE IN THE OVER-RIDDEN CAMBRIAN BEDS

(*After Wills*, Physiographical Evolution of Britain.) *M.S.*, Moine Schists; *C.*, Cambrian; *TOR.*, Torridonian; *L.G.*, Lewisian Gneiss; *G.T.*, Glencoul Thrust; *B.T.*, Ben More Thrust; *M.A.*, Moine Thrust. Inset – details of the imbricate structure.

Cambrian rocks into overlapping *Schuppen*, giving *imbricate structure* (Fig. VII–47). The fault, although affecting chiefly Pre-Cambrian and Cambrian rocks, is believed to be of mid-Palaeozoic age and to belong to the Caledonian orogeny in a broad sense.[1]

Thrust-nappes – Large-scale low-angle thrusts constitute one variety of nappe structure (*thrust-nappes*) of which the Moine Thrust is the classic

[1] The problems associated with structures of the magnitude of the Moine Thrust are such that their elucidation comes only from the work of many geologists over many years. New discoveries and ideas arise continually, especially, it appears, in the heady atmosphere of the Highlands. As an introduction to the complexity of the Moine problems, see the Discussion on McIntyre's paper, 'The Moine Thrust – its discovery, age, and tectonic significance', *Proc. Geol. Assoc. Lond.*, Vol. 65, 1954. The more important primary works are: B. N. Peach and J. Horne *et al.*, 'The Geological Structure of the North-West Highlands of Scotland', *Mem. Geol. Surv. Great Britain*, 1907. B. N. Peach and J. Horne, *Chapters on the Geology of Scotland*, London, 1930. E. B. Bailey, 'The Structure of the South-West Highlands of Scotland', *Quart. Journ. Geol. Soc. Lond.*, Vol. 78, 1922, pp. 82–131; idem, 'The Glencoul Nappe and the Assynt Culmination', *Geol. Mag.*, Vol. 72, 1935, pp. 151–65; idem, 'West Highland Tectonics: Loch Leven to Glenroy', *Quart. Journ. Geol. Soc. Lond.*, Vol. 90, 1934, pp. 462–525.

example. A further account of nappes, including fold nappes, will be found in Chapter XI.

Strike-slip faults

In strike-slip faulting the relative displacement of the blocks is horizontal or approximately so, but an old strike-slip fault plane may, at a later date, be affected by vertical movements due to a different set of forces from that under which it originated, thus making the determination of the original faulting a matter of some difficulty.

Assuming lateral horizontal compression, strike-slip faulting implies that the greatest relief of stress is also horizontal, and the mean stress axis vertical. However, in the majority of examples of strike-slip faulting, one primary set of shear planes and even one single major fault is dominant, indicating that a horizontal force couple was almost certainly involved. However, the movement on transform strike-slip faults (see p. 213) is a simple differential gliding due to the direct movement apart of the edges of interdigitating plates. Such sliding would, however, induce force couples in superincumbent cover rocks above a rigid basement, and the question of stress relationships is accordingly not easily solved. Because strike-slip faults are more readily determinable where they cross the trend of folds, they have been called *transcurrent 'thrusts'* [sic], or, in the Juras where they appear to be formed during the folding, *tear faults*. In the Scottish Highlands they were termed *wrench faults*.[1]

Evidence for strike-slip faulting – Direct evidence of the movement on faults at any time is readily obtained from visible displacements on the faults accompanying earthquakes. With some reservations—firstly that the movement at any one time or place may not represent the average movement of major crustal blocks, and secondly that movements in superficial soft rocks and 'made ground' in cities are generally very irregular—observation not only affords evidence of relative displacement but also, from the comparison of the position of accurately fixed points before and after the earthquake, permits the recognition of the actual movements of blocks.

Well-known examples are the faulting with a lateral shift of 8 ft, of the upper Crystal Springs earth dam, which occurred during the San Francisco earthquake of 1906, and displacements of some 6–8 ft affecting farm fences in New Zealand.

From the mathematical analysis of seismograms 'focal mechanism' solutions can be got, which indicate the direction of movement on faults associated with earthquakes. These reveal that strike-slip movements are occurring on active transform faults,[2] and that there is a large component of

---

[1] Although these various terms may appear more expressive, there is no better general term than *strike-slip* for this class of fault.

[2] B. Isacks, J. Oliver, and L. R. Sykes, 'Seismology and the New Global Tectonics', *Journ. Geophys. Research*, Vol. 73, 1968, pp. 5855–99.

strike-slip connected with many deep-focus earthquakes originating up to 600 km in the earth.[1]

All such observed movements on faults are, however, very small compared with the total displacements postulated after study of the geological evidence, which may indicate tens or hundreds of miles, representing the summation of a great many small movements over a long period of time.

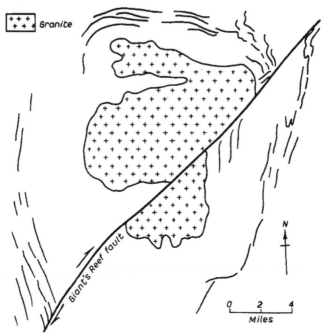

FIG. VII–48. THE GIANT'S REEF FAULT, NORTHERN TERRI-
TORY, AUSTRALIA

A dextral strike-slip fault slicing a granitic massif. The lines show trends of the pre-Cambrian rocks. (From mapping by Walpole *et al.*; courtesy Bureau of Mineral Resources)

The slicing-through of a clearly identifiable igneous massif is seen in the $3\frac{1}{2}$ miles of total displacement on the Giant's Reef Fault in northern Australia (Fig. VII–48). The matching of the Strontian granite on the north of the Great Glen Fault in Scotland with the Foyers granite on the south, now some 65 miles apart, afforded the first strong evidence, along with other data from the region, for a total lateral shift of this amount on the fault[2] (Fig. VII–49), but this appears to comprise a sinistral shift of 83 miles in the Devonian, and a dextral shift of 18 miles during the Cainozoic.[3]

[1] J. H. Hodgson, 'Nature of Faulting in Large Earthquakes', *Bull. Geol. Soc. Amer.*, Vol. 68, 1957, p. 611–44.
[2] W. Q. Kennedy, 'The Great Glen Fault', *Quart. Journ. Geol. Soc. Lond.*, Vol. 102, 1946, pp. 41–76.
[3] N. Holgate, 'Palaeozoic and Tertiary Transcurrent Movements on the Great Glen Fault', *Scottish Journ. Geol.*, Vol. 5, Pt 2, 1969, pp. 97–139.

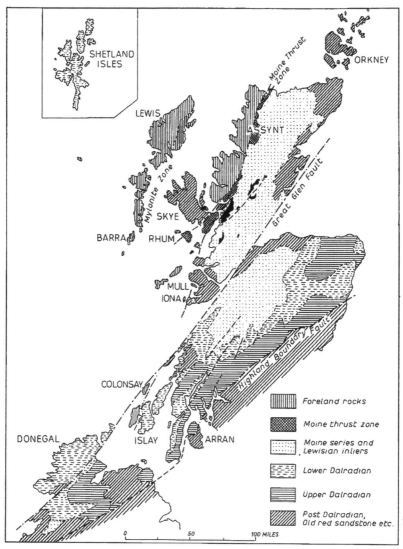

FIG. VII–49. THE GREAT GLEN FAULT

Generalized geological map showing the supposed geography of northern Scotland and Ireland before the lateral displacement on the Great Glen Fault. Other major faults are also shown. (*After Kennedy, and based on a compilation by Dr Janet Watson and Dr G. Wilson*)

The San Andreas Fault in California can be traced for at least 600 miles. It is a dextral strike-slip fault which is still active but which has had a long history. On a map from which younger rocks have been removed, it will be seen that the older rocks have been affected by some 300–350 miles of strike-slip (Fig. VII–50A). The largely chaotic Franciscan rocks appear to

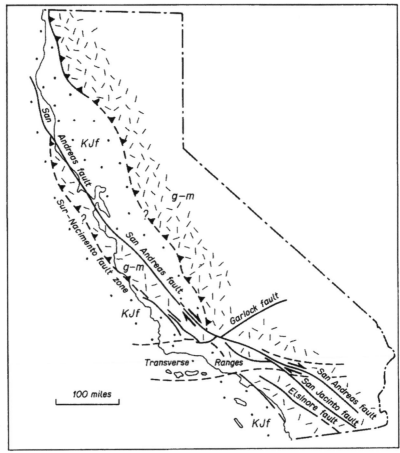

FIG. VII-50A. THE SAN ANDREAS FAULT, CALIFORNIA, AND SOME RE-
LATED FAULTS

KJf. Franciscan rocks. *g–m.* Sierran granite and metamorphic rocks. (*After B. M.
Page*[1] *– partly hypothetical*)

represent oceanic trench accumulations which have been strongly deformed
as they were forced against the American block before the San Andreas
Fault developed, the Mesozoic edge of the continent being represented by the
Sierran–Franciscan boundary and its continuation in the Sur–Nacimiento
Fault. The San Andreas Fault is now regarded as a transform fault[2] originating when in mid-Oligocene time, the American sialic block ('plate') began to
encounter the mid-Pacific plate,[3] but there are difficulties in reconciling age

[1] Fig. VII–50A is based on a map kindly supplied by Professor B. M. Page, this being
modified from a map by P. B. King in *The Evolution of North America*, Princeton, 1959, Fig.
91. I am indebted to Professor Page for discussion about the San Andreas Fault, and for a
statement on which the text herewith is based.

[2] See p. 210, especially Wilson, *op. cit., Nature*, 1965.

[3] D. P. McKenzie and W. J. Morgan, 'Evolution of Triple Junctions', *Nature*, Vol. 224,
1969, pp. 125–33.

relationships and also evidence on land with sea-floor tectonics. The southern part of the San Andreas fault appears to be connected with the opening of the Gulf of California in the Pliocene,[1] while in middle California Lower Miocene rocks indicate some 200 miles of offset, Middle Miocene 125–175 miles, and Upper Miocene about 75 miles.[2] Bolt and others have discussed the seismological evidence in terms of a model.[3]

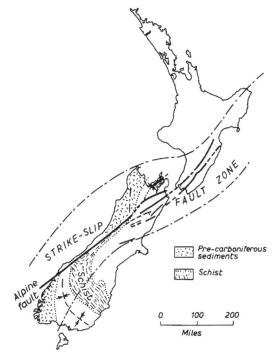

FIG. VII–50B. THE ALPINE FAULT, SOUTH ISLAND OF NEW ZEALAND

(*After Wellman, 1955*)

Similarly for the Alpine Fault in New Zealand[4] (Fig. VII–50B) which, like the San Andreas Fault, is still active, a total lateral shift of 300 miles is postulated since the end of the Jurassic Period, as judged from the tearing apart of a major syncline and other features. It also appears to be a transform fault.

Physiographic evidence for strike-slip faulting includes a wide variety of

[1] D. G. Moore and E. C. Buffington, 'Transform Faulting and Growth of the Gulf of California Since the Late Pliocene', *Science*, Vol. 161, 1968, pp. 1238–41.

[2] Proceedings of conference on geological problems of San Andreas fault system, Stanford, Calif., 1967: *Stanford Univ. Pubs. Geol. Sci.*, Vol. 11, 1968.

[3] B. A. Bolt, 'Seismological Evidence on the Tectonics of Central and Northern California and the Mendocino Escarpment', *Bull. Seismol. Soc. Amer.*, Vol. 58, 1968, pp. 1725–67.

[4] H. W. Wellman, 'New Zealand Quaternary Tectonics', *Geol. Rundsch.*, Vol. 43, 1955, pp. 248–57.

data such as the slicing of river-terraces, the closing of valleys by faulted spur-ends (shutter-ridges) and the offsetting of stream courses.[1]

Many major strike-slip faults have been identified only of recent years,[2] and strike-slip movements on smaller faults are increasingly recognized.[3] Moody and Hill[4] proposed that an array of second- and third-order faults

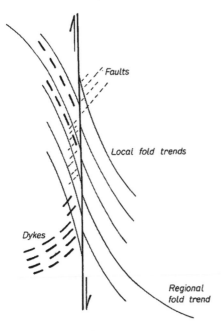

Faults

Local fold trends

Dykes

Regional fold trend

FIG. VII–51

Idealized structural trends associated with a major dextral strike-slip fault. (*After Bishop, 1968*)

and folds might develop in association with major strike-slip faulting, but the theoretical basis of their system has been questioned[5] and in fact the associated faults and folds actually found form a simpler array (Fig. VII–51).[6]

Tear faults – Crossing the broad sweep of the Jura folds at very obtuse angles are several faults with chiefly strike-slip displacement that appear to tear the

[1] C. A. Cotton, 'Geomorphic Evidence and Major Structures Associated with Transcurrent Faults in New Zealand', *Rev. Géogr. Phys. Géol Dynam.*, Ser. (2), Vol. 1, 1957, pp. 16–30.

[2] C. R. Allen, 'Transcurrent Faults in Continental Areas', *Phil. Trans. Roy. Soc. Lond.*, Vol. 258, 1965, pp. 82–9, discusses the Phillipine Fault, Atacama Fault (Chile), Alpine Fault (N.Z.) and San Andreas Fault.

[3] M. Abdel-Gawad, 'New Evidence of Transcurrent Movements in Red Sea Area', *Amer. Assoc. Petrol. Geol.*, Vol. 53, 1969, pp. 1466–79.

[4] J. D. Moody and M. J. Hill, 'Wrench-Fault Tectonics', *Bull. Geol. Soc. Amer.*, Vol. 67, 1956, pp. 1207–46.

[5] J. J. Prucha, 'Moody and Hill System of Wrench Fault Tectonics Discussion', *Bull. Amer. Assoc. Petrol. Geol.*, Vol. 48, 1964, pp. 106–24.

[6] H. Wopfner, 'On Some Structural Development in the Central Part of the Great Australian Artesian Basin', *Trans. Roy. Soc. South Aust.*, Vol. 83, 1960, pp. 179–93. D. G. Bishop, 'The Geometrical Relationships of Structural Features Associated with Major Strike-Slip Faults in New Zealand', *N.Z. Journ. Geol. Geophys.*, Vol. 11, 1968, pp. 405–17.

folds across. Their origin due to differential advance of adjoining segments of the fold arc has been postulated, but they may in fact represent a rejuvenation of more ancient basement faults, as they are relatively straight, and cut as many as 8–9 anticlines.[1] Nevertheless, there are excellent examples of true tear faults on a smaller scale, in which the folds in one block are more tightly compressed and closer spaced than in the other, as shown by Goguel[2] (Fig. VII–52). It is therefore desirable to use *tear fault* (Ger. *Blattverschiebung*) only for strike-slip faults that originate during the folding in the way described.[3]

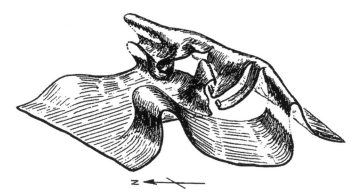

FIG. VII–52. DIAGRAM SHOWING TEAR-FAULTING
(*After Goguel*)

Transform faults – The mechanism and name for transform faults, a type of fault not previously recognized, were proposed by J. Tuzo Wilson in 1965.[4] Growth (or reduction) of the surface area affected is fundamental in transform faulting, as is shown in the simple case of a 'half-shear' in which a terminal rift such as is believed to be represented on the ocean floors by the oceanic ridge system, is filled with dyke injections thus adding to the area (Fig. VII–53A): or on the other hand the formation of a folded arc by the downward plunging of oceanic crust may reduce the area (Fig. VII–53B). Where two such simple transforms are joined end to end, the fault between the moving blocks is a transform fault. Wilson recognised six different types of transform faults, each of which might have either dextral or sinistral

[1] R. Staub, Alpine Orogenese. 'Grundsätzliches zur Anordnung und Entstehung der Kettengebirge', in *Skizzen zum Antlitz der Erde*, Vienna, 1952, pp. 1–46.

[2] J. Goguel, *Traité de Tectonique*: Paris, 1952, pp. 130–1; 152–8. Goguel's representation of tear faulting by means of a deformed rectangular graticule corresponding with the map graticule is particularly expressive.

[3] It may eventually be necessary to restrict the term *wrench faults* to strike-slip faults caused by horizontal force couples, as the word suggests.

[4] J. T. Wilson, 'A New class of Faults and their Bearing on Continental Drift', *Nature*, Vol. 207, 1965, pp. 343–7; 'Transform Faults, Oceanic Ridges and Magnetic Anomalies Southwest of Vancouver Island', *Science*, Vol. 150, 1965, pp. 482–9.

displacements. That shown in Fig. VII–53C is a ridge–ridge type, in which after prolonged movements the ends of the fault become inactive and earthquake activity is confined to the segment between the median axes of the ridges, and to these axes themselves. It will be seen that the apparent displacement of the ridge axis in crossing a transform fault is in the opposite

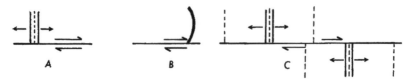

FIG. VII–53. TRANSFORM FAULTING

A. Dextral half-shear, with rift opening along a mid-oceanic ridge.
B. Dextral half-shear, with compressional orogenic arc forming.
C. Ridge–ridge transform fault, after a period of movement; former positions of the rift margins shown by broken lines. (*After J. T. Wilson, 1965*)

sense to the actual strike-slip movements on the active part of the fault. The trace of the fault persists as a prominent linear fracture zone transverse to the axis of the ridge. Rapid advances have been made very recently in our knowledge of these and related fractures of ocean basins and the broader aspects are discussed in Chapter XIV.

Fault-folds (Bruchfalten) – In the majority of great grabens, basement rocks are displayed on the flanks of the graben and the trough is occupied by a considerable thickness of younger sediments, which in most cases were

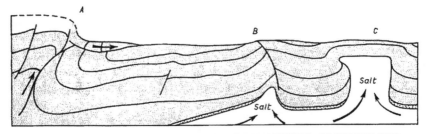

FIG. VII–54. GERMANOTYPE (SAXONIAN) STRUCTURES–BRUCHFALTEN

A. Osning type with high-angle reverse fault and minor low-angle 'nappe'.
B. Dardesheim type, with salt anticline.
C. With salt dome.

deposited in the developing trough and which are absent or much thinner on the upthrown blocks. The structures that are developed where a relatively thick cover of younger rocks lies on a more rigid basement which is then faulted, differ from those of sharply faulted grabens and horsts because of the different reactions of the softer cover rocks. This has been intensively studied by Stille and his school in the mid-German region where thick Mesozoic rocks cover a basement in which pronounced anisotropy is present, due to Hercynian folding and other structural lines (Fig. VII–54). The region includes important salt deposits which influence the structures because of their mobility, so that Saxonian or Germanotype fault-folds (*Bruchfalten*) are more complex than those of comparable regions in which highly mobile beds are absent.[1] In Germany the basement rocks appear to have been faulted chiefly by normal faults along two or more dominant trends, producing either long horsts and grabens, or a mosaic of fault blocks, over the edges of which the cover rocks are draped, so that there remain relatively flat-lying beds in the troughs and on the plateau tops of horsts, but the flanks of these are broadly warped, thus giving anticlines over the horsts and synclines in the troughs. In the synclinal structures the thickness of the younger rocks and the intensity of folding increases, especially in the actual vicinity of salt domes. Local compression is developed probably due to flowage of the mobile beds and to squeezing in front of strong monoclinal warps where small reverse faults may also be formed.

The Coast Ranges of California afford another example of a region exhibiting this type of structure,[2] which Bucher interprets as being caused by alternating tension and compression in an area where thick deposits of sediments are being laid down. According to this view, the pattern of the long blocks bounded by normal faults is determined by tension. The folding of the sediments in the troughs, as well as the minor thrusting, result from later compression and from drag along the faults. Schuh[3] came to much the same conclusion with regard to the Saxonian Bruchfalten. Structures comparable with Bruchfalten are probably much commoner than would appear from the literature, many examples being described in terms of folding. There is a close similarity with folding of the Plains type.

[1] H. Stille, 'Mittledeutsche Rahmenfaltung', *Jahrb. d. niedersachs. Ver. Hannover*, 1910, pp. 141–69; also, with others, in 'Göttinger Beiträge zur saxonischen Tektonik', *Abh. Preuss. Geol. Landesanstalt*, N.F., Vol. 95, 1923–5; Vol. 116, 1930; Vol. 128, 1931. 'Geotektonische Forschungen', *Zur Germanotypen Tektonik I*, Hft. I, edited by H. Stille and F. Lotze, Berlin, 1937.
[2] W. H. Bucher, *The Deformation of the Earth's Crust*, Princeton, 1933, pp. 303–25. B. L. Clark, 'Folding of the California Coast Range Type illustrated by a Series of Experiments', *Journ. Geol.*, Vol. 45, 1937, pp. 296–319.
[3] F. Schuh, 'Die Saxonische Gebirgsbildung', *Kali*, Vol. 16, 1922, Hft. 8, 9, 15, 16.

CHAPTER VIII

Folds

Folds are perhaps the most intriguing of all geological structures.[1] Examples of closely folded beds that may be seen in cliff exposures invite speculation on the mechanism that caused the crumpling of the strata; the interest and beauty of the sweeping curves of great folds drawn in cross-section are aesthetically satisfying in themselves; but above all, the implications of folding with regard to movements in the crust of the earth have always assumed major importance in tectonic thought and consequently in the philosophy of geological science.

Our understanding of the origin and tectonic implications of folding is based on a great many lines of evidence the chief of which are:

  (i) the shapes assumed by folded beds (fold geometry)
 (ii) experiments simulating folding
(iii) the relationships of folds to other structures, especially to faults
 (iv) the relationships of belts of folding to other major tectonic units
  (v) the microstructures developed in folded rocks.

The notion of folding

In geology as in everyday speech the word 'fold' implies that an originally planar object has been acted upon so as to produce some form of plication. Folds result from deformation, so that geological structures due for example to variation in initial dip as in tuffs on the flanks of a volcanic hill, would not be termed folds. In a simple case such as folding a sheet of paper, there is rotation of one or both segments about the median crease, which gives forms commonly seen in folded strata – the anticline or arch, and the syncline or trough fold. Many, if not the majority, of geological folds are formed by a somewhat similar process involving rotation of the flanks or limbs of the folds, but other mechanisms may also produce structures broadly resembling such simple folds. Thus it is usual to describe as a fold any curved or zig-zag

[1] J. G. Ramsay, *Folding and Fracturing of Rocks*, New York, 1967. E. H. T. Whitten, *Structural Geology of Folded Rocks*, Chicago, 1966.

structure shown by the bedding planes of stratified rocks, if this structure has resulted from changes in the attitude of the rocks after deposition. The surfaces of lava flows and their inner flow planes may also be used in field work to reveal folding, and in certain circumstances planar structures in metamorphic rocks may be similarly used, but in such cases some reservations must be entertained until detailed work reveals the true nature and original attitude of the planar structures mapped.

The definition of folding adopted does not imply that all folds are caused by the action of deep-seated or hypogene earth forces, and indeed we must now recognize four important categories of folds:

(i) those folds that are caused primarily by diastrophism, by which is meant the action of deep-seated earth forces which either have a strong horizontal component or which, if acting vertically, are sufficient to cause major uplifts against the force of gravity. Structures due to such processes are examples of *primary tectogenesis*.

(ii) those folds that are due to sliding and flow of great rock masses under gravity, their instability being due to elevation and tilting associated with primary tectogenesis. These are examples of *secondary tectogenesis* and *gravitational tectonics*.

(iii) those folds that are due to local effects rather than to regionally applied hypogene forces, among which are included folds due to compaction (see p. 60), to igneous injections, to injections of salt (see salt domes, p. 275), and the like.

(iv) those folds that are due to the entirely superficial sliding or flow of inadequately supported masses of newly deposited sediments that become detached to form slumps and slides. These have already been discussed in Chapter III.

Fold geometry in cross-section

The geometry of folds must be considered from several aspects. In the first place, we may simplify the subject and treat with folds as shown by one thin bed or chosen horizon plane in a succession of rocks. This is all that is required for the definition and description of certain of the most fundamental facts about folds, whether they are anticlines, synclines, monoclinal folds, and so on. Secondly, we may deal with the mutual geometrical relationships of successive horizon planes in a folded series as seen in cross-section, either the upper and lower separation planes of one bed, or a vertical sequence of beds or selected planes. This is required since many changes may occur in the form of a fold within a succession – even to the extent that an anticline at the surface may overlie a syncline in depth. It is because of such effects that our first consideration, as mentioned above, must be simple and deal with one horizon only.

Folds caused by diastrophism invariably occur in groups, as with waves, and having first discussed individual folds, it will be necessary to deal with *fold-trains* or *fold-systems*. For instance, with the Alpine folding of Europe, a train of genetically related folds may be traced northwards from the Alps themselves, decreasing in intensity until the last are seen in simple structures in southern England, fancifully described as the waves of the Alpine storm breaking on a far-distant shore. Folds due to local effects such as differential compaction over buried hills or to the uprise of salt to form salt domes, are not interrelated in the same way as a train of folds caused by diastrophism.

Finally, time has also to be considered in relation to folding. The recognition of periods of folding or of quiescence belongs rather to historical geology, but since diastrophic folding requires a period of time, many structural effects are possible due to the contemporaneity of folding with deposition or erosion.

Folding of a key horizon

Because of the great significance of the horizontal plane in geology, several fundamental definitions in folding and notably those for the basic terms anticline and syncline relate the attitude of beds to this plane. Ideally an

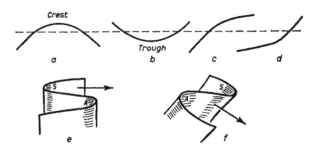

FIG. VIII–1. SIMPLE TYPES OF FOLDS

a. Anticline. *b.* Syncline. *c.* Anticlinal bend. *d.* Synclinal bend.
e. Anticline and syncline plunging at 90°, the arrow showing the stratigraphic top of the succession. *f.* Antiform and synform folds where, if the arrow shows the top of the succession, *A* is in reality a syncline.

anticline is an upwardly convex fold in which a given bed intersects the same horizontal plane in both limbs, dipping away from the crest of the fold on either side. A *syncline* is the reverse of anticline, the beds dipping towards the trough on either side. The highest point at the arch of an anticline is the *crest* of the fold, and the lowest point in a syncline the *trough* (Fig. VIII–1*a*, *b*).

There are many attitudes assumed by folded beds for which, although the folds may be essentially anticlinal or synclinal in type, they do not conform with the ideal forms of such folds. For these types it may be desirable to use anticline for a fold that encloses older beds in its core, and syncline for one that encloses younger beds. Busk [1] suggests *anticlinal* and *synclinal bend* for folds in which the dip of the beds suddenly increases or decreases, but there is no 'turn over', so that any bed intersects a given horizontal plane in only one limb (Fig. VIII–1*c, d*). Again, where folds plunge at 90° or more, the true fold-form is seen in horizontal or near horizontal section, rather than in a vertical section (Fig. VIII–1*e*). There is little objection to the use of the terms anticline and syncline where the stratigraphical succession is known with certainty, but if this is not known, words such as *antiform* and *synform* folds may be used (Fig. VIII–1*f*).

Fold symmetry – The symmetry of folds has traditionally been described according to the dips in either limb, dips being measured by reference to a

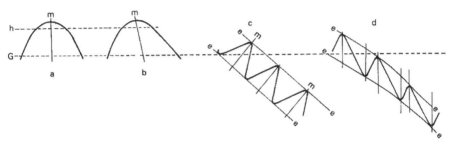

FIG. VIII–2. FOLD SYMMETRY

a. Symmetrical anticline (upright). *b*. Asymmetrical anticline (symmetrical above *h*). *c*. Asymmetrical fold as seen in outcrop at ground level *G*, actually symmetrical about the median line *m* (inclined). *d*. Upright inequant folds, symmetrical at outcrop, asymmetrical as to length of limbs. *e–e*, Fold envelopes.

horizontal plane. According to this usage, in *symmetrical folds* the dip as measured where the limbs intersect a given horizontal plane is at the same angle but in opposite directions in either limb (Fig. VIII–2*a*). In *asymmetrical folds* the corresponding dips are unequal. If a line is drawn that divides the cross-section of a fold into two more or less equal parts, it will be seen that this line is vertical in symmetrical folds, which are accordingly said to be *upright*, and is inclined to the vertical in most asymmetrical folds.[2] Furthermore, in a symmetrical fold the two halves are mirror-images, whereas this is not so in an asymmetrical fold.

It will also be seen that it is necessary to consider the whole of a fold to reveal its symmetry; for example, the fold in Fig. VIII–2*b* would appear symmetrical if mapped only about the crest, but is, in fact, asymmetrical. Some authors have also considered that in symmetrical folds the lengths of the limbs as well as their dips should be equal,[3] so that the folds in Fig. VIII–2*d*

[1] H. G. Busk, *Earth Flexures*, Cambridge, 1929, reprinted N.Y., 1957; see p. 7.

[2] See also pp. 263–5, where this bisecting line is more precisely considered.

[3] B. Stočes and C. H. White, *Structural Geology*, London, 1935, p. 116; F. J. Turner and L. E. Weiss, *Structural Analysis of Metamorphic Tectonites*, New York, 1963.

would be described as upright and asymmetrical. In fact the vast majority of folds are, strictly speaking, asymmetrical, but the direction of inclination of the median line or plane is very significant, and the direction of upward inclination of this plane is now called the *vergence* of the fold. It is true that the relative length of the limbs is an important aspect of fold geometry, for if corresponding limbs in a succession of symmetrical folds are consistently

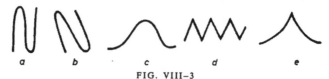

FIG. VIII–3

a. Upright isoclinal folds; *b.* Isoclinal overfolds; *c.* Curvilinear fold; *d.* Angular folds (zig-zag or chevron); *e.* Cuspate fold.

longer, a given bed will rapidly pass beneath the surface as one limb of an anticlinorium or synclinorium. Folds in which one limb is considerably longer than the other may be termed *inequant* (Fig. VIII−2c). If one limb of an asymmetrical fold is overturned the fold is an overfold. In *isoclinal folds* the limbs are folded closely together and are parallel[1] (Fig. VIII−3a, b). *Curvilinear folds* exhibit smooth curves in cross section; but many folds have straight or nearly straight limbs so that they are *angular* (Fig. VIII−3c). The repetition of angular folds gives *zig-zag* or *chevron* folds (Fig. VIII−3d), a

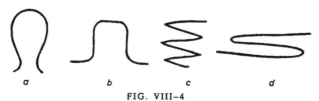

FIG. VIII–4

a. Fan-fold;   *b.* Box-fold;   *c.* Recumbent   zig-zag   folds;
d. Large-scale recumbent folds (nappes).

type well known in the Ruhr and the Belgian coalfields, the Appalachian coalfield, and the Bendigo goldfield, Australia. In *cuspate folds* the beds are smoothly curved in sectors, but adjoining sectors meet in sharp cusps (Fig. VIII−3e). Two fold types that are best known from European examples are the *fan-fold* and the *box-fold* (Ger. *Kofferfalt*). Mont Blanc is the type example of a fan-fold, and box-folds are common in the Juras (Fig. VIII−4). Both types are usually upright folds. By way of contrast, in *recumbent folds* both limbs are horizontal or nearly so. Recumbent folds are represented on a grand scale in *fold-nappes*, first recognized and best developed in the Western European Alps. Nappe structures are more fully discussed below (p. 253).

---

[1] Where the dips in either limb are very steep but not parallel, it is fundamentally incorrect to describe folds as isoclinal.

Orders of folds

Anticlinoria and *synclinoria* are complex large folds of general anticlinal or synclinal form with many minor folds on their limbs (Fig. VIII—5).

Folds of various *orders* are recognized, those of the largest size being termed first order folds and, higher orders being successively smaller, down to minute puckers of microscopic dimensions. The major undulations of an anticlinorium are first order folds, but very large up- or downwarps of the crust are not commonly termed folds but rather *swells*, *sags*, *troughs*, and the like. This is partly, no doubt, because such great structures affect regions whose detailed structure is already complex. Even where a cover of young

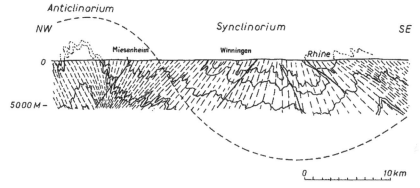

FIG. VIII–5. SYNCLINORIUM AND ANTICLINORIUM IN THE RHINELAND
The dashed lines indicate cleavage. (*After Quiring, 1939*)

sediments is present, it may be difficult to decide to what extent the very gentle, apparently synclinal warping of beds deposited in a broad basin is due to initial dip or to later warping.

In this regard it is well to point out the deceptive appearance given by drawing sections with greatly exaggerated vertical scale. Not only do the dips appear far too high so that even low initial inclinations appear like steep tectonic dips but the fold geometry is distorted because of the dependence of apparent thickness on the dip (Fig. VIII—6), and there are other serious disadvantages that have been discussed by Suter.[1]

Monoclines and terraces

Where horizontal or nearly horizontal beds suddenly dip fairly steeply for a short distance to give a step-like effect which physiographically and structurally is closely analogous with a normal fault (Fig. VIII—7), the structure is called a *monocline* (*monoclinal fold* or *flexure*). Essentially, a monocline

[1] H. H. Suter, 'Exaggeration of Vertical Scale of Geologic Sections', *Bull. Amer. Assoc. Petrol. Geol.*, Vol. 31, 1947, pp. 318–39.

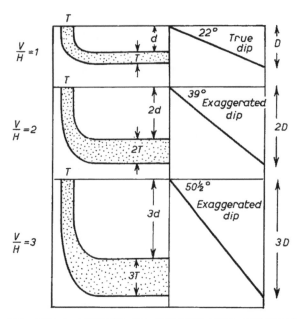

FIG. VIII–6. EFFECTS OF EXAGGERATED VERTICAL
SCALE

A. Apparent variation of thickness of strata in a monoclinal
flexure.

B. Apparent increase in dip with exaggeration of vertical scale.

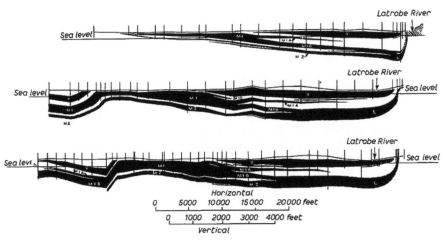

FIG. VIII–7. MONOCLINES IN TERTIARY BROWN COAL, LATROBE VALLEY,
VICTORIA

Vertical lines are bores. The sections are in sequence across the coal-basin. L. Latrobe Seam;
M2. Morwell No. 2 Seam; M1. Morwell No. 1 Seam; Y. Yallourn Seam. Note the great
thickness of the seams. (*By courtesy C. S. Gloe, and the State Electricity Commission*)

forms a link between two blocks of stratified rocks which have been displaced vertically relative to each other. As it is generally clear that the steeply dipping sector is the part that has been rotated, the term *flexure* is particularly applicable and is indeed commonly used. Kelly[1] has given a careful historical account of the various usages associated with the term monocline, and has

FIG. VIII-8. TYPES OF MONOCLINES

A. Typical monocline in otherwise horizontal beds.
B. Monoclines linking tilt-blocks.
C. Monocline composed of anticlinal and synclinal bend.
D. Monoclines facing in opposite directions, producing an anticline.

presented excellent descriptions of the classic region of the Colorado Plateau.

It will be noted that many of the geometrical forms shown as monoclines in Fig. VIII−8 may be completely described as to their geometry in terms of anticlines, synclines and anticlinal or synclinal bends. The essential notion in relation to monoclinal folds is primarily genetic rather than geometrical, and

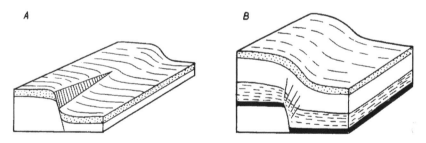

FIG. VIII-9. RELATION OF MONOCLINES TO FAULTS

A. Monocline passes laterally into fault.
B. Monocline passes downwards into fault.

it is open to the geologist to describe a region in terms of folds without implying how the folds originated, or in terms of monoclinal flexures (with which faulting will almost always be associated) if genesis is to be emphasized.

As noted above, monoclines are closely related to faults. Commonly a fault in brittle basement rocks passes up through softer covering strata as a

[1] V. C. Kelly, 'Monoclines of the Colorado Plateau', *Bull. Geol. Soc. Amer.*, Vol. 66, 1955, pp. 789–804.

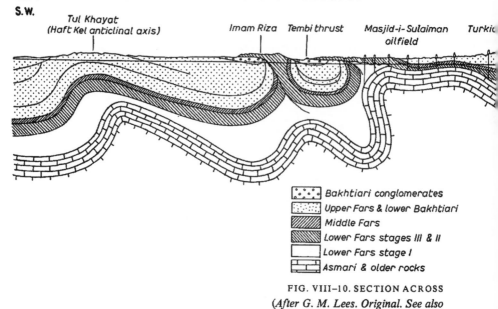

FIG. VIII–10. SECTION ACROSS
(After G. M. Lees. Original. See also

monocline, or again, with a gradual decrease in throw a fault may pass into a monocline either vertically or laterally (Fig. VIII–9).

Although the movement on monoclines generally simulates normal faulting and it is commonly a first assumption that this is the fundamental structure, evidence such as the actual as well as the relative elevation of high blocks, thrusting and other signs of compression may suggest that the faults are in fact reverse faults, perhaps upthrusts.

The use of the term monocline by oil geologists for the structure of a region of broadly uniform dip is currently being replaced by *homocline* for such structures, but it is necessary to be aware of the older usage. There is an important distinction between *regional dip* and *homoclinal dip*. Regional dip is the average dip over a region where there may be marked dip changes or even broad folds. It is thus a calculated figure, whereas homoclines are actual structures observed and measured in the field.

A *terrace* is a local flattening in an otherwise uniformly dipping series of beds, like a monocline in reverse.

GEOMETRICAL RELATIONSHIPS OF SUCCESSIVE STRATA IN FOLDS

Physical principles

The shapes and relative attitudes assumed by successive beds in a folded sequence are widely variable, being determined by the physical properties of the beds and the mechanics of the folding processes.

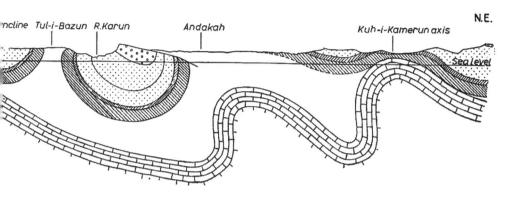

THE IRANIAN OILFIELD
The Science of Petroleum, *Oxford, 1938, p. 79*)

Competent and incompetent beds – If we consider the section across the Persian oilfields region shown in Fig. VIII–10, it will be seen that two types of beds may be recognized according to their geometry after folding. The one group including the Asmari Limestone retains its stratigraphical thickness virtually unchanged during folding; the other (Lower Fars Stage I) shows such great variation in thickness from place to place that its original thickness can only be guessed at. It is clear that these two types of beds have reacted very differently to the folding forces, the one bending rather stiffly, the other flowing into any shape required by the forces operating. Beds of the first type are described as *competent*, and of the second type as *incompetent*.[1]

Competency and incompetency are purely relative properties, implying greater or less rigidity or mobility, as may be judged from the forms assumed by the various beds in any one folded region. Traced to a different region, a relatively competent bed may be found to become more deformable, and more rarely, because of metamorphism, a relatively incompetent formation may become more competent.

In the one region, competency is judged chiefly by the amplitude of folding and by evidence of the preservation of the original thickness of beds, the folds of larger amplitude being characteristic of the more competent beds. The thickness of a formation influences the amplitude of folding as will be shown

---

[1] Originally the words competent and incompetent were used by Willis to express the notion that certain beds might be strong enough, or competent, to lift the superincumbent load in a rising anticline, while others are not; but this idea must be abandoned since C. M. Nevin (*Principles of Structural Geology*, 3rd edn, 1942, New York, p. 50) has shown that no sedimentary rock is sufficiently strong to lift the mass of a great fold.

below, so that in general, thick massive quartzites, sandstones, and limestone are more competent than clays, shales, and tuffs, salt and gypsum deposits being the least competent of all common rocks.

Most theoretical analyses of the generation of folds by buckling of layered materials under compression in the plane of layering (stratification) have been made using either elastic or viscous theory. Relative competency and incompetency of the layers may be treated using viscosity differences, this being regarded as appropriate in relation to slow deformation especially of deeply buried rocks. In a series of publications commencing with the work of Biot in 1961,[1] such studies have given an insight into the relationships between the wave length of rapidly growing ('dominant') folds, layer thickness and viscosity-contrasts.

Dieterich and Carter[2] have applied numerical analysis to the problems of the orientation of principal stress axes in and around folds in layered viscous media, obtaining results which agree with those got from fabric studies of folded rocks, and Dieterich[3] has used the results in relation to cleavage orientations. Price[4] has pointed out that where temperatures and pressures are lower, and folding more rapid, the assumption that competent rock behaves as a viscous material is not applicable. Also elastico-viscous theoretical models in end-on compression yield upright symmetrical folds, which are not common under non-metamorphic conditions. He discusses asymmetrical folds in relation to the rheological model of a solid possessing a yield strength, beyond which it may fail by brittle rupture or deform ductilely in a plastico-viscous manner. Ghosh discusses simple shear of viscous models, which also yield asymmetrical folds.[5]

Bending of an unstratified slab

Neutral-surface folding – If a slab of material of uniform internal structure be bent into a fold, it may be seen by using a square grid, or lines of circles impressed on the side before folding, that there is a regular distribution of tension and compression within the bent slab. The outer, convex side is

[1] M. A. Biot, 'Theory of Folding of Stratified Viscoelastic Media and its Implication in Tectonics and Orogenesis', *Bull. Geol. Soc. Amer.*, Vol. 72, 1961, pp. 1595–1620; idem, 'Theory of Internal Buckling of a Confined Multilayered Structure', ibid., Vol. 75, 1964, pp. 563–8; M. A. Biot, H. Odé, and W. L. Roever, 'Experimental Verification of the Theory of Folding of Stratified Viscoelastic Media', ibid., Vol. 72, 1961, pp. 1621–32.

[2] Biot's work and that of Ramsay, Chapple, and others is briefly reviewed by J. H. Dieterich and N. L. Carter in 'Stress History of Folding', *Amer. Journ. Sci.*, Vol. 267, 1969, pp. 129–54.

[3] J. H. Dieterich, 'Origin of Cleavage in Folded Rocks', *Amer. Journ. Sci.*, Vol. 267, 1969, pp. 155–65.

[4] N. J. Price, 'The Initiation and Development of Asymmetrical Buckle Folds in Non-Metamorphosed Competent Sediments', *Tectonophysics*, Vol. 4, 1967, pp. 173–201.

[5] S. K. Ghosh, 'Experimental Tests of Buckling Folds in Relation to Strain Ellipsoid in Simple Shear Deformations', *Tectonophysics*, Vol. 3, 1966, pp. 169–85.

subject to an extension parallel to the circumference of the fold, while the inner, concave side undergoes compression in this direction. Between the zones of tension and compression one surface suffers no change in length (Fig. VIII–11). This is accordingly known as the *neutral surface*, and the folding as *neutral-surface folding*. The position of the neutral surface is determined by the stress distribution in the slab and is, in fact, variable

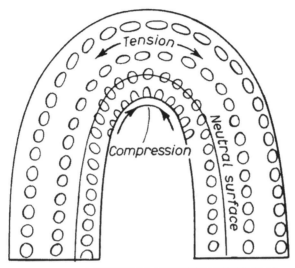

FIG. VIII–11. BENDING OF AN UNSTRATIFIED SLAB
The neutral surface lies between the outer zone of tension and the inner of compression.

according to the physical properties of the material, and the mechanical means employed to bend the slab, which provides an overall set of forces within which the special stresses due to the bending operate (Fig. VIII–12).

The material of the slab yields to the compressions and tensions in several ways, firstly by dilatation (either positive or negative), which will itself consist partly of an elastic expansion or compression and partly of permanent dilatation, this latter being rather readily produced in loosely cemented granular rocks with interstitial pore spaces. In such rocks, readjustments of the packing of grains may accommodate the required strains. Secondly, distortion, either elastic or permanent, will, perhaps most commonly, be induced. Fig. VIII–11 shows the distribution and type of distortion involved. If this occurs without notable dilatation, the ellipses formed from original circles are approximate strain ellipses for elastic strain, and for plastic deformation they are ellipses directly related to the strain ellipses.

Although very few rocks are sufficiently elastic to permit any considerable elastic deformation over a short period, it is conceivable that small elastic strains might, through readjustments of defects in crystal lattices and by actual recrystallization, be dissipated or relaxed over a long time, so that folding might proceed, though very slowly, without passing the elastic limit at any stage.

It is of interest to inquire as to the amount of deformation actually required in neutral-surface folding. For this an approximation will suffice, assuming the neutral surface to be midway within the slab or folded bed, and

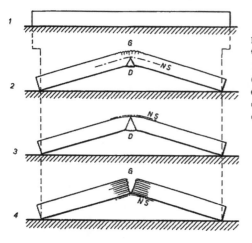

FIG. VIII–12. VARIOUS POSITIONS OF THE NEUTRAL SURFACE
1. Original slab. 2. Neutral surface central. 3. Neutral surface at the extrados. 4. Neutral surface at the intrados. *G*. Tension cracks. *D*. Zone of compression. (*After Seidl, 1934*)

that the folding results in concentric circular arcs. Under these conditions the extension at the outer surface is $\frac{1}{2} . \theta t$, where θ is the angle through which the slab is bent, measured in radians, and t is the thickness of the slab. This is not at all a large deformation for many folds, and is best expressed as a percentage of the length involved. For a 10-ft bed, folded into a semicircular arc a mile wide (radius $\frac{1}{2}$ mile), the extension and corresponding shortening is only 0·2 per cent of the original length of the folded bed, or 0·002 ft per foot of length. The percentage will be much greater for folds of small radius, and it is therefore to be expected that considerable distortion as well as dilatation will be involved in closely folded thick beds. Where distortion is involved, the position of the neutral surface is affected as the folding proceeds, and the shape of the slab, with regard to the parallelism of its upper and lower surfaces, may also be affected.

In experiments it will be found that the compression in the core causes lateral expansion along the length of the slab, the free edges turning up at the top (Fig. VIII–13). These effects indicate that the distribution of stress and strain with strong folding are not at all simple, and experimental and actual folds exhibit a number of special strain effects in bent slabs, which are discussed below.

Special strain effects in bending

(i) *Dilatation and bulk flow* – Evidence for the internal rearrangement of grains in unmetamorphosed beds in the cores of folds [1] has not received much attention and is not easily demonstrated. However, sandstones folded into sharp curves yield a good deal of evidence indicative of inter-granular movements, more especially in regions where the beds were water-bearing two-phase systems at the time of folding. In a sandstone without any trace of cleavage, folded isoclinally, the quartz grains may show no more evidence of strain at the sharpest part of the flexure than do other quartz grains from the limbs. There can in such a case be no doubt that the local strains in folding were performed chiefly by relative displacements of the grains, and it is probable that this would include some compaction as well as the rolling of grains by external rotation.

(ii) *Crumpling of intrados* – Crumples due to compression on the concave side of a bent slab [2] are formed between kink-planes or small faults.[3]

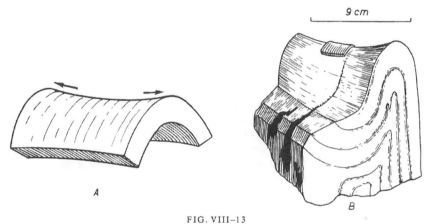

9 cm

FIG. VIII–13
A. BENDING OF AN ELASTIC SLAB
(The experiment is simply made with a rubber eraser.)

B. FOLD IN GNEISS, SHOWING SADDLE
(*After Koenigsberger*, Natwiss., *Jahrg. 12, Hft. 28, 1924, p. 568*)

(iii) *Tension joints at the extrados* – With brittle materials, tension cracks form at right angles to the extension, and thus radially in the fold, in the zone above the neutral surface. As these form, the slab loses coherence and can no longer transmit tensional stress. Accordingly, the effective upper surface at

---

[1] This section does not refer to the crumpling of thin-bedded rocks in the core of major folds, which is discussed below.

[2] It is convenient, following Goguel, to refer to the concave side of a fold as the *intrados* and the convex side as the *extrados*, the terms being derived from architecture in describing arches.

[3] The majority of illustrated examples are in well-stratified rocks and are the result of disharmonic folding.

FIG. VIII–14A. FORMATION OF SLIP-BANDS AT EXTRADOS AND
INTRADOS OF BENT METAL ROD
(*After Seidl, 1934*)

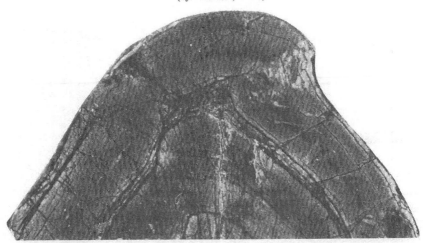

FIG. VIII–14B. FOLDING IN CAMBRIAN CHERT, NEAR HEATHCOTE, VICTORIA
Compression in the core of the upper bed gives rise to small thrust faults dipping towards the
hinge. The crest is lifted, and the space beneath filled with the highly plastic rock of the thin bed
below. The whole is comparable with a faulted anticline and saddle reef, or with folds in coal.
(Natural size)

any instant is the surface joining the tips of the growing cracks. The effective
thickness is thus reduced, the neutral surface moves down, and eventually
may pass to the lower surface while the slab cracks completely across. This
affords partial explanation of the radial tension joints characteristically found
in brittle rocks folded in a near-surface zone without high confining pressure.[1]

(iv) *Faulting at extrados and intrados* – Experiments with metals illustrate

[1] See, however, the discussion of jointing in Chapter IX.

very finely the formation of slip bands in the '45°' position both above and below the neutral surface (Fig. VIII—14A). In rocks these may be represented by faults, which will be normal (tensional) faults on the extrados and reverse (compressional) faults on the intrados. Crest-faulting is discussed more fully under Domes (p. 273); core-faulting is illustrated in Fig. VIII—14B in which the neutral surface is at the top of each of the faulted beds.

(v) *Recrystallization in neutral-surface folds* — A very fine experimental analogue in a recrystallized steel plate is given by Seidl[1] (see Fig. VIII—15).

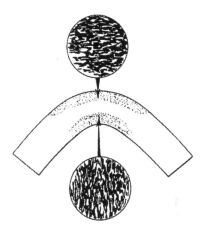

FIG. VIII–15. RECRYSTALLIZATION IN
BENT STEEL BAR
Note the grain elongation is circumferential
at the extrados and radial in the intrados
of the arch. (*After Seidl, 1934*)

Not only is the neutral surface, with a superior zone of tension and an inferior zone of compression very well illustrated, but it is seen that the long axes of the recrystallized grains are circumferential in the outer surface and radial in the core of the fold.

The external geometry of neutral-surface folds is variable in detail, but, broadly regarded, there is no excessive change in the thickness of the folded strata except at quite sharp flexures, where the bed thickens, while in the limbs between anticline and syncline there is less distortion and at the point of inflexion on any limb there is no distortion or dilatation.[2]

Because the distortion required is proportional to the thickness of the slab, and inversely proportional to the radius of curvature of the fold, it will in general be true that thick massive formations will, under given stress conditions, yield folds of large radius and hence will be recognized as competent beds. On the other hand, very thin beds may fold with negligible distortion and thus with negligible resistance to folding arising from internal strain, and

[1] E. Seidl, *Bruch und Fliessformen in der technischen Mechanik und ihre Anwendung auf Geologie*, Bd. 5, Krümmungsformen, Berlin, 1934. Seidl anticipates much later work. H. Ramberg has described similar recrystallization in a folded quartzite in 'Strain Distribution and Geometry of Folds', *Bull. Geol. Inst. Univ. Uppsala*, Vol. 42, 1963, pp. 1–20.
[2] E. L. Ickes, 'Similar, Parallel and Neutral Surface Types of Folding', *Econ. Geol.*, Vol. 18, 1923, pp. 575–91.

these may therefore fold more readily and into more numerous folds of smaller radius.[1]

Bending with external neutral surface — A slab that is subject to overall lateral tension or compression and at the same time to a bending movement, may be stretched or squeezed as a whole while it bends, so that there is no neutral surface within it. It is possible to conceive a neutral surface to exist

FIG. VIII-16. FOLDING OF BANDED IRON-FORMATION WITHOUT TENSIONAL EFFECTS

Laminae within the core exhibit features of rheomorphic (flow) folds. (Natural size)

outside the slab, either above its convex or within its concave side (Fig. VIII-12) but the bending of the slab itself can no longer be considered as neutral-surface folding, and the total deformation, either by stretching or by compression, may be much greater than has so far been considered.

Folding combined with tension is illustrated by an experiment of H. Cloos in which a clay cake rests on a rubber balloon which is slowly inflated (Fig. VII-29B). The stretching of the balloon transmits tension to the clay, and its curvature induces bending. As with most of Cloos' experiments the consistency of the clay is such that it yields by numerous complementary shears,

[1] M. Smoluchowski has discussed the question for elastic strains. 'Über ein gewisses Stabilitätsproblem der Elastizitätslehre, etc.', *Bull. Acad. Sci. Cracovie*, 1909.

which represent miniature normal faults. The experiment reproduces very faithfully the features of broadly domed shield areas, but also represents beds stretched and domed over salt plugs (p. 281) and over buried hills (p. 259).

In folding combined with compression there may be no tensional zone in the specimen, and the possible effects, including bulk distortion, folding, or shearing, are shown in Fig. VIII−16. In geology such conditions, in large folds, are likely to occur only in beds that are overlain by a considerable thickness of other strata, or which are incompetent at shallow depths. It would not be possible to distinguish the effect from compression arising in the core of a neutral-surface fold, unless the section could be fully reconstructed with knowledge of the structures in all horizons at the time of folding, which

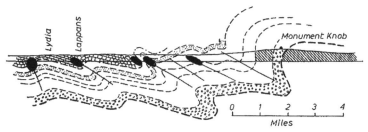

FIG. VIII−17. SOUTH MOUNTAIN FOLD, MARYLAND
Black ellipses show oölite-deformation in Cambrian limestone (white).
(*After E. Cloos, 1947*)

is rarely possible. The South Mountain fold, Maryland, described by E. Cloos, for which the deformation of oölite grains affords a most intimate picture of the strain, may however be considered as an example. If a neutral surface existed, it must have lain far above the upper surface of the Cambrian limestones (Fig. VIII−17).[1]

FOLDING OF A STRATIFIED MASS

Flexural-slip folding

In folding a pile of very thin sheets of relatively strong material, which may freely slide over each other like the pages of a book when opened or closed, it may be assumed that although each sheet exhibits neutral surface folding, the distortion of any small segment in each is negligible because of its extreme thinness. The distortion whereby folding is achieved in the pile of sheets depends on the mechanical means employed to produce the folds, but we may first assume a simple arch to be produced with the central part held while the limbs freely rotate to make an arcuate fold (Fig. VIII−18). Since in the idealized case each sheet retains its original length and thickness unchanged,

[1] E. Cloos, 'Oölite Deformation in the South Mountain Fold, Maryland', *Bull. Geol. Soc. Amer.*, Vol. 58, 1947, pp. 843–918.

the geometry requires that the sheets shall slip over each other, the slip relative to the upper sheet being proportional to the distance from this sheet.

Considering a succession of folds of this type, it is seen that between any two points where the dip is zero the lengths of the sheets are the same, the slip in one sense on an anticline being compensated in each bed by equivalent slip in a syncline. It follows that if it be assumed that the anticlinal and synclinal

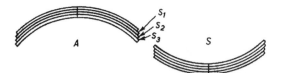

FIG. VIII–18. SLIP IN THIN SHEETS FOLDED INTO
AN ANTICLINE (A) AND INTO A SYNCLINE (S)
Note that the slip in *A* is compensated by that in *S*.

axes remain fixed during the folding, all the slip is concentrated in the fold limbs. This may be simply demonstrated by bending a paper pile across which lines have been ruled normal to the sheets (Fig. VIII−19). It will also be clear that so long as the sheets remain in contact the total thickness must under all conditions of curvature remain constant, so that the fold consists of parallel curves.

FIG. VIII–19. CONCENTRATION OF SLIP IN FOLD
LIMBS
A pile of paper sheets is clamped at the ends. Lines were ruled at right angles across the edges of the pile. (*After Seidl*)

We have now seen that there are two possible ways in which folds consisting of a family of parallel curves may be formed, the one as a result of neutral-surface folding, the other by the slipping mechanism discussed above. Each requires a certain type of local strain within the folded object and the geometrical treatment specifies only this. If, for instance, differential gliding along surfaces parallel with the folded surface could occur in an unstratified mass rather than in a pile of rather stiff sheets, the geometrical requirements for parallel folding could also be satisfied. In fact, the stress distribution in an isotropic slab during bending may be gauged from neutral surface folding,

and in this we have seen that shear planes are necessarily formed in the 45°
position, relative to the principal stress axes in the slab. The parallel slip that
occurs in the paper experiment is due to the lack of adherence between the
sheets, and to their strength which, under end-on push, permits them to be
forced one over the other.

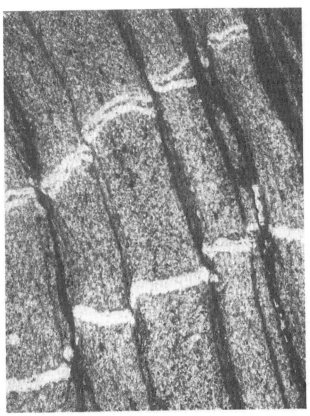

FIG. VIII–20. SANDSTONE LAMINAE SEPARATED BY
SHALE AND SHOWING BEDDING-SLIP BY THE DIS-
PLACEMENT OF QUARTZ VEINLETS

An anticline is situated to the left of the illustration. ×15.

It is therefore clear that folding of the type discussed for the sheets depends
entirely on the original lack of isotropy, with low adherence between sheets of
relatively rigid material in which stress may be transmitted without excessive
plastic yielding. This may be translated into geological terms to mean that
the adherence of the grains between two adjacent beds must be so low,
and within the beds themselves so high, that the beds act as *gliding units* (Fig.
VIII–20). Folds of this type are known as *flexural-slip* folds. A sharply
defined sandstone–shale or limestone–shale junction may act as a gliding surface, but within one graded bed, although stratification may be well-

defined, the argillaceous component gives sufficient adherence across the stratification planes to inhibit gliding along them.

Parallel folds

The conditions just discussed are essentially those visualized by van Hise when he proposed to distinguish *parallel folds* from other types.[1] Parallel folds are often also called concentric folds, since, according to the view that

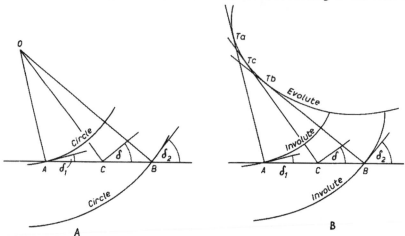

FIG. VIII–21. CIRCULAR ARCS AND INVOLUTES

A. Reconstruction of parallel fold between dip readings as shown by a circular arc of origin O.

B. The same, reconstructed as an involute. (*After Mertie*)

any smooth curve may be resolved into a series of circular arcs, the beds of any one such arc are folded about one centre or *origin*.[2] However, Mertie[3] has shown that the geometrical relations may also be satisfied by regarding the curves as parallel involutes, in which case the term concentric is inapplicable[4] (Fig. VIII–21).

Many formations of limestone or sandstone in beds some few feet thick exhibit essentially parallel folding with clear indications of slip along the separation planes (*bedding-slip*).[5] This is shown by slickensiding and by the

[1] C. R. van Hise, 'Principles of North American Pre-Cambrian Geology', *U.S. Geol. Surv.*, 16th Ann. Rept., Pt 1, 1896, pp. 581–843.

[2] H. G. Busk, *Earth Flexures*, Cambridge, 1929, reprinted N.Y., 1957.

[3] J. B. Mertie, Jr, 'Stratigraphic Measurements in Parallel Folds', *Bull. Geol. Soc. Amer.*, Vol. 51, 1940, pp. 1107–34; idem, 'Delineation of Parallel Folds and Measurement of Stratigraphic Dimensions', ibid., Vol. 58, 1947, pp. 779–802.

[4] While it seems desirable, following Mertie, not to use the term concentric for these folds, the relative simplicity of using circular arcs for constructing parallel folds has great advantages in practice.

[5] H. Cloos and H. Martin, 'Der Gang eine Falte', *Fortschr. Geol. u. Pal.*, Bd. 11, 1932 (Deecke-Festschr.), pp. 74–88. R. Hoeppener, 'Faltung und Klüftung im Nordteil des

displacement of early-formed mineral veins (calcite, quartz) from bed to bed across the fold (Fig. VIII – 20). In any fold of anticlinal type the upper of any two adjacent beds slides relative to the lower towards the crest of the anticline, a rule that has many applications in structural geology (see drag-folds

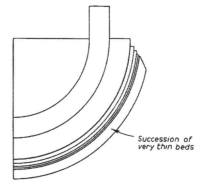

FIG. VIII-22. SLIP WITH BEDS OF DIFFERENT
THICKNESS
(*After Nabholz*)

Succession of
very thin beds

and cleavage). Where beds of different thickness are present in a fold of this type, the amount of relative slip depends on the thicknesses of the beds (Fig. VIII – 22).

Limits in construction of parallel folds – When a parallel curvilinear fold is constructed geometrically it will be seen that, starting from a sinusoidal fold,

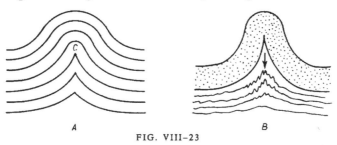

A   B

FIG. VIII-23

A. Construction of parallel folds showing cusp and decreasing dips in depth.
B. Construction of closed parallel fold in thick sandstone, with squeezing of incompetent beds out of the fold core.

the radii of curvature become progressively larger both upwards and downwards, so that the dips and the 'intensity' of the folding decrease; eventually the beds would approximate to horizontality. In this construction a certain volume of beds below the level at which the anticlines become cuspate is unaccounted for (Fig. VIII–23A). This implies that, below that particular

Rheinischen Schiefergebirges', *Geol. Rundsch.*, Vol. 41, 1953, pp. 128–44. E. S. Hills, 'Examples of the Interpretation of Folding', *Journ. Geol.*, Vol. 53, 1945, pp. 47–57. W. K. Nabholz, 'Untersuchungen über Faltung und Klüftung im nordschweizerischen Jura', *Eclog. Geol. Helv.*, Vol. 49, 1956, pp. 373–406. (Nabholz's explanation of joints rotated during folding in plastically deformable interbedded strata is not adhered to in this book.)

FIG. VIII–24. PARALLEL FOLD IN CARBONIFEROUS VARVES, NEW SOUTH WALES, WITH SQUEEZING OF SOFT ROCK FROM THE CUSPATE CORE

stratigraphical horizon, the construction should no longer be continued according to the rules for parallel folds. What in fact occurs is that any bed or formation which does fold as a competent unit may continue to yield closer parallel folds. The underlying incompetent beds are squeezed downwards in anticlines,[1] adapting themselves to the form assumed by the competent beds (Fig. VIII–23B). Thus in drawing sections, beds or formations that are judged from field data to be competent may be shown in parallel folds. Interbedded more mobile beds which yield plastically are virtually only a medium in which the competent beds are flexed, and in regions where these incompetent beds include salt deposits, gypsum, soft clays, or shales, ideal examples of parallel folding in more competent limestones and sandstones may be found, virtually floating in these yielding rocks (Fig. VIII–10).[2]

Various graphical methods have been proposed to aid in section-drawing assuming parallel folding to obtain. The most widely used method is that developed by Busk,[3] in which it is assumed that the fold is composed of a number of tangential circular arcs. Between successive dip-readings at the surface, chosen to form the basis of the construction, a series of circular arcs

---

[1] Unless they break through the fold upwards forming diapiric injections (q.v.).

[2] See, however, the discussion of salt tectonics, p. 292, where it is pointed out that synclines in beds resting on thick salt may actually sink into the salt until they reach an underlying supporting bed.

[3] H. G. Busk, *Earth Flexures*, Cambridge, 1929.

is drawn, the two dip-readings thus controlling the reconstruction of the fold in the sector between them. Each sector is bounded by limiting radii to which the dips are normal, the centre of curvature for each being known as its *origin*. An example is shown in Fig. VIII–25, but for further expansion of more complex examples Busk's book should be referred to.

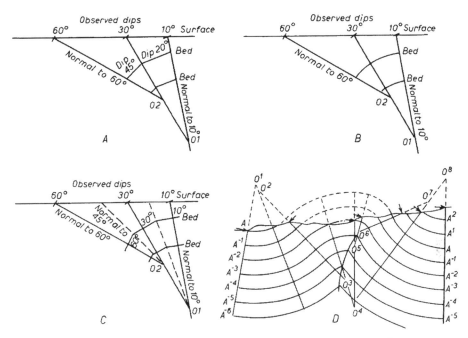

FIG. VIII–25. METHODS OF RECONSTRUCTION OF PARALLEL FOLDS

A. Method of chords. B. Method of circular arcs. C. Method of tangents (*after Coates*). D. Method of circular arcs applied to an asymmetrical anticline (*after Busk*)

Special methods are required to deal with asymmetrical folds in which the steeply dipping limb is thinned. Gill, in addition to Busk and other authors, has given special attention to this.[1]

Similar folds

Van Hise designated *similar folds* as the only important alternative type to parallel. The geometrical criteria for similar folds are that successive beds in a folded sequence take on similar curves, so that the form of the fold remains the same in a vertical direction.

Geometrical constructions representing similar folds can be made in two

[1] W. D. Gill, 'Construction of Geological Sections of Folds with Steep-limb Attenuation', *Bull. Amer. Assoc. Petrol. Geol.*, Vol. 37, 1953, pp. 2389–406. See also J. Coates, 'The Construction of Geological Sections', *Quart. Journ. Geol. Min. & Met. Soc. India*, Vol. 17, 1945, pp. 1–11.

ways, both of which simulate possible conditions for rock folding. Firstly, if the length of lines representing the beds in cross-section is kept the same, similar folds are produced by shortening the section while the beds fold, each sector rotating about planes (P) (Fig. VIII–26A) and each bed in any one sector rotating the same amount. This simulates folding by lateral compression. Secondly, if the length of the section of folds be maintained the same after as before folding, the beds are stretched by differential upward and downward displacements across the bedding planes along a direction (S) (Fig. VIII–26B). This may be brought about most simply by differential shearing in this direction, as may be illustrated by drawing a line across a pack of cards, and pushing up the centre of the pack. Although the effect is purely one

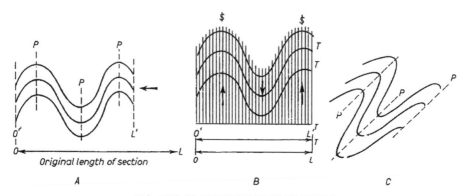

FIG. VIII–26. TYPES OF SIMILAR FOLDS
A. Due to lateral compression, with rotation about planes 'P'. B. Due to shear, without lateral compression. C. Same as A but with oblique planes 'P'.

of micro-faulting, an appearance of folding results known as *shear-folding*, where the shear planes are so closely spaced as to be microscopic.

With similar folding of types A and B, the geometry may be varied so that the lines P or S make some angle other than $90°$ with the bedding planes, as shown in Fig. VIII–26C.

It is a notable geometrical property of all the constructions that the distance T representing the thickness of the section measured parallel to the direction P or S is constant throughout. In the case B, T represents the original thickness of the strata measured in the direction of S, but in the case A, the limbs are thinned and the crests and troughs thickened relative to the original thickness, and thus no section measured normal to the bedding in the conventional way will afford a measure of the original stratigraphic thickness. Mass displacement of material from the limbs to the crests and troughs goes on, as described by van Hise,[1] and this is a characteristic of incompetent beds in certain circumstances.

[1] C. R. van Hise, 'Principles of North American Pre-Cambrian Geology', *U.S. Geol. Surv.*, 16th Ann. Rept., Pt 1, 1876, pp. 581–843.

These various effects can be matched closely in geological examples on all scales.

Similar folding with lateral compression – Examples are best seen where competent formations are interbedded with incompetent. Fig. VIII–27 shows a small-scale example from Mount Isa, Queensland, in the lead ore-body. The competent bands exhibit parallel folding, but the overall pattern is that of

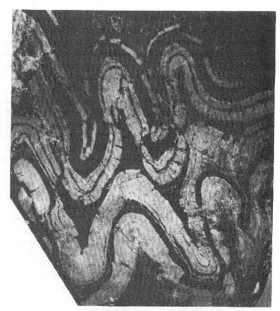

FIG. VIII–27. SMALL-SCALE COMPOSITE FOLDS OF SIMILAR HABIT, LEAD ORE-BODY, MT ISA, QUEENSLAND

Natural size. (*From R. Blanchard and G. Hall*, Proc. Aust. Instit. Mining & Metall., *No. 125, 1942, Fig. 11*)

similar folding with marked evidence of differential flow in the incompetent layers, which adjust themselves to the predetermined fold form. These are, strictly speaking, *composite folds of similar habit.*

Shear-folding (Gleitbrett folds)

Many beautiful examples of small-scale shear folds are shown by sandy laminae in slates (Fig. VIII–28), and by different coloured layers in certain schists. Most rocks exhibiting shear folding also show what have been called *Gleitbretter.*[1] Although the whole rock may be traversed by minutely spaced shear planes which in slates are the cleavage planes, the shearing movements are greater along certain more or less evenly spaced narrow zones between which are bands of relatively small deformation, so that the rock appears to be sliced (Fig. VIII–29). The geometry of the resultant small folds depends

[1] W. Schmidt, *Tektonik und Verformungslehre*, Berlin, 1932. See also B. Sander, *Einführung in die Gefügekunde der geologischen Körper*, Vienna, 1948–50, 2 vols. E. B. Knopf, 'Structural Petrology', *Geol. Soc. Amer.*, Mem. No. 6, 1938.

FIG. VIII–28. SHEAR FOLDING IN SLATE FROM 'CENTRE
COUNTRY' (CREST OF FOLD), BENDIGO
Natural size.

on the angle between the shear planes and the bedding, the sense of the slip in
relation to the bedding (right-hand- or left-hand-side down) and the properties
of the beds involved (Fig. VIII–30). In general (but not invariably) the
Gleitbretter are wider in thicker beds than in thinner, of the same
composition.[1]

It is important to note that in asymmetrical folds formed by the Gleitbrett
mechanism, the thinning of the steeply dipping limb and its eventual tearing

[1] See also discussion of Cleavage, Chapter X.

FIG. VIII–29. *Gleitbretter* IN SMALL-SCALE SHEAR-FOLDS IN SLATE,
DONNELLY'S CREEK, VICTORIA. ×2

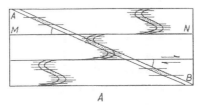

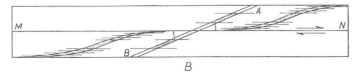

FIG. VIII–30. *Gleitbretter* MECHANISM

The spacing of shear planes affecting bed AB in the direction of shear MN
varies. Dotted sectors of beds remain parallel to their original position.
The shape of folds depends on the relative attitude of bed in relation to
shear direction.

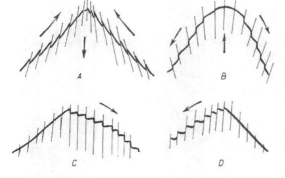

FIG. VIII-31. SENSE OF SHEAR
IN SHEAR-FOLDS

A, B. Orthorhombic symmetry.
C, D. Monoclinic symmetry.

FIG. VIII-32. COMBINATION OF FLEXURAL FOLD IN SAND-
STONE LAMINAE, AND SHEAR-FOLDING IN SLATE. ×½.

The fold exhibits monoclinic symmetry. Fig. VIII–20 illustrates
bedding-slip in the sandstones of this specimen.

apart is not due to the faulting or shearing of a pre-existing fold, but rather that the folding and the shearing develop together.

It is in small-scale examples in slates and schists that shear-folding is most readily demonstrable, and the extent to which major folds may be of this origin remains conjectural. The chief features, especially the sense of the shearing (right-hand-side down or left-hand-side down) in relation to the fold may be judged from the illustrations (Fig. VIII – 31) in which it is seen that in some folds the shearing is symmetrical about the axial plane, whereas in others it remains constant in sense of movement in the two limbs of the fold. In general the relationships near the axial plane itself are complex, with variable sense of displacement.

Where beds of different physical properties alternate, some may exhibit purely shear folding while others, generally the coarser and thicker beds, show limited shear, combined with rotation (Fig. VIII – 32). In section-drawing, the beds that are shear-folded are best constructed keeping the thickness constant in the direction of the shear planes (cleavage in slates); the coarser and thicker beds may show some apparent thinning in the steeper limbs, due to shear, but because they have also been rotated, their thickness in the fold limbs approximates more closely to the original thickness.

Angular folds and kinking

Folds with straight or nearly straight limbs, and sharply curved or even pointed crests and troughs are variously called *zig-zag*, *chevron*, or *accordion* (concertina) folds, and an additional type termed *mitre-folding* is here recognized. The first mechanism by which they may be formed is by the rotation of

FIG.   VIII–33.   CONCERTINA-FOLDS
DUE TO ROTATION

A. Symmetrical. B. Asymmetrical, with
fault at apex.

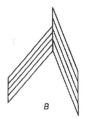

the limbs about median planes, such as may be imitated with two packs of cards in contact along an edge (Fig. VIII – 33). These are strictly similar folds, but, if minor effects at the crests and troughs are for the moment ignored, the volume remains constant and the beds retain their original thickness. Bedding-slip is essential as the beds must slide over each other, but distortion is restricted to the crests and troughs, the limbs merely rotating. It is obviously unlikely that folds of this type could form in thick competent strata, and it is found in nature that they are characteristic of thin-bedded rocks, especially where there is rapid alternation of more rigid beds such as sandstone, with interbedded shales, or other incompetent rocks which aid bedding-slip. Accordingly they are well exhibited in the Flysch deposits marginal to the European Alps and also in the Ruhr and the Belgian coal basins, but they are perhaps known in greatest detail from detailed mapping and mining in the goldfields of central Victoria, especially at Bendigo and Ballarat, where there is a very thick succession of flysch-type sandstones (greywackes) and shales (now slates) (Fig. VIII–34).

If angular folds formed by the above-described mechanism are asymmetrical about the median plane, the beds cannot 'match' at the crests and

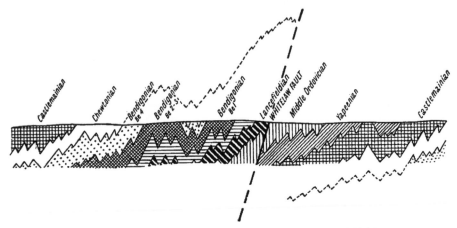

FIG. VIII–34. CONCERTINA-FOLDING, BENDIGO, VICTORIA
Scale, 1 in. = 2 miles. (*After Thomas*)

troughs since the steeper limbs occupy a greater vertical thickness. The differential movement will generally be accommodated by a fault. However, it would appear to be in general easier for a symmetrical attitude to develop, and in fact many angular folds are symmetrical about the median plane, although where this plane is inclined the folds must be described as asymmetrical (Fig. VIII–35).

The sharp bending of foliated materials about certain planes is called in German *Knickung* and the planes, *Knickungsebene*. In English the planes are best termed *kink planes*. Kinking in crystals has already been referred to (p. 119), but recently the analogous deformation of foliated rocks such as slate, phyllite, and thin-bedded sediments has been the subject of a good deal of field and experimental study.[1]

[1] M. S. Paterson and L. E. Weiss, 'Experimental Deformation and Folding in Phyllite', *Bull. Geol. Soc. Amer.*, Vol. 77, 1966, pp. 343–73; idem, 'Folding and Boudinage of Quartz-Rich Layers in Experimentally Deformed Phyllite', ibid., Vol. 79, 1968, pp. 795–812; J. F. Dewey, 'Nature and Origin of Kink-Bands', *Tectonophysics*, Vol. 1, 1965, pp. 459–94; idem, 'The Origin and Development of Kink-Bands in a Foliated Body', *Geol. Journ.*, Vol. 6, Pt 2, 1969, pp. 193–216; see also Proc., Conference on Research in Tectonics, *Geol. Surv. Canada*, Paper

Ideally, kinking results from the rotation of layers or foliae about planes oblique to the layering, in such a way that the kink planes bisect the angle between the undisturbed material and that within a *kink-band*, which latter lies between two parallel kink planes. It is fundamental that kink-bands are of finite width, so that kink planes are essentially different from shear planes; nevertheless, conjugate kink-bands develop, like Luders' lines in metals, symmetrically about the principal stress axes and parallel to surfaces of maximum

FIG. VIII–35. ASYMMETRICAL CONCERTINA-FOLDS,
SYMMETRICAL ABOUT THE KINK PLANES
Kanmantoo Group, Scott Cove, South Australia. (*By courtesy M. F. Glaessner*)

shear-stress. Continued deformation leads to the development of complex arrays of kink-bands, and the foliae or layers are thus folded (Figs. VIII–35, 37); or again, microscopical kink-bands may develop as crenulate cleavage, post-dating an earlier cleavage. There are many similarities, too, with large-scale folds, especially with chevron-folds and box-folds.[1]

Angular folds with inclined kink planes are common and because of their distinctive features are here termed *mitre-folds*.[2] Consideration of the form of many of the Jura folds clearly demonstrates the significance of kink planes in types such as box-folds, for which it is indeed more appropriate to recognize

68–52, 1968; P. Matte, 'Les Kink-Bands – Exemple de Déformation Tardive dans L'Hercynien du Nord-Ouest de L'Espagne', *Tectonophysics*, Vol. 7, 1969, pp. 309–22.

[1] R. T. Faill, 'Kink Band Structures in the Valley and Ridge Province, Central Pennsylvania', *Bull. Geol. Soc. Amer.*, Vol. 80, 1969, pp. 2539–50; J. F. Dewey, 'Nature and Origin of Kink-Bands', *Tectonophysics*, Vol. 1, 1965, pp. 459–94.

[2] The term mitre refers both to the 45° position of the kink planes and of the peculiar resemblance of the resulting folds to a bishop's mitre.

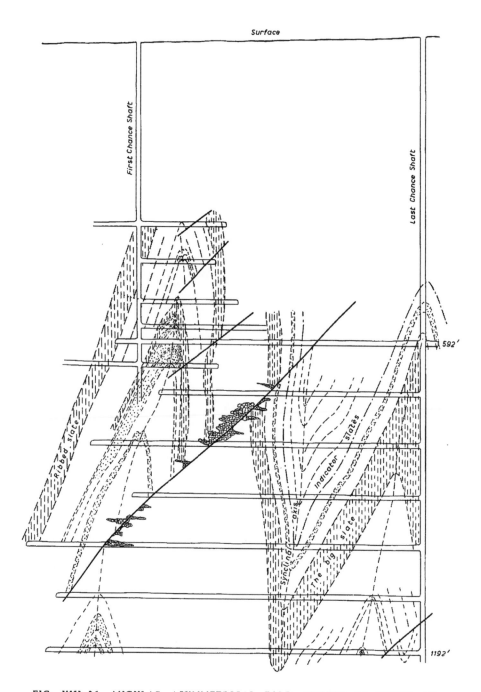

FIG. VIII–36. ANGULAR ASYMMETRICAL FOLD, MAPPED IN DETAIL IN
DEPTH AT BALLARAT EAST, VICTORIA

Note attenuation of the steep limb, probably due to shearing after initial kinking about
axial planes. Note also the fault with quartz veins in *en echelon* tension gashes. (*After
Baragwanath*, Geol. Surv. Vict., *Mem. No. 14, 1923, Fig. 56*)

FIG. VIII-37. KINK PLANES IN FOLDING

A. Upright. B. Inclined. C. Inclined in two directions (conjugate folds).

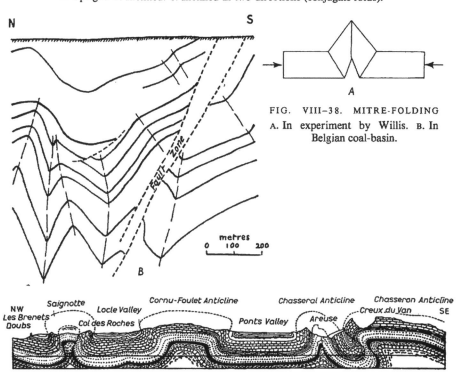

FIG. VIII-38. MITRE-FOLDING

A. In experiment by Willis. B. In Belgian coal-basin.

FIG. VIII-39A. KINK PLANES IN JURA FOLDS AND FAULTS

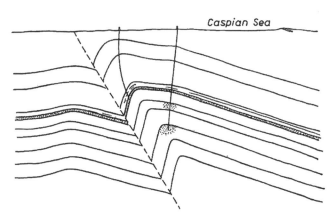

FIG. VIII-39B. KINK PLANES IN FOLDS, DAGNESTAN, EASTERN-FORE-CAUCASUS

(After I. O. Brod, 17th Geol. Congr., U.S.S.R., 1937, p. 161)

the lateral or inclined kink planes than to speak of a median axis of folding (Fig. VIII–39A). In section drawing it is particularly important to recognize any sudden change in dip that may indicate the presence of a kink plane. Examples are shown in Fig. VIII–39B.

FIG. VIII–39C. KINK PLANES IN SLUMPED BEDS
(*After A. Heim*, Neues Jahrb. f. Min., *etc.*, *Jahrg. 1908, Bd. II, Pl. 13*)

Features of the hinge of angular folds

The plane about which the limbs rotated in angular folds is a plane of weakness in all rocks. As previously noted, a fault upthrust along the steeper limb is generally developed if the folding is asymmetrical about the plane. This plane of weakness was recognized long ago by Dunn at Bendigo, where thin Tertiary monchiquite dykes follow closely along it, the plane itself having developed during Palaeozoic folding [1] (Fig. VIII–40).

Under conditions favouring fracturing on the extrados of the folds, brittle beds are broken apart, as is sometimes seen in thin limestones embedded in shales. At greater depths, however, the folds are generally rounded for a short distance about the crests and troughs, and their geometry may then be resolved in terms of the principles already discussed.

Saddle reefs – Where relatively competent sandstones are interbedded, the rounded crests and troughs of each bed show approximately parallel folding; consequently there is a tendency for an opening to form between the base of one bed and the top of that beneath. Such potential openings may be occupied

[1] E. J. Dunn, 'The Bendigo Goldfield', *Mines Dept. Vict.*, Special Repts., Nos. 1 and 2, 1896. The dykes in fact often deviate several feet from 'centre country' due no doubt to inhomogeneities in the rocks.

by 'saddle reefs', which are best known at Bendigo and in Nova Scotia, Stillwell[1] has shown that the Bendigo reefs are in part replacements of slates and sandstones, but the potential spaces formed during folding must be regarded as the primary control in their formation. Bendigo saddle reefs

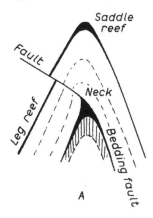

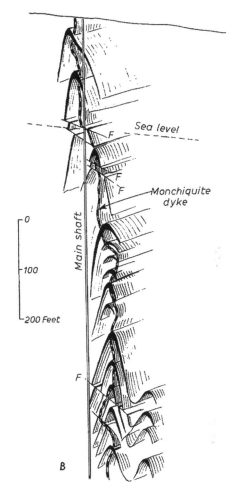

FIG. VIII–40. BENDIGO GOLDFIELD
A. Saddle reefs and 'neck', Bendigo, Victoria. (*After Stillwell, 1917–19*)
B. Section at the Great Extended Hustlers Shaft, Bendigo, to illustrate similar folding (partial block-diagram). Saddle reefs solid black.

commonly possess a prolongation upwards from the top of the saddle, known as the *neck* of the reef (Fig. VIII–40). This follows a thrust fault which, commencing on a bedding plane on one limb of an anticline, passes across the crest and, deviating downwards, continues for some feet in the rocks of the other limb. It is not known whether the curvature is due to continuation of the folding after the fault was formed, or simply to shearing of the more sharply

[1] F. L. Stillwell, 'The Factors Influencing Gold Deposition in th Bendigo Goldfield', *Bulls. No. 4, 8, 16, Advisory Council of Science and Industry*, Melbourne, 1917–18–19. Idem, 'Replacement in the Bendigo Quartz Veins and its Relation to Gold Deposition', *Econ. Geol.*, Vol. 13, 1918, pp. 108–111.

dipping limb at an appropriately competent thick sandstone. Similar small thrusts also occur in places where there is no saddle reef.

It is of interest especially in relation to suggestions as to the origin of saddle reefs to note the close analogy of the necks to the *queues anticlinales* found in the Belgian coalfields at sharp anticlinal arches [1] (Fig. VIII−41). The coal is apparently squeezed along a fault which passes into the opposite limb at the fold axis, as in saddle reefs. Also as with saddle reefs, *queues* are more

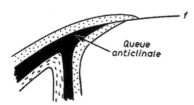

FIG. VIII−41. *Queue anticlinale* IN COAL, BELGIAN COALFIELD

(*After Kaisin*)

numerous at anticlines than at synclines. The implication is that the corresponding structures are of mechanical origin, and hence that the features of saddle reefs are due less to replacement than to folding processes.

Rheomorphic folds

Many folds in highly incompetent beds that have undergone very considerable plastic deformation and translation do not comform with the relatively simple geometrical types already discussed; they display exaggerated thinning on the limbs and thickening at the crests and troughs of folds, convolutions due to folding of the first-formed folds themselves, and similar effects that reflect the great mobility of the beds, so that the resultant folds may be termed *rheomorphic*.[2]

The geometry of rheomorphic folds may be accounted for by a combination of shearing and bulk distortion, with thinning normal to the shear direction and elongation parallel to it (Fig. VIII−42). The shear direction is clearly that of mass transport corresponding with the flow planes in a fluid, as is clearly shown in the rheomorphic folds of salt domes where the shape of the salt intrusion permits adequate knowledge of the direction of injection of the salt. This is mainly upwards, but also somewhat lateral near the top, where an 'overhang' is often present (Fig. VIII−42). (See Salt Domes, p. 275.) Folds due to the push of glacier ice on soft brown coals and clays also have the characteristics of rheomorphic folds and in recrystallized Pre-Cambrian rocks such as have been mapped in detail at Broken Hill, New South Wales,

[1] F. Kaisin, 'Quelques réflexions sur la dysharmonie', *Bull. Soc. belge Géol.*, Vol. 45, 1936, pp. 258−71. O. Stutzer and A. C. Noé, *Geology of Coal*, Chicago, 1940, p. 344.

[2] S. W. Carey, who has given an excellent account of folds of this type, terms them *rheid*-folds, but since he attributes a particular significance to the word rheid, it is thought preferable to use the more descriptive term rheomorphic for these fold types. 'The Rheid Concept in Geotectonics', *Journ. Geol. Soc. Aust.*, Vol. 1, 1954, pp. 67−117.

similar deformation is to be attributed to crystal plasticity synchronous with metamorphism and transfusion.[1] In the Westport area, Ontario, the folding of metamorphic rocks including gneisses, marbles, and quartzites has been attributed by Wynne-Edwards[2] to non-uniform laminar flow. Anticlines represent advancing parts and synclines are retarded, while folded anticlines are due to synchronous unsteady flow.

The extent to which the apparent thickening of folds of this type at crests and troughs is to be attributed to flattening (compression with concomitant extension at right angles) must in all cases remain uncertain unless critical data permit its evaluation, for in most cases simple shear-folding, with thinning of the limbs, will produce an identical fold geometry. It also must be pointed out that uncritical reference of the complexities of folds in schists and

S W.　　　　　　　　　　　　　　　　　　　　　　　　**NE.**

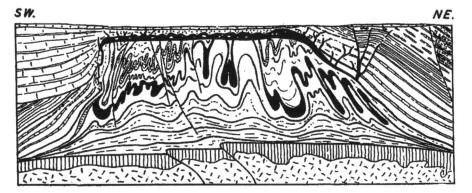

FIG. VIII–42. RHEOMORPHIC FOLDS IN SALT DOME NEAR HANOVER
The salt squeezed from left and right into the dome. (*After Seidl*)

gneisses to 'rheomorphic folding' may lead one to overlook the possible effects of several phases of folding, which result in a complex fold geometry (see Chapter IX).

Fold nappes on the scale in which they are represented in the western Alps, are rheomorphic folds, affording the most complex regional structures known to geology.

Nappe structures – Thrust nappes have been mentioned in Chapter VII, but the grandest concept in nappe tectonics is that of many large recumbent overfolds, or *fold nappes*. The word *nappe* (Fr.; Ger. *Decke*) refers to any thin extensive sheet of rocks that acts as a cover, and this includes gravel beds or

[1] S. W. Carey, 'The Rheid Concept in Geotectonics', *Journ. Geol. Soc. Aust.*, Vol. 1, 1954, pp. 67–117; for primary references on Broken Hill, see J. K. Gustafson, H. C. Burrell, and M. D. Garretty, 'Geology of the Broken Hill Deposit, Broken Hill, N.S.W.', *Bull. Geol. Soc. Amer.*, Vol. 61, 1950, pp. 1369–437. B. P. Thomson, 'The Geology of the Broken Hill District', in *Geology of Australian Ore Deposits*, 5th Empire Min. and Met. Congr., Melbourne, 1953, pp. 533–77.
[2] H. R. Wynne-Edwards, 'Flow Folding', *Amer. Jour. Sci.*, Vol. 261, 1963, pp. 783–814.

flows of lava as well as tectonic nappes. In English, however, the word is used only for tectonic structures, and this usage is increasing in French and German. It refers to a sheet or slab of rocks of large dimensions that has been translated from its original situation for some distance, in extreme cases of the order of miles or even tens of miles, moving forward over the rocks beneath and in front of it and finally covering them as a cloth covers a table. In a complex structural environment each nappe retains its individuality as a tectonic unit.

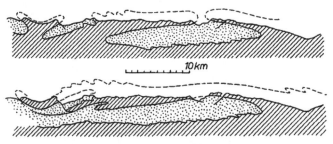

FIG. VIII–43. GLÄRUS 'DOUBLEFOLD' AND ITS REINTER-
PRETATION BY M. BERTRAND AS A NAPPE
(*After Goguel*, Traité de Tectonique, *Masson, Paris, 1952*)

The Western Alps of Europe are the classic area for fold nappes. Large-scale recumbent folding and important low-angle thrusting is obvious in many Alpine sections, but earlier notions of the major movements were conservative. In the famous example of the Glärus structure, folds overturned in two directions were postulated (the double fold of the Glärus), the hypothesis of a vast unilateral movement in one great fold coming later (Fig. VIII–43). Indeed the evidence for translation of rock-masses was derived rather from facies geology than from purely structural mapping, when it became known that the movements had brought into juxtaposition rocks originally deposited in sedimentary zones lying in the reverse order from their present situations[1] (Fig. VIII–44). The combination of structural and facies mapping provided

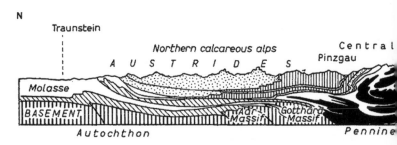

FIG. VIII–45. GENERAL SECTION ACROSS

[1] See F. Heritsch, *The Nappe Theory in the Alps* (trans. P. G. A. Boswell), London, 1929.

the basis for reconstructing the *allochthonous nappes* especially the Pennine nappes for which no point of origination can be seen, as the zone of roots plunges steeply down and is much attenuated (Fig. VIII–45). The origin of *parautochthonous* and *autochthonous nappes* stripped from crystalline basement wedges in the Calcareous Alps (Helvetic Nappes) is however clear.[1] The

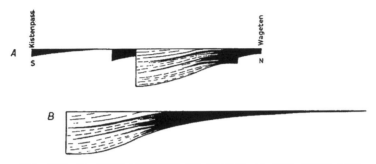

FIG. VIII–44. DISPLACEMENT OF FACIES AS THE BASIS FOR NAPPE-INTERPRETATION

A. Arrangement of facies as now found from Kistenpass to Wageten.
B. Original arrangement of facies in the sedimentary trough. (*After A. Heim*, Geologie der Schweiz, *Leipzig, 1921, p. 26*)

'travel' of the Pennine overfolds, the over-riding of one nappe by others supposedly originating in succession farther to the south, and the 'plunging' of the prows into the soft rocks of the Molasse trough to the north of the Alpine structures afford a most stimulating image of *tectonique en mouvement*, but it may be that the western Alps are unique in the grandeur of scale and the complexity of folding of their nappes. In many other regions of alpine tectonics, slicing along low-angle reverse faults is more prominent than recumbent folding on the scale seen in the Pennine nappes.

In the Carpathians, control of 'napping' by the inhomogeneities of the geosynclinal sediments is notable. Thus not only does facies aid in revealing

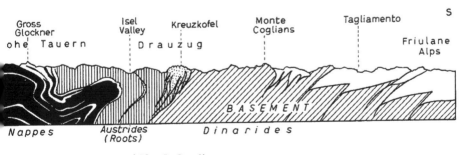

THE EUROPEAN ALPS (*After R. Staub*)

---

[1] R. Trümpy, 'Die helvetischen Decken der Ostschweiz', *Eclogae Geol. Helv.*, Vol. 62, 1969, pp. 105–42.

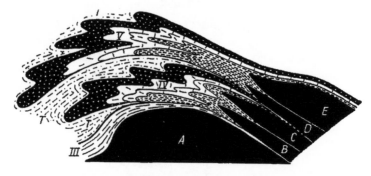

FIG. VIII–46. NAPPES AND CRYSTALLINE WEDGES

A, Basement with sedimentary cover *III*; *IV*, *V* nappes with crystalline
cores *B*, *C*, *D*, *E*; *T*, Tertiary between the nappes. Neritic calcareous facies,
solid black; Marl, shale and mudstone, broken lines. (*After Collet*)

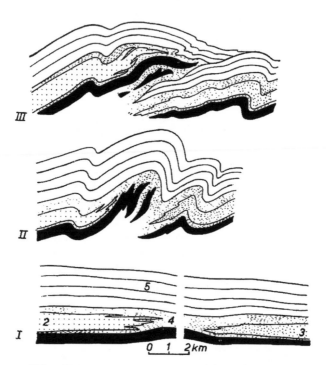

FIG. VIII–47. RELATIONSHIP BETWEEN NAPPES
AND ROCK-MASSES IN THE CARPATHIANS

The diagrams show the origin of thrusts in the axial region
of the Flysch zone. In stage I (undeformed) the middle of
the section is omitted to reduce the length. II shows compression of the axial zone, with a thick cover of Krosno
shales and micaceous sandstones. III shows the same with
thinner Krosno cover and thicker sandy complexes of the
Upper Cretaceous and Eocene beneath. 1, Lower
Cretaceous; 2, M. & U. Cretaceous and Eocene sandy
facies; 3, the same in Inoceramian facies; 4, the same in
shales and marls; 5, Krosno Beds. (*After Ksiazkiewicz*)

nappes, but it also tends to influence their structural evolution due to the presence of competent and incompetent masses (Fig. VIII−47).[1]

The details of the structure of nappes are highly involved, as may be seen by reference to the various illustrations. Most of the complexities derive from

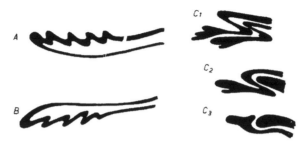

FIG. VIII−48. TYPES OF NAPPES AND COMPLEXITIES
A. Rising nappe. B. Plunging nappe. $C_{1, 2, 3}$. Types of involutions involving two nappes. (*After A. Heim*, Geologie der Schweiz, *Bd. II, 1921*)

the high mobility of nappes, especially those of the European Alps, and the action of nappes as plungers, whereby they form folds in front of the prow as this pushes against soft sediments (Fig. VIII−46A), or again push against other nappes either from above or from below, forming *involutions* (Fig. VIII−48). Where part of a superficial nappe is left isolated by erosion it is

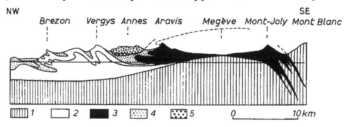

FIG. VIII−49. ORIGIN OF THE MORCLES NAPPE FROM CRYSTALLINE BASEMENT WEDGES, WITH RELATIVELY SMALL TRANSLATION (PARAUTOCHTHONOUS NAPPE)
In front of the nappe is autochthon, folded by the push of the nappe.
1, Hercynian massifs. 2, Autochthonous sediments. 3, Morcles nappe. 4, Tertiary. 5, Nappe-outlier (Klippe) of the Annes.
(*After Ed. Parejas, from Collet, 1927, Fig. 17*)

called a *Klippe*, and where erosion through a nappe reveals the underlying structural units, the area is called a *window* (Ger. *Fenster*).

The *root* of a recumbent fold nappe is the core of the anticline where it is seen in its relation to an older basement (Fig. VIII−46B) or, in the case where

---

[1] M. Ksiazkiewicz, 'Geology of the Northern Carpathians', *Geol. Rundsch.*, Vol. 45, 1956, pp. 369−411.

the basement is too deeply buried to be seen, the core is vertical or nearly so and the nappe appears to be 'rooting' in depth.[1]

Because of the translation of rocks in nappes, it is important to recognize various conditions as follows: *autochthonous* beds and folds – those that have not been translated; *parautochthonous* beds and nappes – those that have undergone relatively small translation and which can be traced back to their origin or roots (see Morcles Nappe, Fig. VIII–49). Nappes which have been translated for great distances are sometimes called *allochthonous*.[2]

FOLDS DUE TO VERTICAL FORCES

The role that vertically acting forces play in folding has been variously emphasized since the early stages of the development of geological thought on the origin of fold mountain chains. While it was at first natural to assume that the elevation of these chains was due to vertical forces, later emphasis on tangential forces and lateral compression in folding, and evidence for the uplift of fold mountains after rather than during folding, subordinated the role of vertical forces in the folding itself. Vertical components of primary tangential forces are, indeed, concerned in the elevation of practically all anticlines, at least in relatively superficial rocks, but what is usually implied by vertical forces is that the major features of the folding appear to be due to forces that elevate large anticlines or domes against the ever-present force of gravity. To exclude downward movements due to gravity is, however, not entirely logical, except perhaps for the reason that in slumping and large-scale gravitational sliding the movements are chiefly lateral due to a resolute of gravity acting on an inclined plane. On the other hand folds due to compaction (p. 60) are due directly to gravity and must be considered in the present context along with those due to upward push from beneath.

Supratenuous folds

The geometry of anticlines or domes formed by compaction over buried hills is described as *supratenuous*, because the formations are thinner at the crest of an arch than in the neighbouring troughs.[3] Folding of the Plains Type, well known in the Mid-Continent petroleum fields of the U.S.A., is typically supratenuous and has characteristics as follows:[4] (1) the net result of the folding is a

[1] For further details, see L. W. Collet, *The Structure of the Alps*, London, 1927. L. Kober, *Bau und Entstehung der Alpen*, Vienna, 1955. R. Staub, 'Bau der Alpen', *Beitr. Geol. Karte Schweiz*, N.F., Vol. 52, 1924.

[2] A. Tollmann has given a useful review of the nomenclature of nappes and associated structures in 'Die Grundbegriffe der deckentektonischen Nomenklatur', *Geotekt. Forsch.*, Hft. 29, 1968, pp. 26–59.

[3] C. M. Nevin and R. E. Sherrill, 'Studies in Differential Compaction', *Bull. Amer. Assoc. Petrol. Geol.*, Vol. 13, 1929, pp. 1–22.

[4] S. K. Clark, 'The Mechanics of the Plains-Type Folds of the Mid-Continent Area', *Journ. Geol.*, Vol. 40, 1932, pp. 46–61.

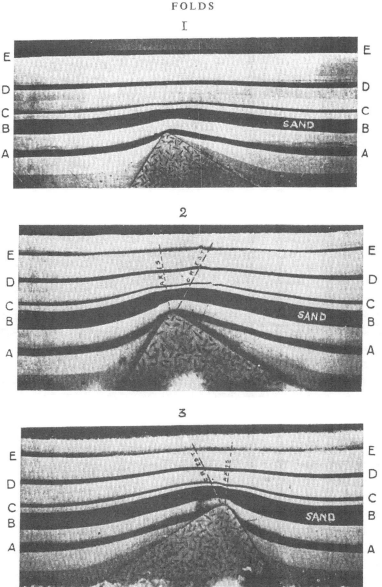

FIG. VIII–50. EXPERIMENTS IN DIFFERENTIAL COMPACTION

Except where indicated, the sediments are clay. Numbers 2 and 3 are opposite
sides of the same hill. (*After Nevin*)

relative local elevation without a corresponding depression, so that there are
only anticlines and domes rising above the regional dip; (2) the folds become
more pronounced in depth, and there is a thinning of the strata above the
crests of the folds; (3) the folds are usually asymmetrical, and the crestal
plane dips towards the steeper limb of the fold (Fig. VIII–50). The deposition

of thicker sequences in the basins and thinner on the flanks of buried hills may accentuate the supratenuous character of folds of this type (Fig. VIII−51).

The extent to which faulting of the basement beneath folds of the Plains Type may have contributed to the elevation of the anticlines or domes is uncertain, but evidence for such movements is growing.[1] G. M. Lees has also drawn particular attention to this mechanism in relation to folding of the forelands fronting deep geosynclinal zones,[2] and structures substantiated by drilling demonstrate the reality of the process. The stretching of many domes (see p. 281) is best accounted for by push from beneath, but this is not

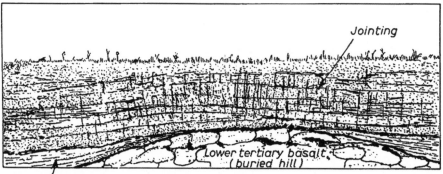

Jointing

Lower tertiary basalt.
(buried hill)

Lensing beds on flank of hill

FIG. VIII–51. SMALL BURIED HILL OF BASALT COVERED BY MARINE TERTIARY ROCKS, ROYAL PARK, VICTORIA

Note lensing of lower beds against the flanks of the hill, and anticline in overlapping red sandstones, with radial jointing.

necessarily an indication of faulting in the basement, as it may also be due to the sub-surface movement of incompetent rocks, which is admirably seen in salt domes.

In inferring the forces that have caused folding, it is necessary to consider first the immediate cause, whether this involves movements of basement blocks directly reflected in folds, secondary sliding from upwarps, regional compression, or some other cause that may be deduced from local geological data. In a second stage of the study one must then consider the forces that give rise to the features to which the folds are related. For instance, it will later be shown (p. 340) that folds due to local vertical movements may arise within the zone of deformation along a major strike-slip fault. In a third stage, the origin of this fault-zone must itself be related to regional forces that may have operated.

[1] See pp. 58–9.
[2] G. M. Lees, 'Foreland Folding', Anniv. Address, *Quart. Journ. Geol. Soc. Lond.*, Vol. 108, 1952, pp. 1–34.

CHAPTER IX

Tectonic analysis of folds

Folds have so far been considered chiefly in cross-section, and it is now necessary to deal with fold geometry in three dimensions. Although much use is made in geological reports of cross-sections of folds, their properties along the regional trend are more immediately obvious in regional mapping and map study.

Tectonic elements of folds

The *crest* of an anticline, as defined by a particular bed, is the line joining the highest points of the fold along its trend, and the *crest plane* is the surface[1] that joins the crests of successive beds or formations in depth. The crest is

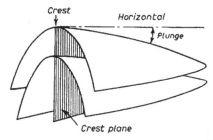

FIG. IX–1. CREST OF PLUNGING ANTICLINE

generally determinable with particular precision in the field, since it is the line along which the direction of dip reverses to form the fold. A corresponding nomenclature is used for synclines, for which the *trough* and the *trough plane* may be recognized for the lowest points of the fold.

The inclination from the horizontal of the crest or trough of a fold as defined by a given bed, is the *plunge* of the fold for that bed[2] (Fig. IX–1). With long folds it is generally found that the plunge varies so that the crest undulates (Fig. IX–2), and indeed in strongly deformed rocks the changes in

[1] Terms such as crest *plane* and axial *plane* are used in this book in preference to crest surface and axial *surface* which some authors use. No one, however, talks of 'axial-surface cleavage', and indeed it is not inappropriate to use *plane* for approximately planar portions of gently curving natural surfaces.

[2] Plunge of folds was formerly called *pitch* and as such is shown on all older maps.

plunge may be very strong. Overturned plunges are also known, but these may represent incongruous drag folds (see p. 294) or effects due to super-posed folding (p. 283).

When any fold is followed along its trend it will eventually be found to die

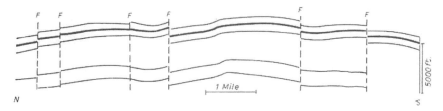

FIG. IX–2. UNDULATING CRESTS OF SUCCESSIVE BEDS IN A SIMILAR FOLD, BENDIGO, VICTORIA, SHOWN BY LONGITUDINAL SECTION OF SADDLE REEFS. (*After Herman,* '*Structure of Bendigo Goldfield*', Bull. Geol. Surv. Vict., *No. 47, 1923*)

out. With an anticline the crest plunges downwards and the limbs decrease in length until they both pass into a zone of regional dips, so that the fold has the form of a very much elongated dome. The bowing·of crests and troughs along the trend of folds as seen in longitudinal section itself gives an appearance of folding but this does not indicate cross-folding, being rather an inherent

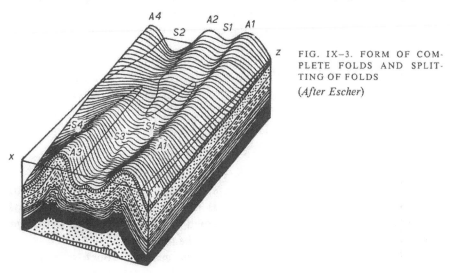

FIG. IX–3. FORM OF COM-PLETE FOLDS AND SPLIT-TING OF FOLDS
(*After Escher*)

property of all folds.[1] Where a fold bifurcates so that, for example, a syncline appears either on the flank or at the crest of an anticline, the newly appearing folds differ in plunge or in trend from the original single fold (Fig. IX – 3).

Although plunge is a primary tectonic element of folds, it is as well to point out that, quite apart from cross-folding of synchronous or later date, simple

[1] Cross-folding is discussed on pp. 283–8.

warping or tilting as in block faulting changes the primary plunge and the effects of such displacements must be deducted from the measured angles to arrive at the primary fold geometry.[1]

Normal cross-section of a fold – Although it is usual to draw geological cross-sections in the vertical plane for the purpose of illustrating regional geology, it must be realized that a vertical section across a plunging fold gives a false impression of the true fold geometry, the more so, the steeper the plunge. This is readily seen by considering a number of sections of a semi-cylindrical

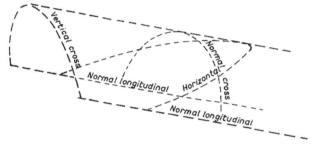

FIG. IX-4. SECTIONS AT VARIOUS ANGLES ACROSS A
PLUNGING CYLINDRICAL FOLD
(*After Challinor*)

plunging anticline (Fig. IX−4).[2] In the discussions that follow it will be assumed that 'cross-section' (or 'profile') means a section drawn normal to the plunge of a fold.[3]

Axial plane

Despite a somewhat variable usage for the term *axis*, most structural geologists agree on the recognition of the *axial plane* as a surface that approximately bisects a fold. From the discussion of the formation of folds it is clear that there is a plane in each fold that is significant in its formation and which can be recognized from the geometry of the fold in cross-section. In general this plane passes through the line in each successive bed about which the dip changes most rapidly, which line does not necessarily correspond with the crest or trough, and is termed the *hinge* of the fold (Ger. *Gelenk, Scharnier*) (Fig. IX−5).

In asymmetrical similar folds this plane does not, in fact, bisect the fold,

[1] J. Challinor, 'The Primary and Secondary Elements of a Fold', *Proc. Geol. Assoc. Lond.*, Vol. 56, 1945, pp. 82–8.

[2] J. Challinor, op. cit.

[3] See particularly the discussion of tectonic analysis of folds, below. See also M. J. Fleuty, 'The Description of Folds', *Proc. Geol. Assoc.*, Vol. 75, Pt 4, 1964, pp. 461–92; S. Simpson, Discussion, *Geol. Mag.*, Vol. 102, 1965, pp. 179–80.

but it corresponds with the direction of shear in shear folds and with the plane of flexing in angular folds, and is equally as important as the hinge in parallel folds. Even where lateral kink planes are present in mitre-folds it is generally possible to recognize a median hinge in addition, but in folds that are circular

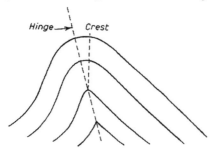

FIG. IX–5. ASYMMETRICAL FOLD, WITH HINGE AND CREST CONVERGING TO THE CUSP

arcs and in box-folds there is no hinge. Nevertheless, a median plane will generally be significant either from the geometry of beds beneath the core, or from the similitude of successive folds across the strike (Fig. IX–6). Accordingly, the recognition of the axial plane depends on the fold geometry itself. It

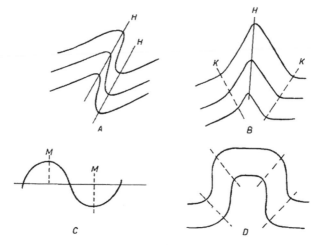

FIG. IX–6. THE HINGE OF FOLDS

A. Similar folds with asymmetrically located hinges.
B. Mitre-fold with median and lateral hinges (conjugate folds).
C. Semicircular folds. Median planes M recognized from similitude.
D. Box-fold without median hinge (conjugate folds).

is either a bisecting plane, or a plane passing through the hinges in successive beds of a folded sequence.

The axial plane extends along the length of a fold, but it is a fact that its extension holds little interest, for in practice it is the trace of this plane as seen in cross-sections that is of particular value in fold geometry. This trace was formerly commonly called the fold 'axis',[1] and indeed 'axial plane' can only be defined by reference to this axis as it is traced along and through a fold. Therefore in this book, the trace of the axial plane in the profile of a fold is recognized as one of the tectonically significant axes of a fold, and is named the *transverse axis*.[2]

Longitudinal fold axis

In current practice, many schools of tectonic analysis following Wegmann recognize only one axis in folds, this being defined as a line which, if moved parallel with itself, generates the fold.[3] This axis is obtained by statistical

FIG. IX–7. TRANSVERSE AND LONGITUDINAL FOLD AXES, AND AXIAL PLANE
(Shaded)

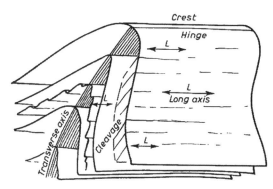

examination of the attitude of bedding, cleavage, minor folds, lineations, and the like within a small region throughout which the fold structures are homogeneous. The axis so determined is not a single entity but a direction which pervades the folds. It is here regarded as preferable to identify this axial direction as the *longitudinal* (or *long*) *axis* of folds, but in most recent works it is referred to simply as the *fold axis*. Folds that conform with the definition of this axis given above are described as *cylindrical* (Fig. IX–7), and successive small domains of most folds may be treated as cylindrical even though the axis may vary in plunge from domain to domain. Some folds are, however, demonstrably conical,[4] and conical folds may also result from superposed folding.

The Significance of plunge – Where folds plunge at a more or less constant

[1] J. D. Dana, *Manual of Geology*, New York, 1880, pp. 93, 95. A. Geikie, *Textbook of Geology*, 4th edn, 1903, Vol. 2, p. 675.

[2] The adoption of this term would avoid the ever-recurring circumlocution 'trace of the axial plane', which is a particularly unfortunate term because one never actually sees an axial plane, but only its 'trace'.

[3] C. E. Wegmann, 'Beispiele tektonischer Analysen des Grundgebirges in Finnland', *Bull. Comm. Geol. Finl.*, Vol. 13, No. 87, 1929, pp. 98–127.

[4] F. Moseley, 'Conical Folding and Fracture Patterns in the Pre-Betic of Southeast Spain', *Geol. Journ.*, Vol. 6, Pt 1, 1968, pp. 97–104.

angle, the completion of profiles (drawn at right angles to the plunge) is most properly carried out by projecting data along the direction of the long axes to meet the chosen plane of the profile. If the plunge changes along the trend, the region may be divided into a number of sectors for the purpose of profile drawing, a statistical 'average' for the plunge being arrived at for each sector, and the attitude of the profile being modified accordingly at different levels (Fig. IX–8). The method depends upon the recording of a great many field observations of the attitude of beds and other structures, rather than the somewhat haphazard application of Pumpelly's rule to only a few drag-folds (see p. 293). Some of the methods used are treated below, but it may also be

FIG. IX–8. USE OF PLUNGE FOR PRO-
JECTION IN SECTION-DRAWING

mentioned that a useful method of map study, known as 'looking down-structure' depends on the same principle.[1] In this the map is turned so that the observer looks along the fold trend and down the plunge, the map being held at an appropriate tilt to give an effect of foreshortening. With practice the observer obtains a useful visual impression of the profile.

π- *and* β-*diagrams* – In homogeneous cylindrical folds there is one long axis, which may be determined by plotting on an equal area projection the poles of a large number of bedding planes whose dip and strike have been measured, to make a π-diagram.[2] With an ideal fold of the type under discussion, these poles will all fall on one great circle (the π-circle) (Fig. IX–9), and the normal to this is β, the long axis, whose attitude may accordingly be read off from the projection.[3] In practice, the poles will 'spread' rather widely and the π-circle may then be obtained by first contouring the π dots for density of spacing, and then drawing an appropriate π-circle through the maxima. Many more observations will in general be made on the fold limbs than near the

---

[1] J. H. Makin, 'The down-structure method of viewing geological maps', *Journ. Geol.*, Vol. 58, 1950, pp. 55–72.

[2] Refer to p. 146, Chapter VI.

[3] β symbolizes the B-tectonic axis of Sander with which the long axis of most folds conforms. See Chapter XIII.

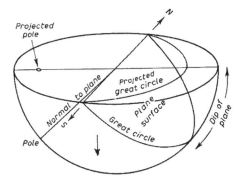

FIG. IX-9. SPHERICAL PROJECTION (LOWER HEMISPHERE) AND THE PROJECTION OF THE GREAT CIRCLE AND THE POLE REPRESENTING AN INCLINED PLANE STRIKING N-S

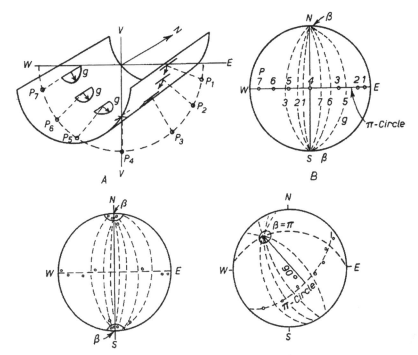

FIG. IX-10. π AND β DIAGRAMS

A. A syncline striking N–S, with horizontal trough, showing diagrammatically how the poles P_1–P_7 of dipping surfaces are projected to an imagined sphere, and lie on a great circle (π-circle). g shows imagined great circles (on the lower hemisphere) of spherical projections about dip readings in one limb of the fold.

B. Poles and great circles on the Schmidt Net, corresponding with dip readings 1–7 in A. The fold and the readings are geometrically precise. In actuality, slight deviations occur and errors are made in measurement.

C. Shows the same fold, with dip readings deviating from the ideal. Where the great circles intersect represents the axial trend of the fold. This is shown by the centre of the contoured maximum of intersection points (β), which also corresponds with the normal to the π-circle.

D. Shows the plotting of π and β for a plunging fold.

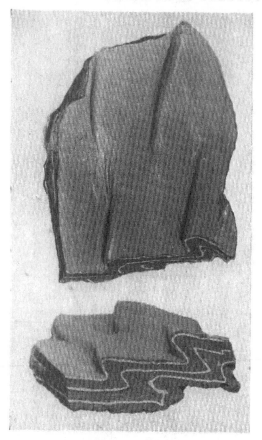

FIG. IX–11. *En echelon* FOLDS
IN SCHIST

(*After Campbell*)

hinges, and in strongly inequant folds the prevalence of observations on the
long limbs will reveal the inequant character of the structure.

β-diagrams are constructed by plotting the plane of the bedding[1] at each
observed point as a great circle on an equal area net. The strike at each
observation point is the diameter of the great circle, and it is seen (Fig.
IX–10) that the intersection of great circles representing beds in either limb,
represents the point of emergence of the long axis, termed β. Again, it is found
in practice that there is a great number of intersections of the great circles,
which must be contoured for density of distribution, β being then represented
by the centre of the maximum so found.[2]

Homoaxial and *en echelon* folds – The methods outlined above are applicable

---

[1] In addition other planar structures believed to be related to the folding, especially cleavage,
may be incorporated, but the discussion here refers only to bedding. The student should also
refer to Chapter XIII and to the discussion of cleavage, schistosity, and lineation.

[2] Examples of the use of these diagrams will be found in the following works: L. E. Weiss, 'A
study of tectonic style – Structural investigation of a marble-quartzite complex in Southern
California', *Univ. Cal. Pub. Geol. Sci.*, Vol. 30 (1), 1954, pp. 1–102; idem, 'Geometry of
Superposed Folding', *Bull. Geol. Soc. Amer.*, Vol. 70, 1959, pp. 91–106.

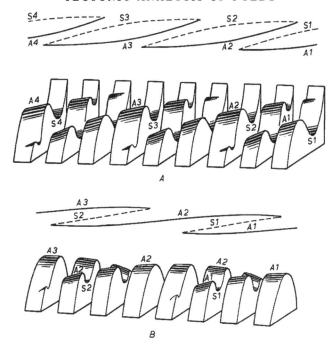

FIG. IX–12. TWO TYPES *en echelon* FOLDING

A. Dissected block diagram of folds (below), with plan projection above. An elliptical pattern of right-hand *en echelon* folds.
B. Zig-zag pattern of right-hand *en echelon* folds. (*After Campbell*)

in a relatively simple way to individual folds, and to fold zones in which all the structures, major and minor folds, cleavage, and the like are related genetically to one longitudinal axial direction, and may thus be described as *homoaxial*. It has been noted above in relation to the bifurcation of folds that the plunge of the offshoots differs from that of the original single fold, and further complexities arise in connection with cross-folding, but *en echelon* folding is perhaps of greater general significance. In this a parallel series of short folds is present, whose long axes are oblique to some major structure such as folds of a lower order, or the trend of strike-slip faulting or regional cleavage (Fig. IX–11). Campbell has demonstrated that *en echelon* folding is represented in many Australian Pre-Cambrian and Palaeozoic fold systems.[1] He describes those in which the 'overlap' of adjacent folds is towards the right as dextral, and those overlapping towards the left as sinistral (Fig. IX–12). Where synclines link the ends of adjoining anticlines the

[1] J. D. Campbell, 'En echelon folding', *Econ. Geol.*, Vol. 53, 1958, pp. 448–72. F. Mendelsohn further discusses *en echelon* folding: *Econ. Geol.*, Vol. 54, 1959, pp. 505–15; also 'The Structure of the Roan Antelope Deposit', *Bull. Inst. Min. Met.*, London, Vol. 68, 1958–9, pp. 519–33.

pattern is *zig-zag*, but where each syncline is linked with its own anticline the pattern is *elliptical*. The geometrical relationships of cleavage and lineation in major and minor folds are not described in these examples, which were elucidated by intense formational mapping for the purposes of mining geology.

Where mappable formations are lacking, or the formations are very thick and have no marker beds within them, as is often found in greywackes and their slaty or schistose derivatives, some indications of the fold characteristics may be obtained from π- and β-diagrams.

Firstly, a diagram exhibiting many more readings relating to one set of fold limbs than to the other indicates that the limb of an anticlinorium (or synclinorium) is represented. Further, marked maxima on either limb with a gap representing the absence of low dips, points to angular folding. The presence of a subsidiary β-point additional to the predominant β but close to it, indicates the presence of some fold deviating rather sharply from the main trend, as, for example, with minor folds arranged *en echelon* on a major fold. Where there are two opposed plunges, a spread of β-points between the two maxima indicates broad arching of the fold crests, but a hiatus in the β-points indicates sharply opposed plunges such for instance as arise from the splitting of folds. Other considerations of β- and π-diagrams may be suggested from the actualities observed in the field at good exposures, and a good discussion is given by Weiss.[1]

Domes and basins

We have seen that anticlines eventually 'plunge out' at either end and thus have the form of elongated domes, and in fact there is every gradation in type from anticlines tens of miles long to sub-circular domes. The essential feature of a dome is its *closure* which is brought about by the dips radiating from the highest point or *culmination*, and where an anticline, due to plunge changes, has a local structure of this type it may be identified and named as a local dome (Fig. IX–13). It is because of the great importance of domes in oil geology that many more examples have been intimately studied than is the case for basins, and indeed in geological usage the word *basin* has many more and less precise connotations than its antonym 'dome'. Any region of structural depression with a more or less complete basin shape may be termed a basin, whether of the order of a mile or several hundred or even thousands of miles in width as with coal basins, artesian basins, sedimentary basins, and ocean basins.

In the same way as a short anticline (or elongated dome) may be termed a *brachy-anticline*, a short anticlinorium with the general shape of a complex

[1] L. E. Weiss, 'Geometry of Superposed Folding', *Bull. Geol. Soc. Amer.*, Vol. 70, 1959, pp. 91–106.

dome is called a *brachy-anticlinorium*. The structure of the thick Lower and Middle Palaeozoic geosynclinal deposits of central Victoria, including many of the well-known goldfields such as Bendigo and Castlemaine, consists of brachy-anticlinoria and brachy-synclinoria in which the individual folds (often with as many as six anticlines in a distance of a mile) preserve generally straight and parallel trends (Fig. IX–14). The beds are a flysch-like succession of thin and impersistent sandstones and slates, and determination of the regional structure depends entirely upon the mapping of palaeontological zones, chiefly in the graptolitic Ordovician rocks.[1]

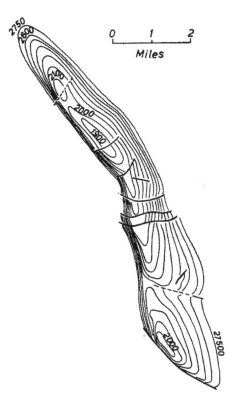

FIG. IX–13. ANTICLINE WITH LOCAL DOMES, SUMATRA.

(After Barnwell, from Lees, Quart. Journ. Geol. Soc. Lond., *Vol. 108, 1952, Fig. 11)*

Closure of domes – The amount of closure for any given bed in a dome is measured by the vertical distance from the culmination to a horizontal plane tangential to the same bed at the highest point where the dip is reversed, which point marks the place where a fluid may escape. Reversal of dip forms either a *saddle* on the crest, or a syncline on the flank of the dome. The geometry of a dome is best represented cartographically by means of *structure contours*, which are contours relative to some horizontal datum plane (generally sea-level), drawn on a particular geological horizon in the dome.

[1] D. E. Thomas, 'The Structure of Victoria with Respect to the Lower Palaeozoic Rocks', *Mining & Geol. Journal*, Mines Dept. Vict., Vol. 1, No. 4, 1939, pp. 59–64.

Generally these are *stratum contours*, as they are drawn on the top (or more rarely on the bottom) of a bed or formation that is significant for the geologist. They often represent a key horizon that is readily identified in bore samples or by geophysical means. The approximate measure of

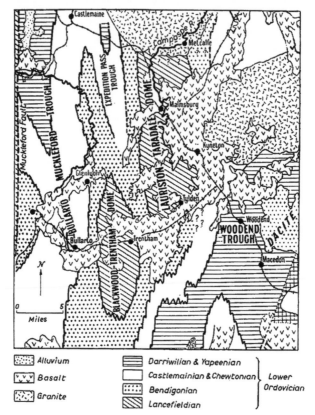

Alluvium		Darriwilian & Yapeenian	
Basalt		Castlemainian & Chewtonian	Lower
Granite		Bendigonian	Ordovician
		Lancefieldian	

FIG. IX–14. BRACHY-ANTICLINORIA AND BRACHY-
SYNCLINORIA, CENTRAL VICTORIA
(*After Thomas, 1939*)

closure, subject to the limitations of the contour-interval and the probable accuracy of the contouring, may readily be estimated from such a map (Fig. IX–15).

What is termed 'subsurface mapping' is of great significance in oilfield and mining geology and several important works have been devoted to it.[1]

---

[1] Margaret S. Bishop, *Subsurface Mapping: Preparation and Interpretation*, New York, 1959. L. W. LeRoy and J. W. Low, *Graphic Problems in Petroleum Geology*, New York, 1954. Other works on related topics are W. L. Donn and J. A. Shimer, *Graphic Methods in Structural Geology*, New York, 1958. P. C. Badgley, *Structural Methods for the Exploration Geologist*, New York, 1959.

Faulting of domes and basins

Where faults are geometrically related to individual folds the prima facie hypothesis that they are genetically related must be entertained. The stretching and normal faulting of rocks domed over salt plugs has already been mentioned (p. 275), but stretching is common in most anticlines and domes, more especially, but not exclusively, where the folding may be attributed to vertically acting forces.

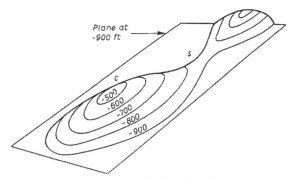

FIG. IX-15. DOME SHOWING CLOSURE
C, culmination; *S*, saddle.

Stretching across the fold in the direction of the dip often gives rise to a longitudinal *crestal graben* or fault zone with numerous associated normal faults (Fig. IX–16). There is a marked tendency for the graben to divide dichotomously towards each end of a long dome, a condition that is well matched by experimental analogues formed by upward push on a rectangular slab (Fig. IX–17).[1]

Cross-faulting of domes and basins may, as with cross-folding (p. 283), be either geometrically connected with the folding process, or entirely subsequent and due to a different diastrophic episode. Cross-faults of the latter type will not be treated here. The commonest arrangement of genetically related cross-faults consists of a number of normal faults transverse to the long arch of the fold, which represent the effects of tension along the length of the arch, or in a basin, along the trough (Fig. IX–13, 18). Again, where a horizontal force couple is involved in the folding, the faults may be oblique to the long axis, and they may include strike-slip as well as normal faults. Radial normal faults are also well displayed on many rounded domes (Fig. IX–19).

In many domes, faulting genetically connected with the formation of the dome is limited to a belt along the crest at certain horizons, while the limbs

[1] E. Seidl, *Bruch und Fliessformen in der technischen Mechanik und ihre Anwendung auf Geologie*, Bd. 5, Krümmungsformen, Berlin, 1934.

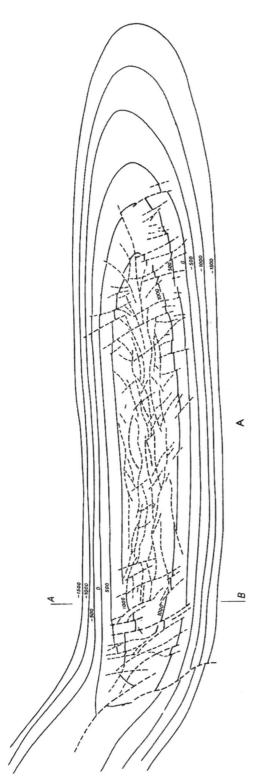

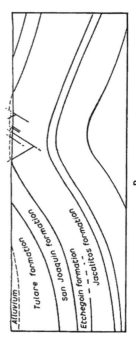

FIG. IX–16. DOME WITH CRESTAL (EPI-ANTICLINAL) GRABEN, KETTLEMAN HILLS DOME, CALIFORNIA

A. Structure contours drawn on *Littorina* zone of Etchegoin Formation (North Dome of Kettlemen Hills). Note longitudinal and transverse faults.

B. Section along *A–B*. (*After Woodring, Stewart and Richards,* U.S. Geol. Surv. *Prof. Paper 195, 1940*.) Length of Structure – 16 miles.

FIG. IX–17. EXPERIMENTAL DOME IN CLAY, SHOWING CRESTAL GRABEN AND BIFURCATING PATTERN OF GRABENS ON THE NOSE

Isohypses are contours of the clay surface, with dip of surface as shown. (*After H. Cloos*, '*Hebung* – *Spaltung* – *Vulkanismus*' Geol. Rundsch., *Vol. 30, 1939, pp. 401–527*)

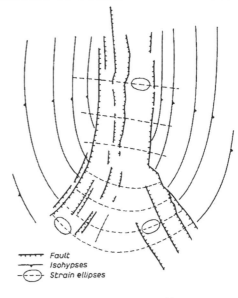

⊥⊥⊥⊤ *Fault*
⌄⌄⌄ *Isohypses*
⊸(⊂⊃)⊸ *Strain ellipses*

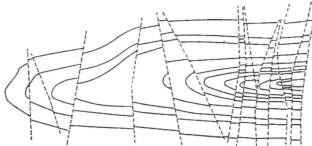

FIG. IX–18. COAL-BASIN WITH NORMAL CROSS-FAULTS, RUHR.

(*After Stegemann, in* Geology of Coal, *by O. Stutzer and A. C. Noé, Chicago, 1940, Fig. 93*)

remain unfaulted. This led Irwin to call such faulting *epi-anticlinal*.[1] The Kettleman Hills structure affords the best-known example[2] (Fig. IX–16).

Salt intrusions and salt domes

Salt domes are so named from the doming of strata over salt intrusions which have the form of plugs or stocks resembling intrusive igneous masses. In plan the intrusions range from circular to elliptical or elongated, and, rarely, they

[1] J. S. Irwin, 'Faulting in the Rocky Mountain Region', *Bull. Amer. Assoc. Petrol. Geol.*, Vol. 10, 1926, pp. 105–29. T. A. Link, 'The Origin and Significance of "Epianticlinal" faults as revealed by Experiments', ibid., Vol. 11, 1927, pp. 853–66.

[2] W. P. Woodring, R. Stewart, and R. W. Richards, 'Geology of the Kettleman Hills Oil Field, California', *U.S. Geol. Surv.*, Prof. Paper 195, 1940.

are very irregular (Fig. IX–20). The more elongated intrusions give rise to anticlines (*salt anticlines*) rather than to domes in the overlying beds. The structure of the surrounding rocks is known in detail in many places because of intensive exploration by drilling and by geophysics in the search for oil,

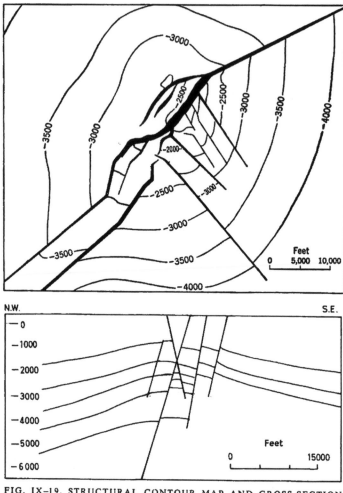

FIG. IX–19. STRUCTURAL CONTOUR MAP AND CROSS-SECTION OF THE VAN OILFIELDS, ILLUSTRATING EPI-ANTICLINAL FAULTS.

(*After Nevin*, Principles of Structural Geology, *New York, 1942, Fig. 73*)

and again, where the salt itself is mined, the internal structure of the salt intrusion is known in detail, and the sinuous flowing folds may be exhibited to the public as objects of wonder and beauty.

Salt intrusions occur in contrasting geological environments ranging from regions such as the Gulf Coast oilfields of Texas and Louisiana where the regional geology shows beds that are flat-flying or nearly so, to strongly folded regions as in Roumania and Algiers.

The essential features of a salt intrusion and salt dome are shown in Fig. IX—20; they include strong upward drag of the sediments adjoining the salt, the doming above, and, in the salt itself, most involved folding. A *cap rock* consisting of calcite, gypsum, anhydrite, and sulphur in various proportions is generally present, and the top of many salt plugs spreads laterally to make a mushroom-shaped *overhang*. Typically associated with salt domes intrusive into flat-lying beds is a sub-circular area surrounding the dome itself, where

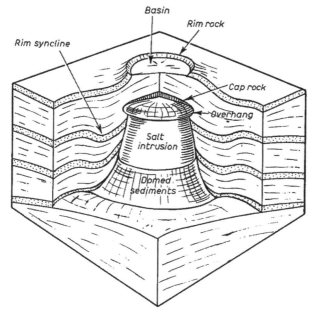

FIG. IX–20. BLOCK DIAGRAM SHOWING THE FEATURES
OF A SALT PLUG WITH DOME, RIM-SYNCLINE AND DRAG

the sediments sag gently downwards, producing, in combination with the upward drag along the dome, a *rim-syncline*. The term *peripheral sink* applies to the space evacuated by the salt that rises to form a salt intrusion. The rim-syncline thus occupies the peripheral sink.[1] Because of the differential erosion of hard and soft rocks in the dome, the topographic expression of the structure is often a basin, ringed around with one of the resistant beds of the succession which forms the *rimrock*. Salt domes have been located in oilfields by noting such topographic features.

Origin of salt intrusions – While the injection of the salt is indisputable in all examples, the motive forces have been a subject of much discussion. Students of the Roumanian salt plugs have naturally suggested that orogenic forces

[1] C. H. Ritz, 'Geomorphology of Gulf Coast Salt Structures, and its Economic application', *Bull. Amer. Assoc. Petrol. Geol.*, Vol. 20, 1936, pp. 1413–38.

were mainly responsible for the salt injections, but some other explanation is clearly required where the beds are flat-lying. The uprise of a salt-mass in such an environment is believed to be due almost entirely to hydrostatic

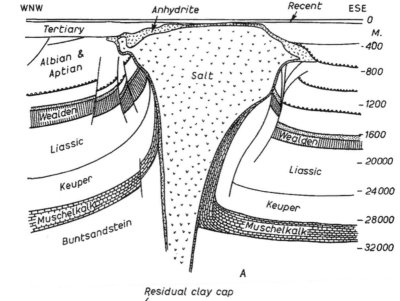

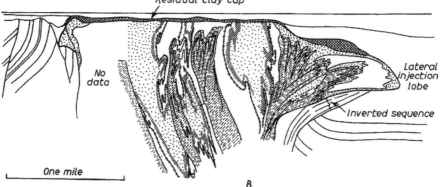

FIG. IX–21. SALT DOMES

A. Section of the salt intrusion at Wienhausen-Eicklingen, showing the 'overhang'. (*After Bentz, in Erdöl und Teknotik, Hannover-Celle, 1949, p. 10, Fig. 2*)

B. Section of the salt intrusion at Wathlingen-Hänigsen showing the 'overhang', folding and faulting in lobate salt injections. (*After Bentz, Pl. 2*)

forces that arise from the lower density of rock-salt (Halite 2·2, Sylvine 1·9 – excluding Anhydrite 2·9) as compared with normal sediments (average 2·5) that overlie the parent salt-bed. This suggestion was first made for the Hanoverian salt intrusions, but is perhaps even better applied to those of the

Gulf Coast region, since the German examples show a clear regional pattern of distribution related to major structure lines (Fig. IX–22).[1]

The idea that salt domes may rise through sediments because of the lower density of the salt has been greatly strengthened by model experiments, especially those carried out in accordance with the theory of scale models,

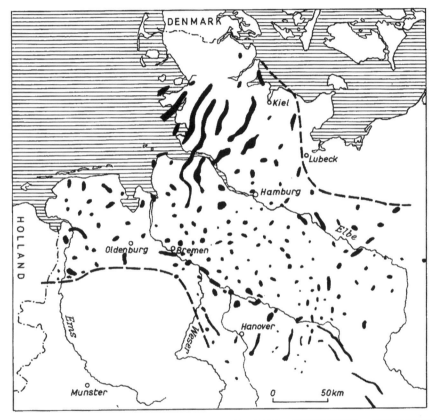

FIG. IX–22. DISTRIBUTION OF PERMIAN SALT INTRUSIONS, NORTHWEST GERMANY, SHOWING ARRANGEMENT ON TECTONIC LINES
(*After H. Cloos*, in Erdöl und Tektonik, *Hannover-Celle, 1949, p. 293, Fig. 1*)

convincingly demonstrating not only the ability of buoyancy forces to cause the rise of mobile material of low density through normal overburden, but suggesting also an origin for the rim-syncline by deflation of a source bed due to the concentration of salt into the intrusion. The drag along the walls of the intrusion, the doming above, and the faulting of the domed strata are all

---

[1] O. Heermann, 'Der tektonische Nordrand des Hannoverschen Beckens', *Erdöl und Tektonik in Nordwestdeutschland*, Amt für Bodenforschung, Hannover-Celle, 1949, pp. 56–68. The book affords a great amount of interesting detail about salt domes.

reproduced in these experiments[1] (Fig. IX–23). Calculation suggests that salt will flow under overburden of the order of 1000 ft, and the elevation of active salt domes in Persia above the surrounding countryside is in agreement with the figure that may be calculated from the known densities and thicknesses of the local rocks.[2] The suggestion is made that too great a thickness of overburden may inhibit the further rise of a dome, but in southern Persia the salt, of Cambrian age, has risen through 20,000 ft of sediments. The implication may be that its rise has been conditioned at each period by progressive

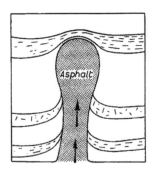

FIG. IX–23. EXPERIMENTAL ANALOGUE OF SALT PLUG, WITH ASPHALT RISING THROUGH MUD

(After Parker and McDowell)

sedimentation. Gansser has drawn attention to the influence of highly saline pressure-water in promoting the mobility of rock-salt in Persia.[3]

According to the hypothesis of uprise under hydrostatic forces, a slight updoming of the source bed is required to initiate the flow of salt towards the high point. For the Hanoverian salt-domes the probability is that such elevations were tectonically produced, since there is a clear geometrical pattern related to tectonic lines in the distribution of the salt intrusions.

O'Brien[4] has recently pointed out the possible influence of volcanic necks in localizing the salt intrusions of Southern Persia originally described by Harrison.[5] The Persian salt plugs carry up innumerable 'exotic blocks' of dolerite and trachyte, and O'Brien argues that the salt made its way up and

[1] L. L. Nettleton, 'History of Concepts of Gulf Coast Salt-Dome Formation', *Bull. Amer. Assoc. Petrol. Geol.*, Vol. 39, 1955, pp. 2373–83. T. J. Parker and A. N. McDowell, 'Model Studies of Salt-Dome Tectonics', ibid., pp. 2384–470, and earlier papers especially by M. B. Dobrin, 'Some Quantitative Experiments on a fluid salt-dome model and their geologic implications', *Trans. Amer. Geophys. Union*, Vol. 22, Pt 2, 1941, pp. 528–42 and L. L. Nettleton, 'Fluid Mechanics of Salt Domes', *Bull. Amer. Assoc. Petrol. Geol.*, Vol. 18, 1934, pp. 1175–204. See also M. T. Halbouty, *Salt Domes, Gulf Region, United States and Mexico*, Houston, Texas, 1967.

[2] G. M. Lees, 'Salt-Dome Depositional and Deformational Problems', Symposium on Salt Domes, *Journ. Inst. Petrol. Technol. Lond.*, Vol. 17, 1931.

[3] A. Gansser, 'Über Schlammvulkane und Salzdome', *Vierteljahrsschr. Naturf. Ges. Zürich*, Jhg. 105, 1960, pp. 1–46 (with excellent illustrations).

[4] C. A. E. O'Brien, 'Salt Diapirism in South Persia', *Geol. en Mijnb.* N.S. Nr. 9, Jhg. 19, 1957, pp. 357–76.

[5] J. V. Harrison, 'The Geology of Some Salt-Plugs in Laristan (Southern Persia)', *Quart. Journ. Geol. Soc. Lond.*, Vol. 86, 1930, pp. 463–552.

around volcanic diatremes or pipes. It is not claimed that the hypothesis is applicable to all salt intrusions, although it is true that many such are localized in volcanic regions.

In Roumania and Algiers, where salt intrusions occur in strongly folded rocks, it is certain that diastrophic forces are to a large extent responsible for their development and structure.[1]

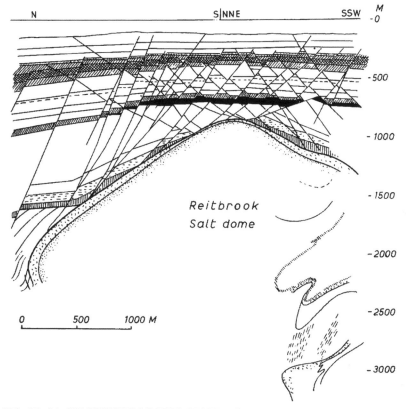

FIG. IX–24. STRETCHING OF BEDS ABOVE A SALT PLUG, WITH NUMEROUS NORMAL FAULTS

(*After Behrmann, in* Erdöl und Tektonik, *Hannover-Celle, 1949, p. 198, Fig. 5*)

The rocks of the dome over salt intrusions afford very fine examples of broad arching with associated tensional faults, as discussed for domes (p. 275) (Fig. IX – 24). In the salt itself the folds are characterized by features indicating strong plastic flow (rheomorphic folds). Balk's careful mapping has shown that the folds at Grand Saline, Texas, are very steeply plunging, as if a

[1] E. Fulda, 'Salztektonik', *Zeitschr. deutsch geol. Gesellsch.*, Abt. 79, 1927, pp. 178–96. A. Pustowka, 'Über rumänische Salztonaufbrüche', *Neues Jahrb. f. Min.*, etc., Beil.-Bd. 61, Abt. B, 1929, pp. 317–98. J. V. Harrison, 'Comments on Salt Domes', *Verh. Kon. Ned. Geol. Mijnb. Gen.*, Dl. XVI, 1956, pp. 1–8.

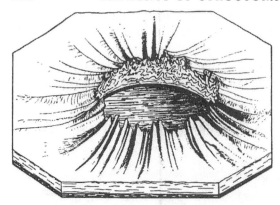

FIG.   IX–25.   FORM   OF
FOLDS IN SALT PLUG AT
GRAND SALINE
(*After Balk*)

sheet had been pushed through a small orifice (Fig. IX – 25).[1] This is well
reproduced in the experiment of Escher and Kuenen.[2]

Diapiric structures

The mobility of salt or gypseous clay and other geological materials
(including coal) may be such that in folded and faulted terrains these rocks
are squeezed, and actually injected into weak zones or potential spaces of all

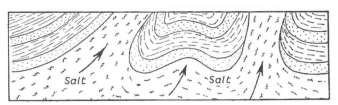

FIG. IX–26. DIAPIRIC FOLDS, SHOWING SALT DEPOSITS
INJECTED UPWARDS IN THE DIRECTION OF THE ARROWS,
AND CUTTING THROUGH ANTICLINES IN THE OVERLYING
BEDS

kinds. These effects are known as *diapiric*, and the phenomenon as diapirism.[3]
Salt domes are diapiric structures, and the use of the term is also extended to
include certain magmatic injections (see Chapter XII). In diapiric folds
(*Injektivfalten*) the mobile beds are injected chiefly through the cores of
anticlines (Fig. IX – 26), but where faulting is associated, the injections may
be highly irregular, especially with rock salt.[4] The 'active' extrusion of salt at
the surface forming salt glaciers or maintaining the elevation of salt plugs

[1] R. Balk, 'Structure of Grand Saline Salt Dome, Van Zandt County, Texas', *Bull. Amer.
Assoc. Petrol. Geol.*, Vol. 33, 1949, pp. 1791–1829.

[2] B. G. Escher and Ph. H. Kuenen, 'Experiments in Connexion with Salt Domes', *Leidsche
Geol. Med.*, Dl. 3, 1929, pp. 151–82.

[3] *Diapirism and Diapirs*, Symposium, *Amer. Assoc. Petrol Geol.*, Mem. No. 8, 1968.

[4] J. Bertraneu, 'Les Diapirs Triasiques du Bou Taleb Occidental (Algèrie)', *Geol. en Mijnb.*,
N.S. Nr. 9, Jhg. 19, 1957, pp. 377–82.

despite solution by rain, are indicative of contemporaneous upward movement, but, as explained above, this is not necessarily due to diastrophic forces.

Cross-folding and folded or superposed folds

When the plunge of folds is studied on a regional basis it is invariably found
that the long axes of folds are themselves bowed both upwards and downwards, this being the normal form of folds. It may, moreover, become
apparent that the culminations and depressions of the plunge are themselves
systematically arranged on lines that cross the fold axes so that two directions
of folding are represented, giving *cross-folds*. Such a structural pattern may
develop during the one folding episode (simultaneous cross-folding), or it may

FIG. IX–27. AXIAL PLANE
FOLDING
(*Based on Scotford*)

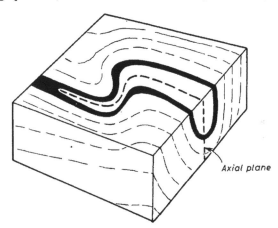

result from the folding of an already folded series (subsequent cross-folding).
In the latter case we have to deal with *folded folds* exhibiting double, or in the
general case multiple folding.[1] Folded folds need not, however, show crossing
of the axes as the later may be *coaxial* with the earlier, and indeed it is clear
from the many examples of the folding of early-formed folds seen in salt
domes that complex structures of this type may form simply as a later stage
(though not as a distinct episode) in the folding of highly mobile rocks. The
frontal involutions of nappes (Fig. VIII–48) also demonstrate this.

A special case of double-folding is afforded where the second folds are
chiefly in a horizontal plane, so that the axial planes of the first folds are bent
(Fig. IX–27). This has been referred to as *axial plane folding*.[2]

Simultaneous cross-folds – These may be formed firstly by irregularities
in the surface of the basement on which a folded series lies. This is a
'passive' influence of the basement on the folds, but the irregularities may

---

[1] The term *refolding* is used, but is not recommended as it implies some previous loss of fold
structures.

[2] D. M. Scotford, 'Metamorphism and Axial Plane Folding in the Poundridge Area, New
York', *Bull. Geol. Soc. Amer.*, Vol. 67, 1956, pp. 1155–98.

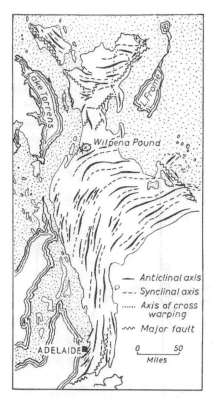

FIG. IX–28. FOLD TRENDS, FLINDERS RANGES, S. AUSTRALIA

(*After Sprigg*)

Legend within map:

— Anticlinal axis
--- Synclinal axis
..... Axis of cross warping
∿∿ Major fault

0 ___ 50
Miles

FIG. IX–29. WILPENA POUND, A DISSECTED BASIN IN QUARTZITE, SOUTH AUSTRALIA, AT THE INTERSECTION OF DIVERGING TRENDS

(*By courtesy Mines Department, South Australia*)

be determined for instance by ancient faults, so that there is a pattern in the buried surface. The effect has been suggested particularly for the alpine nappes, which moved forward over such a basement. The axes of cross-folds in nappes divide them into *tectonic segments* of alternate synclinal and anti-clinal type along the length of the nappe.[1] Secondly, 'active' faulting or warping in the basement, accompanying and perhaps causing the folds in the overlying beds, may give rise to cross-folds, or again these may arise from secondary stresses in a thick geosynclinal sequence, such that local domes

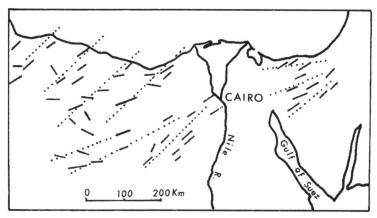

FIG. IX–30. LONG AXES OF DOMES IN NORTHERN EGYPT

Dotted lines indicate regional trend of Cretaceous and early Tertiary deformation. Note domes arranged *en echelon*. (*After R. Said*, Geology of Egypt, *Amsterdam, 1962*.)

and basins develop in them. These latter are commonly formed, but a regular arrangement of cross-fold axes is not always to be seen (see Fig. IX–14). Cross-folding in the North Flinders Ranges, South Australia, is due to the intersection of trends at the junction of two divergent fold belts (Fig. IX–28). It gives rise to the striking topographic feature of the Wilpena Pound formed in the basined Pound Quartzite (Fig. IX–29).

Where a series of brachy-anticlines or domes lies *en echelon* along a major tectonic line, the crests of the domes are oblique to the major trend and appear as cross-folds along it. This is an example of second order tectonics due very probably to primary fault movements on the major trend, and is well shown in northern Egypt and Palestine[2] (Fig. IX–30).

[1] E. Argand, 'Sur la Segmentation Tectonique des Alpes Occidentales', *Bull. Soc. Vaudoise Sci. Nat.*, Vol. 48, Ser. 5, 1912, pp. 345–56. 'Cross-folding', by N. Rast and J. I. Platt, *Geol. Mag.*, Vol. 94, 1957, pp. 159–67; R. Staub, Alpine Orogenese, 'Grundsätzliches zur Anordnung und Entstehung der Kettengebirge', in *Skizzen zum Antlitz der Erde*, Vienna, 1953, pp. 1–46 (*Querfalten*, p. 31).

[2] L. Picard, *Structure and Evolution of Palestine*, Jerusalem, 1943. A. Shata, 'New Light on the Structural Developments of the Western Desert of Egypt', *Bull. Inst. Désert Egypte*, Vol. 3, No. 1, 1953, pp. 101–106.

Subsequent cross-folds – Subsequent cross-folds result in superposed folds, and these may be formed either by the direct influence of folding forces on an already folded series or by the influence of basement faulting and warping on overlying folds. Understanding of the geometry of superposed folds is aided

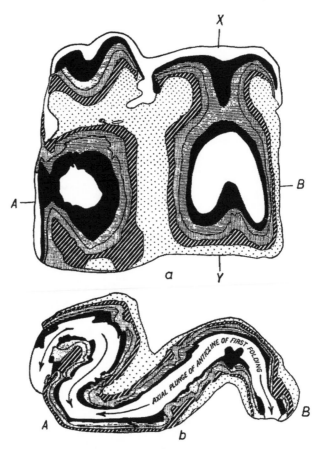

FIG. IX–31A. FOLDED FOLDS

Plasticine model, sliced off to simulate erosion. *A–B*, axial direction of first folding; *X–Y*, axial direction of cross-folding. (*a*) shows surface outcrops after slicing (note the 'prongs'); (*b*), section, before slicing, along *A–B* of (*a*).
(*After Holmes and Reynolds*)

by the construction of graphical or real models for idealized types with which actual examples may be compared. Reynolds and Holmes in their study of cross-folding in County Donegal, Ireland,[1] made a model in plasticine (Fig. IX–31) of the case in which original recumbent folds are folded by a second

[1] Doris L. Reynolds and A. Holmes, 'The Superimposition of Caledonoid Folds on an Older Fold-System in the Dalradians of Malin Head, Co. Donegal', *Geol. Mag.*, Vol. 151, 1954, pp. 417–44. Reynolds and Holmes point out that the 'down-the-plunge' or 'down-structure' method of map study is inapplicable in structures of this type.

event whose fold axes are at right angles to those of the first folds. It will be seen that the outcrop pattern has special peculiarities, in this case with bilaterally symmetrical closures having a mushroom or tuning-fork shape.

O'Driscoll[1] has used models made from thin card laminae to represent the effects of subsequent shear-folding on earlier folds, Ramsay[2] constructed block diagrams, and Weiss[3] has used the stereographic projection. The resulting interference patterns of folds are complex even for idealized types but

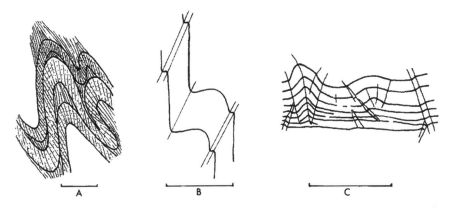

A   B   C

FIG. IX–31B. FOLDS OF SUCCESSIVE PHASES, CO. MAYO, EIRE
A. Fold style of D_1 deformation. B. D_2-fold style. C. D_5 folds (D_3 folds are major folds; D_4 produced strain-slip and crenulation cleavages). The scales represent 6 in. (*After Dewey, 1967*)

Carey[4] has shown how some shear folding may be unravelled from outcrop patterns, while Ramsay and Tobisch[5] illustrated how the shapes of minor folds may be modified by internal movement during subsequent folding.

Phases of deformation and folding – Major superposed folding is one aspect only of the general subject of repeated deformations affecting a folded zone, whether due to entirely separate orogenic cycles, to successive stages in one cycle, or to local constraints which result in cross folding in one fold event. Detailed study of areas with complex folds and related structures has revealed that in many instances a sequence of phases or stages of deformation can be identified. As many as five such phases, excluding faulting, have been recognized, as for instance in the lower Palaeozoic rocks of County Mayo,

[1] E. S. O'Driscoll, 'Cross Fold Deformation by Simple Shear', *Econ. Geol.*, Vol. 59, 1964, pp. 1061–93.

[2] J. G. Ramsay, 'Interference Patterns Produced by the Super-Position of Folds of Similar Type', *Journ. Geol.*, Vol. 70, 1962, pp. 466–81.

[3] L. E. Weiss, 'Geometry of Superposed Folding', *Bull. Geol. Soc. Amer.*, Vol. 70, 1959, pp. 91–106.

[4] S. W. Carey, 'Folding', *Journ. Alberta Soc. Petrol. Geol.*, Vol. 10, 1962, pp. 95–144.

[5] O. T. Tobisch, 'Large-Scale Basin-and-Dome Pattern Resulting from the Interference of Major Folds', *Bull. Geol. Soc. Amer.*, Vol. 77, 1966, pp. 393–408.

Eire,[1] four in the Proterozoic of central Europe,[2] and at least three, possibly even five, in the Northern Highlands of Scotland.[3] Not all deformation phases are of equal importance in modifying earlier structures, and in general, the *style* of successive phases is different, as may be seen in the amplitude of folds of different generations, the genesis of folds by different mechanisms, the

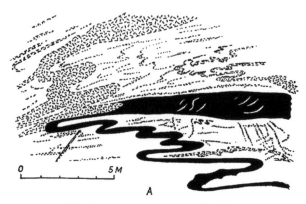

FIG. IX–32A. PTYGMATIC FOLDS
Folded basic dyke originally straight. (*After Sederholm, from Wilson*, Geol. Mag., *Vol. 89, 1952*)

flexing or fracturing of cleavage or schistosity, and the like; but if a subsequent phase involves strong metamorphism, only relics of earlier structures may remain and these are difficult to identify.

In deciphering the deformation history of a region it will be necessary to identify folds, cleavage, and lineations formed at various stages, and it is convenient to use a labelling system such as D_1, D_2, D_3, etc. (or F_1, F_2, F_3, etc.) for deformation (or folding) stages, S_1, S_2, etc., for cleavage or foliation developed during D_1 (F_1), or D_2 (F_2) stages, L_1, L_2, etc., for lineations, and SS for bedding.[4] B_1, B_2, etc., are generally used for fold axes of successive generations, and stages of recrystallization may also need to be identified. Examples of fold styles of successive stages are shown in Fig. IX–31B.

Correlation of stages of deformation between different parts of an orogenic belt, based on intensity and style of deformation, is hazardous because the

---

[1] J. F. Dewey, 'The Structural and Metamorphic History of the Lower Palaeozoic Rocks of Central Murrisk, County Mayo, Eire', *Quart. Journ. Geol. Soc. Lond.*, Vol. 123, 1967, pp. 125–55.

[2] J. Holubec, 'Structural Development of the Geosynclinal Proterozoic and its Relations to the Deeper Zones of the Earth's Crust', *Českoslov. Akad. Věd., Mat. a. Přiod. Věd.*, Vol. 78, Pt 8, 1968, pp. 1–77.

[3] See review in E. H. T. Whitten, *Structural Geology of Folded Rocks*, Chicago, 1966, Chapter 11.

[4] Schemes of labelling such as that used by Holubec (op. cit. supra) in which the subscripts do not correspond in numbering with the deformation stages are likely to lead to confusion, and can be avoided by a combination of sub- and superscript numerals.

attributes of any one phase may change considerably along the strike and also across the belt as well as in depth.

Ptygmatic folds — The curious meandrine contortions of quartzo-felspathic veins typically seen in gneisses especially in injection complexes and transfused rocks are known as ptygmatic structures or ptygmatic folds [1] (Fig. IX–32).

Structures of very similar geometry are found in other rocks, including slumped sediments, quartz veinlets in slate, enterolithic structures in gypsum and clay, rheomorphic folds in salt and folds in tectonically deformed rocks including softened and folded basalt dykes in gneisses, but a notable feature of ptygmatic folds in gneisses is that two or three veinlets crossing each other may exhibit such folding along different axes, or an apparently unfolded vein may be crossed by a folded vein. Furthermore, the quartz and felspar crystals are unstressed. It is clear that examples of ptygmatically folded basalt dykes are due to post-crystallization softening, and plastic flow by flattening of the host rock and the dyke, and experiments simulating such conditions reproduce ptygmatic folds reasonably well.[2] In such instances the ptygmatic folds should show at least a general correspondence in pattern with the shearing and flow of the host rock.

FIG. IX–32B.
PTYGMATIC FOLDS
Stages in formation of ptygmatic folds in aplites, possibly due to the loss of volume during transfusion and granitization. (*After Bradley, 1957*)

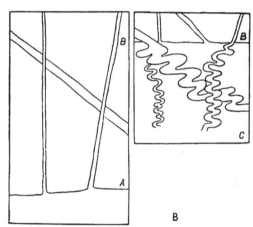

On the other hand, Wilson adduces clear evidence in several cases that only the rock adjacent to a ptygmatic vein yielded along with the vein substance as this folded, so that bulk compression is ruled out. He suggests an origin before the crystallization of the vein, by forceful push of the injected vein substance against an obstacle. Bradley[3] connects the folding with loss of

[1] An excellent review of the literature with a new theory of the formation of ptygmatic folds is given by G. Wilson, 'Ptygmatic Structures', *Geol. Mag.*, Vol. 89, 1952, pp. 1–21.

[2] Ph. H. Kuenen, 'Observations and experiments on ptygmatic folding', *C.R. Soc. Géol. Finlande*, No. 12, 1938.

[3] J. Bradley, 'Geology of the West Coast of Tasmania, Part III, Porphyroid Metasomatism', *Proc. Roy. Soc. Tas.*, Vol. 91, 1957, pp. 163–90.

volume by the host rocks during transfusion, but in fact no hypothesis is adequate to account for all the features of folded quartzo-felspathic veinlets.

Jointing in folded rocks

The jointing of folded rocks is subject to many local variations, but a broad geometrical classification into three types is often possible for joints that appear to be genetically related to the folds.

1. *Longitudinal joints* are normal or nearly normal to the bedding, and intersect it along lines parallel or nearly so with the longitudinal fold axis. In transverse section these joints are radially arranged in the fold (Fig. IX–33) and they are often ascribed to tension consequent on the bending of the beds. In general, however, it appears that such joints form at a late stage in the folding, and, again they are present equally on straight fold limbs as on the

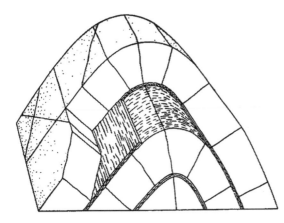

FIG. IX–33. TYPES OF JOINT-ING IN A SIMPLE FOLD
(*After de Waard, 1955*)

curved sectors of folds. Accordingly, it appears that they may be a result of stress-relaxation, or, in folds with straight limbs, they may represent complementary shear directions or Griffith fractures, operative after the fold form was achieved.

2. *Transverse*, or *cross joints* are usually clean-cut fractures that traverse several beds with little or no deviation in direction. They are approximately normal to the longitudinal axis of folds, and thus may serve as a rough indication of the plunge. Cross joints clearly have some connection with the deformation of rocks along the strike of a folded belt. In cleaved rocks they may traverse the cleavage which was formed during plastic flow, and in such instances, they may be ascribed to stress-relaxation, or to the longitudinal extension of the folded beds. More rarely kinking or shearing may be associated with cross-joints.

3. *Diagonal joints* are of two main types, viz., those that are restricted within each bed, and those that cut across the whole folded structure, being diagonal

master-joints. These latter may be assumed to have some genetic connection with the folding where this appears to be caused by horizontal shearing movements. The restricted type of diagonal jointing is little understood. It is best seen in interbedded harder and softer beds such as sandstones and mudstones, or coal and shale, but appears to be much less prominent below the zone of weathering, suggesting that differential expansion and contraction due to weathering processes, or unloading by erosion, may account for certain examples. This may apply particularly where the beds are polygonally jointed, with two or three sets of oblique joints normal to the bedding, these having different angular relationships in adjoining beds (Fig. VI−13). Master-joints that cut across several beds in folds are not necessarily unrelated to the folding processes. Shear joints may be related to tear faults of synchronous origin, and in the final stages of any folding, the continued application of stress may be expected to yield both shear and tension joints, as well as faults.[1]

Disharmonic folding

The notion expressed by the term *disharmonic folding* is that the folds of one bed or group are unrelated to, or out of harmony with those of an adjoining bed or beds. Some variation in the form of folds in successive beds is normally to be expected, and the idea of disharmony must not be pressed too far lest essential relationships be overlooked. The section of the Persian oilfields shown in Fig. IX−34 exhibits for instance many examples of correspondence between surface structure and that in depth, due to the influence of the highly mobile Lower Fars Stage 1, but the chief effect is a strong lateral shift of the folds. The essential feature of disharmonic folding is that one formation exhibits more numerous and smaller folds than the more competent beds enclosing it, as shown in Fig. IX−35. On a grand scale, a *décollement* such as that of the Juras, over a rigid basement, is an example of what Bertrand has called *dysharmonisme en grande* rather than of disharmonic folding in the conventional sense of this term. Disharmonic relationships may also be shown in faulting affecting adjacent formations (Fig. VII−22).

Disharmonic folding is due essentially to the interbedding of competent and incompetent beds or formations, which yield differently to stress and among which secondary stresses may be generated both by the competent beds squeezing the incompetent as in fold cores, and by the mass translation of incompetent beds which must then push against the competent. Indeed it is likely that many small disharmonic folds in incompetent beds are due rather

[1] For examples of joint studies in folded rocks, see D. de Waard, 'The Inventory of Minor Structures in a Simple Fold', *Geol. en Mijnb.* N.S., Jhg. 17, 1955, pp. 1–11; H. Cloos, *Einführung in die Geologie*, Berlin, 1936, pp. 272–9; W. K. Nabholz, 'Untersuchungen über Faltung und Klüftung im nordschweizerischen Jura', *Eclog. Geol. Helv.*, Vol. 49, 1956, pp. 373–406. A. Williams, 'A structural History of the Girvan District S.W. Ayrshire', *Trans. Roy. Soc. Edinb.*, Vol. 63, Pt 3, 1958–9, pp. 629–67.

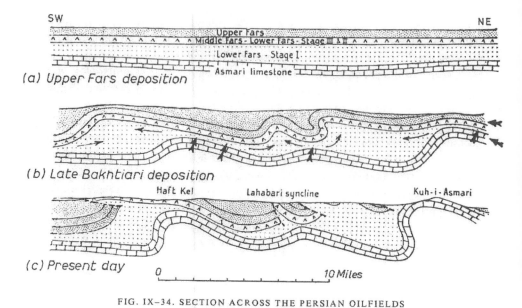

SW   NE

Upper Fars
Middle Fars - Lower Fars - Stage III & II
Lower Fars - Stage I
Asmari limestone

(a) Upper Fars deposition

(b) Late Bakhtiari deposition

Haft Kel   Lahabari syncline   Kuh-i-Asmari

(c) Present day 0   10 Miles

FIG. IX–34. SECTION ACROSS THE PERSIAN OILFIELDS
(*After O'Brien,* Geol. en Mijnb., *1957*)

FIG. IX–35. DISHARMONIC FOLDING IN THE MT ISA, QUEENSLAND, LEAD
ORE-BODY

(*After R. Blanchard and G. Hall,* Proc. Aust. Inst. Min. & Metall., *No. 125, 1942.*) Length of
view, 8 ft.

to the resistance offered by adjoining competent rocks than to the active push or drag of these competent beds on the more mobile rocks. Disharmonic relations are common in the vicinity of the hinges of angular folds, where they are caused by local stresses associated with the hinge or core effects discussed above (p. 229).

'Drag-folds' – Although widely used by 'hard-rock' geologists, the term *drag-fold* in fact lacks precision, and is commonly applied to several different types of structures. The first is where an anticline and syncline of small amplitude lie immediately adjacent to each other in the limb of a larger fold (Fig. IX–36A); the second usage is for small folds in incompetent beds disharmonically related to adjacent competent beds in fold limbs (Fig. IX–36B); a third

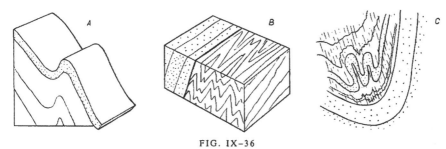

FIG. IX–36

A. An isolated drag-fold in a competent bed on a fold limb.
B. Incompetent beds showing numerous minor folds (disharmonic folds) (section near Bacchus Marsh, Victoria).
C. Crumpling in fold core (section at Chewton, Victoria)

is for examples in the cores of folds (Fig. IX–36C), and again, shear flexures and flexures due to 'drag' along a fault have been termed drag-folds, although this latter usage is not widely accepted. The concept common to all examples in folded rocks is that more competent beds transmit shearing stress to interbedded less competent beds, as a result of bedding-slip. If, however, the incompetent beds have numerous small folds throughout them this mechanism is in fact an unlikely one, especially when such thin beds are enclosed among incompetent beds which can accommodate to differences in fold shape by flowage.[1] For these reasons it has been thought preferable to refer to small folds on the limbs of larger, as parasitic folds rather than as drag-folds. The geometrical similitude of small folds to larger, expressed as Pumpelly's Rule that 'the degree and direction of the pitch of a fold are often indicated by those of the axes of the minor plications on its sides',[2] in fact suggests that such

[1] H. Ramberg, 'Fluid Dynamics of Viscous Buckling Applicable to Folding of Layered Rocks', *Bull. Amer. Assoc. Petrol. Geol.*, Vol. 47, 1963, pp. 484–505; idem, 'Evolution of Drag Folds', *Geol. Mag.*, Vol. 100, 1963, pp. 97–106; S. K. Ghosh, 'Experimental Tests of Buckling Folds in Relation to Strain Ellipsoid in Simple Shear Deformations', *Tectonophysics*, Vol. 3, 1966, pp. 169–85.
[2] R. Pumpelly, J. E. Wolff, and T. N. Dale, 'Geology of the Green Mountains in Massachusetts', *U.S. Geol. Surv.*, Mon. No. 23, 1894, p. 158.

small folds are often caused by direct tectonic deformation rather than by secondary effects due to drag or constraint between rock masses moving relative to each other (*tectonique d'entrainement*).

Pumpelly's Rule is indeed basic to the use of minor folds in the construction of π- and β-diagrams, and it is accordingly important to note that many minor folds deviate strongly from the regional tectonic pattern and have been termed *incongruous* (or *independent*). Van Hise and Leith noted that in the Vermillion district the axes of 'drag-folds' lie in any direction in the plane of bedding, and in the Silurian rocks of Victoria incongruous 'drag-folds' plunge at angles up to 90° (Fig. IX−37), in a region [1] where the general plunge is

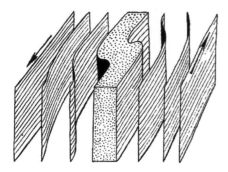

FIG. IX−37. INCONGRUOUS DRAG-FOLD IN A SANDSTONE BED AMONG SHALES

FIG. IX−38. DRAG-FOLDS AND FACING
A. In normally dipping beds.
B. In overturned beds.
The major anticlinial axis lies to the left in each case.

A

B

about 15°. This is doubtless due to horizontal shearing along already vertically dipping beds in these strongly folded rocks.

To summarize, it may be said that the term drag-fold is best applied to folds such as are shown in Fig. IX−36A, which presumably result from the general shearing component developed in regional folding, rather than to the numerous minor folds of incompetent beds, and certainly not to the minor plications in schists and slates, or to examples due to squeezing in fold cores.

Congruous drag-folds may be used to determine the facing in highly inclined strata, as their transverse axes are approximately parallel to that of the major fold (Fig. IX−38).

[1] C. R. van Hise and C. K. Leith, 'The Geology of the Lake Superior Region', *U.S. Geol. Surv.*, Mon. No. 52, 1911, p. 123. D. R. Derry, 'Some Examples of Detailed Structure in Early Pre-Cambrian Rocks of Canada', *Quart. Journ. Geol. Soc. Lond.*, Vol. 95, 1939, pp. 109–34. E. S. Hills, 'Silurian Rocks of the Studley Park District', *Proc. Roy. Soc. Vict.*, Vol. 53, 1941, pp. 167–91.

CHAPTER X

Cleavage

Rocks that possess *cleavage* may be split into thin sheets along parallel or sub-parallel *cleavage planes* that are of secondary origin and are formed by the combined effects of metamorphism and deformation.[1] In any one bed, cleavage is *independent of the bedding*, which it may intersect at any angle, but because it is influenced by the physical properties of rocks it may change its attitude in passing from one bed to another (Fig. X–1).

Cleavage planes are surfaces of low cohesion including: (i) mechanically induced planes such as shear planes, kink planes, or planes of flattening of mineral grains, oöids or other particles in plastically deformed rocks; (ii) the plane, other than bedding, to which flaky or elongate mineral grains formed by recrystallization tend to lie parallel or nearly parallel. The grade of metamorphism required to yield slaty cleavage is low, and accordingly many original sedimentary grains of quartz and felspar or flakes of mica are also preserved. Nevertheless, there are always present in slates minerals such as graphite, sericite, chlorite and others formed by metamorphism. Where cleavage is developed in limestones, recrystallization of calcium carbonate, as well as plastic deformation of the original grains of calcite is found.

Phyllites and schists exhibit many structures that are closely analogous with slaty cleavage, and indeed the chief distinguishing features of these rock types as compared with slates is their more complete recrystallization and consequent larger size of the component mineral grains. This permits the ready recognition of the parallel orientation of newly formed flaky minerals, which occurs along the original bedding as well as in or near the direction of the cleavage, whereas in slates this is not so obvious. Thus in phyllites and schists it is usual to distinguish *bedding fissility*, *foliation*, or *schistosity* as well as the foliation *transverse* to the bedding, which latter is analogous with certain types of cleavage.[2]

'Flow' cleavage and 'fracture' cleavage – It has been usual, following C. K. Leith, to distinguish two chief types of cleavage, viz., *flow cleavage* and

---

[1] According to this definition, very closely spaced joints are not regarded as cleavage since they do not involve metamorphism.

[2] For further discussion of schistosity, see Chapter XIII.

FIG. X–1. REFRACTION OF CLEAVAGE IN SLATE WITH
INTERBEDDED SANDSTONE LAMINAE, BENDIGO,
VICTORIA. ×3.

fracture cleavage.[1] Broadly regarded, the distinction depends on the extent to which recrystallization pervades the whole rock giving '*flow cleavage*', as compared with the mechanical effects of shearing and slipping along individually recognizable discrete surfaces which are prominent in '*fracture cleavage*'. Where recrystallization is dominant, the mineral orientation largely determines the cleavage direction and the rock may be split along any chosen plane, whereas when individual shear, kink or flexure planes are present the rock splits along these. The distinction is, however, not a rigid one, for both effects are combined in the plastic deformation that gives rise to cleavage, and in slates that possess excellent cleavage, individual microscopic planes of shearing or flexuring may generally be identified, although these are so closely

[1] C. K. Leith, 'Rock Cleavage', *U.S. Geol. Surv.*, Bull. 239, 1905.

spaced that the rock was clearly deformed by flowage, and may now be split along any chosen macroscopic surface parallel with the cleavage (Fig. X–2). Such cleavage may itself be deformed by subsequent micro-shearing and flexing, giving rise to what is often called 'fracture cleavage', but this second cleavage generally itself involves minute flexing of the original cleavage and is indeed rarely an effect of fracture but rather of limited plastic flow.[1] Accordingly the use of the terms 'flow' and 'fracture' cleavage is not recommended.[2]

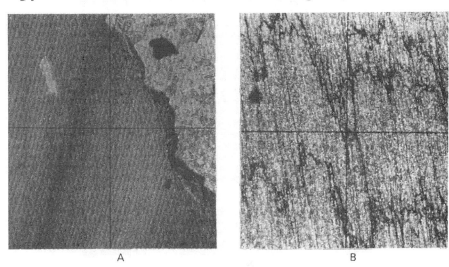

A   B

FIG. X–2. SLATY CLEAVAGE

A. Cleavage in slate at contact with felspathic sandstone, Bendigo, Victoria. ×30.
B. Cleavage in slate with fine carbonaceous laminae, showing microscopic shear planes, Bendigo, Victoria. ×30.

Fracturing gives rise to jointing rather than to cleavage, but in some rocks close-spaced fractures divide the rock into irregular thin lenticles or sheets and may be termed *close-joints cleavage*. Fiedler[3] gives an excellent account of such cleavage in limestones. The discussion which follows applies to true or slaty cleavage.

Mechanical effects in cleavage-formation

Two mechanical effects of deformation most generally associated with cleavage in slates are, on the one hand, micro-shearing or shear-flexuring along

[1] J. L. Talbot, 'Crenulation Cleavage in the Hunsrückschiefer of the Middle Moselle Region', *Geol. Rundsch.*, Vol. 54, 1964, pp. 1026–43.
[2] The use of the term flow cleavage for bedding schistosity is also not recommended since this is not, strictly speaking, cleavage. The term *false cleavage*, being of no special significance, is now outmoded.
[3] K. Fiedler, 'Strukturgeologische Untersuchungen zur Querplattung (Sigmoidalklüftung) in den Oberkreide-Kalken des Osnings', *Mitt. Geol. Staatsinst. Hamburg*, Hft. 34, 1965, pp. 5–125.

FIG. X–3. ROTATION OF PORPHYROBLASTS

A. Rotation of calcite porphyroblasts, with sigmoidal lines of
   residual inclusions, Castlemaine, Victoria. ×30.
B. Syntectonic garnet porphyroblast, showing rotation during
   growth, with sigmoidally aligned inclusions of quartz. Schichallion complex. ×15. (*Photo: N. Rast*)

individual cleavage planes or in narrow zones, and, on the other, elongation of
the elemental parts of the rock in the general direction of the cleavage dip,
combined with shortening at right angles to the cleavage. Micro-shearing is
revealed by displacements of the bedding in laminated slates (Figs. X–2, 12),

or by the rotation of porphyroblasts (Fig. X–3), especially where these contain lines of inclusions. In this case, the growth of a porphyroblast and its simultaneous rotation cause the lines of inclusions to assume a sigmoidal curve. This is most pronounced in *snowball garnets* in schists, which have rotated sufficiently to produce a spiral arrangement[1] (Fig. X–3B, 4).

Effects such as these imply differential plastic flowage which, while combining many intimate displacements in a variety of directions (gliding, twinning, and rotation for individual crystals, shearing on complementary planes for the whole rock) averages these movements to a dominant direction determined by the local strain-distribution, which itself is determined by the forces and constraints locally affecting the rock (Figs. X–5, 10 give a visual impression). Differential plastic flow along one prominent direction, although necessarily involving also microscopic displacements in other directions,[2] will be

FIG. X–4. 'SNOWBALL GARNET' IN
PHYLLITE, ALP PIORA

(*After Mügge*, Neues Jahrb. f. Min.,
Vol. 61, B.B., Abt. A, 1930)

referred to as unilateral shearing. Where two shearing directions are equally developed, the rock is elongated parallel to one bisectrix of the shears, and shortened parallel to the other.

Elongation parallel to the cleavage and shortening normal to it are revealed by pressure fringes around resistant porphyroblasts, by many described examples of distorted fossils, and by deformed oöids in oölitic limestone such as have been discussed in Chapter V. These effects are known as 'flattening', which occurs in the AB plane of a strain ellipsoid. The connection between cleavage which is a unilateral shear direction and that which lies in the plane of flattening is not simple, as it involves both the primary formation of cleavage, and the effects of secondary movements on a pre-existing cleavage.

Where flattening occurs, it is clear that we may expect effects due both to the flattening itself, and to the presence of the complementary shears (S′ and S″), which must also be present. This is demonstrated in sandstones interbedded with slates, in which the shear directions may be seen on suitably weathered surfaces, while the fold geometry demonstrates the AB plane. The

[1] N. Rast, 'Metamorphic History of the Schichallion Complex (Perthshire)', *Trans. Roy. Soc. Edinb.*, Vol. 63, 1958, pp. 413–31.
[2] These are revealed by petrofabric analysis, Chapter XIII.

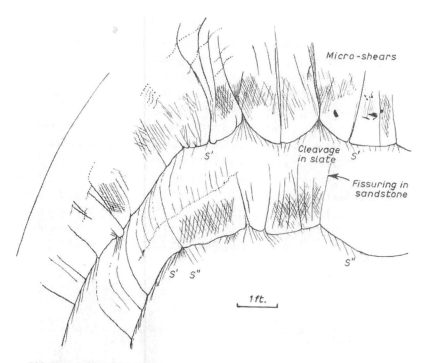

FIG. X-5. DETAILS OF THE NAPOLEON ANTICLINE, BENDIGO, VICTORIA
S', S" micro-shear planes pervading sandstone; 'fissuring' planes (coarse cleavage) in sandstones, and slaty cleavage in slates.

FIG. X-6. DETAIL OF 'FISSURING' IN SANDSTONE AT
NAPOLEON ANTICLINE
(*Photo by courtesy Geological Survey of Victoria*)

details are shown in Fig. X−5. The sandstones are intimately pervaded by complementary microshear planes so that the rock is uniformly cleavable, but in addition it is sliced by prominent bands of strong cleavage-development, which weather out to form open fissures. Such rocks have been termed 'fissured sandstones'[1] (Fig. X−6). Near the fold hinges, fissuring develops on either S′ or S″ or some intermediate plane, but in the limbs it more often approximates to one of these shear directions, although there is considerable variability due to inhomogeneity of the rocks and consequently in the stress-distribution. Fissuring, as contrasted with the intimate shearing of the rock, represents strong localized shearing along certain zones, in response to the bending of the bed, and most probably develops at a late stage, post-dating the

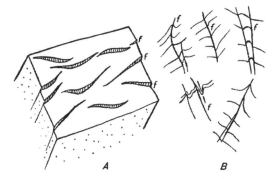

FIG. X−7. MICRO-FAULT DISPLACEMENTS ON FISSURE-PLANES

A. Seen on exposed bedding plane in sandstone, Castlemaine.
B. Various types of displacements of stratification on fissure planes in slate with sandy laminae, seen in section at Chewton, Victoria.

first intimate shearing. The final movements on fissures in fold limbs are shears with a displacement of an inch or so, which have the effect of increasing the dip (Fig. X−7A) in the limbs of the fold.

While these movements proceed in the sandstones, the interbedded slates develop cleavage and related structures. In the fold illustrated in Fig. X−5, the cleavage dips outwards in each limb, but exhibits 'fanning' at kinks and bulges in the overlying sandstone near the hinge, the cleavage-dip ranging between two directions which correspond broadly with S′ and S″. The geometry suggests that S′ is represented in the left-hand limb, and S″ in the right-hand limb of the fold, but cleavage lies in any position between these, in the kinks and bulges of the fold. Cleavage varies from a complementary shear direction to the median plane of flattening between S′ and S″, with unilateral shearing effects prominent in the fold limbs.

[1] E. J. Dunn, 'The Bendigo Goldfield', *Mines Dept., Vict.*, Special Repts., Nos 1 and 2, 1896.

Regional development of cleavage parallel to the median plane of flattening is best known from E. Cloos' study of the deformation of oölitic limestones in Maryland.[1] Microscopical study of orientated thin sections shows that the

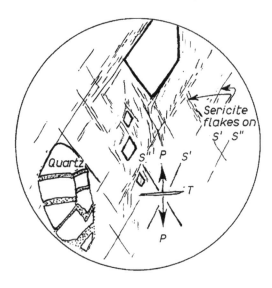

FIG. X–8. CLEAVAGE, SHEARING AND TENSION CRACKS IN SCHISTOSE QUARTZ PORPHYRY, LLANBERIS, WALES

S', S" complementary shears, PP plane of flattening and cleavage, T tension gashes.

(*After A. Harker*, Metamorphism, *Fig. 79A, and slide 6146SM, Cambridge*)

long axes of the deformed öoids, which approximate to triaxial ellipsoids, lie parallel with the dip of cleavage observed in the field. Many additional examples from various rock-types are figured by Harker,[2] especially from sheared quartzites and quartz-prophyries, in which the equal development of

FIG. X–9. CLEAVAGE AVERAGING IN THE PLANE OF FLATTENING PP, BUT RANGING FROM S' TO S"

Slate with sandy laminae, Chewton, Victoria.

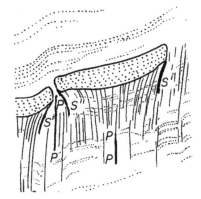

S' and S" is readily seen, sometimes with tension gashes due to stretching in the direction of the A deformation axis (Fig. X–8). In the experimental deformation of marble it has been shown that the grains of calcite become

[1] E. Cloos, 'Oölite Deformation in the South Mountain Fold, Maryland', *Bull. Geol. Soc. Amer.*, Vol. 58, 1947, pp. 843–917.
[2] A. Harker, *Metamorphism*, London, 3rd edn, 1952, Figs 73, 79, etc.

flattened normal to the axis of maximum compressive stress, or parallel to the axis of elongation in a tensile test.[1]

Thus both observation and experiment indicate that cleavage may form in the plane of flattening normal to the maximum compressive stress. It may, however, also deviate from this plane towards the complementary shear directions according as one or other of these is the more dominant (Fig. X–9). Nevertheless as we are dealing with rocks rather than fluids, the second, subordinate shear direction should also be present. Under the microscope, a slate is seen to be divided into lozenge-shaped sectors, due to the inter-action of the two shear directions (Fig. X–10). The angle between these

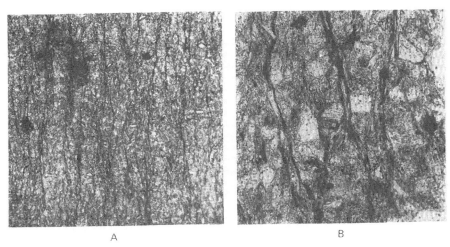

A   B

FIG. X–10. SHEARING IN SLATE
A. In clay–slate, with bedding shown by horizontal mica flakes.
B. In sandstone (with quartz and felspar).

is always acute about the vertical or near-vertical deformation axis, ranging from about 25° to 30° in fine-grained slates up to some 40° in sandstones.[2]
Rotation in fold limbs – We have seen that near the fold-hinges the coarse cleavage known as fissuring in sandstones lies in or between S′ and S″, with the AB deformation plane parallel to the transverse axis of the fold. In the fold limbs, however, fissuring remains nearly normal to the bedding and accordingly dips at low angles in steeply dipping limbs (Fig. X–11). As this occurs even where the limbs are relatively straight, it cannot be regarded as a local effect of compression in the fold core, but appears rather to be due to rotation of the fissuring along with the beds in the fold. Shearing movements do not, however, cease during rotation of the bed, and in the final stages of folding the displacements of an inch or so, mentioned above, take place.

[1] F. J. Turner *et al.*; see Chapter V, p. 111.
[2] It is probable that the angle is somewhat reduced by rotation of early-formed shear planes during the process of cleavage development, especially in fine-grained highly plastic rocks.

These effects are best seen in thicker sandstones which acted as relatively competent beds during folding. The results of continued folding on fine-grained slates interbedded with sandstones are different. In these, cleavage retains its original attitude, but marked unilateral shearing occurs especially

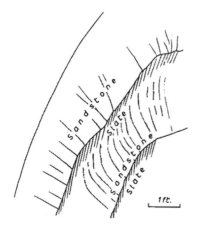

FIG. X–11. FISSURING IN STEEPLY DIPPING FOLD LIMBS, WITH VARIABLE AND EXCESSIVE REFRACTION OF CLEAVAGE FROM SLATES TO SANDSTONES

in the fold limbs. As in fissuring, the whole rock is pervaded by micro-cleavage, but in addition it is divided by narrow zones, generally dark coloured, of stronger movement. If the displaced slices are conceived to be returned to their original positions so that sheared parts of beds are in contact

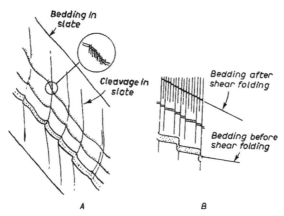

FIG. X–12. VARIATION IN SPACING OF MICRO-SHEARS IN SLATE AND SANDSTONE, CHEWTON, VICTORIA

with each other, it becomes clear that the folding of such slates is predomin-antly shear-folding, and that the dip is largely the result of the displacement of slices of the slate along shears parallel with the cleavage (Figs. VIII–30, 31). Accordingly, the initiation of cleavage must have occurred at an early stage in the formation of the fold. The parallelism of the slaty cleavage from the fold hinge to the limbs shows clearly that its attitude (dip) was little affected by rotation of the limbs, but rather that the dip of the beds is increased by sliding on the cleavage. Drag between the slates and the enclosing sandstones causes the ends of cleavage planes adjacent to the sandstone to be bent over towards anticlinal hinges, producing a sigmoidal curve in the cleavage.

Shear-flexuring associated with the displacements on the cleavage is directly related to the mineral composition and thickness of the laminae involved. In fine-grained slate at the top of a graded bed the spacing of shear planes is so close that an appearance of uniform dip is presented by sandy

FIG. X–13. PUCKERING OF SANDSTONE LAMINAE AT FOLD AXIS, CASTLEMAINE, VICTORIA

laminae one or two sand grains thick, but the distance apart of the shear zones increases in thicker sandstone layers, and the segments of these, between shear zones, preserve a dip that indicates approximately their attitude before macro-shearing commenced [1] (Fig. X–12).

As the hinge of a fold is approached the shearing displacements become irregular, some being right-handed throws and some left-handed, and, combined with the effects of flexing along the shears, the result is complex puckering at the fold axis, which is useful in slate belts as an indication of the presence of a fold which might otherwise be overlooked (Fig. X–13).

[1] E. S. Hills and D. E. Thomas, 'Deformation of Graptolites and Sandstones in Slates from Victoria, Australia', *Geol. Mag.*, Vol. 81, 1944, pp. 216–22. E. S. Hills, 'Examples of the Interpretation of Folding', *Journ. Geol.*, Vol. 53, 1945, pp. 47–57.

Refraction of cleavage – Where cleavage is developed in beds of different physical properties in the limbs of folds its attitude changes and it is said to be *refracted* in passing from one bed to another. Refraction does not occur at the hinges of folds, and the amount of refraction increases with the dip in the limbs (Fig. X–5), attaining nearly a right angle between sandstones and slates where the beds dip vertically. At the hinge of the fold shown in Fig. X–5 it is

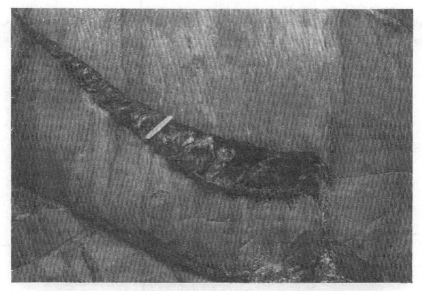

FIG. X–14. REFRACTION BETWEEN FISSURING IN SANDSTONE AND CLEAVAGE CRACKS IN EARLY-FORMED QUARTZ VEIN, CASTLEMAINE

clear that the dihedral angle between the cleavage directions S′ and S″ in the sandstones is little different from that in the slates, so that the refraction in the limbs must be attributed largely to the rotation of the shear planes in the sandstone as the fold developed while the slaty cleavage maintained approximately its original dip. The violent refraction seen in steeply dipping beds may thus be accounted for as being due chiefly to rotation of the fold limbs, rather than entirely to the difference in dihedral shearing angle of the different beds.

In graded beds, the macroscopic cleavage gradually curves as the composition of the bed changes, due to the passage from shear-folding in the slates, to a cleavage attitude affected by increasing rotation at the base of the bed, in the sandstone. The intimate shearing of thin sandstones interbedded in slates and clayey sandstone near the base of a graded bed is, however, complex, as might be expected in beds that are simultaneously rotating and shearing (Fig. X–1).

Refraction between pre-cleavage quartz veins and enclosing slates or sandstones is very marked (Fig. X–14).

Metamorphism and cleavage

The shearing phenomena so far described appear to be entirely of mechanical origin, but although the effects in sandstones are closely similar to those seen in experimentally deformed rocks under high confining pressure, no experiment has succeeded in reproducing slaty cleavage with all its accompanying minutiae. This may be attributed to two effects, firstly to the essential role played by metamorphism, and secondly to the influence of water in the system.

The original clay minerals and hydrolysates of argillaceous sediments recrystallize during the formation of slate. Loosely attached water is driven off, sericite and other micas, chlorite, and hematite are formed and organic matter is changed to graphitic derivatives. However, quartz and sedimentary micas are, in general, only mechanically deformed. The specific gravity of slate is notably higher than that of the parent material, partly because of a reduction in the voids and partly because of the loss of combined water during recrystallization. North [1] gives the average specific gravity of Welsh slates as 2·86, which is considerably higher than shales and sandstones and even more so than wet clay.

Reduction in volume is, in all probability, essential for the development of cleavage. Purely mechanical shearing involves an increase in volume due to

FIG. X–15. DILATATION DUE TO SHEARING WITHIN CONSTRAINING BEDS
The amount of dilatation is reduced with closer spacing of the shear planes.

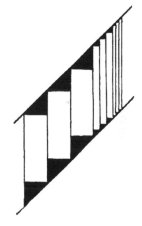

three main causes. These are, firstly, the breaking of atomic bonds and consequent separation of the parts of crystals; secondly, the effect of 'overriding' of close-packed particles as shown in Andrade's analysis of dilatancy (Chapter V), and thirdly, due to the tendency for openings to form where shear planes spaced a finite distance apart, abut onto a non-shearing surface, such as a slate–sandstone contact (Fig. X–15).

[1] F. J. North, *The Slates of Wales*, Cardiff, 1925.

It will therefore come about that shearing is aided by an inherent reduction in the volume of a rock, which compensates in some measure for the expansion due to shearing, and accordingly the stage of recrystallization is that at which slaty cleavage is most likely to develop. This affords an explanation for the absence of cleavage in closely folded sandstones and shales deformed at shallow depths in the crust, or under conditions in other ways not favourable to recrystallization.

Influence of water – The expulsion of water from the original mineral constituents and from the voids in slates implies that at the time of cleavage-formation rocks are two-phase systems containing aqueous solutions.[1] Direct

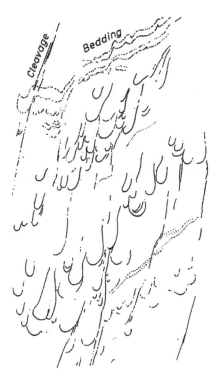

FIG. X–16. LOBATE BEDDING IN CLEAVED SANDSTONE, CHEWTON, VICTORIA

Natural size

evidence for this is available in several ways. The 'eyes' of pressure fringes are filled with quartz, chlorite, and carbonates; dark-coloured prolongations due to staining by carbonaceous matter extend for several inches into shear zones in sandstones overlying black slates, and the bedding of cleaved sandstones may exhibit downward projecting lobes indicative of the movement of a mobile constituent (water) upwards along the bounding shear zones. This may

[1] J. G. Maxwell considers the water to be exclusively the pore water normally found in shales, 'Origin of Slaty and Fracture Cleavage in the Delaware Water Gap Area, New Jersey and Pennsylvania', *Petrologic Studies*: A Volume to Honour A. F. Buddington, *Geol. Soc. Amer.*, 1962, pp. 281–311.

be termed *festooned bedding* (Fig. X−16),[1] but where the channels for upward movement of water or water-sediment mixtures were less numerous, isolated breakthroughs are seen, resembling chimneys.

The observation that the obtuse angle between the complementary shearing planes S′ and S″ in slate, rather than the acute angle, encloses the normal to the axial planes of the folds and presumably therefore the axis of maximum compressive stress may perhaps be partly accounted for by the presence of water in the voids, which introduces pore pressures. However, the growth of new minerals may also be involved in establishing this relationship, and again it may be influenced by rotation after shearing at some originally different angle.

Bulk distortion and flow in slate − The effects of flattening, elongation, and unilateral shear demonstrable in slates are related to the bulk distortion and plastic flow of the rock under the influence of regionally applied forces and the local constraints due to competent and incompetent beds in folds. Cleavage is always best developed in rocks in which intimate bedding-slip on the stratification planes is inhibited by the adherence of clay particles and other colloidal matter across the stratification planes. The rocks appear to have been permeated by a fluid medium, and also to have undergone reconstitution and recrystallization during cleavage formation. The attitude of cleavage planes in relation to local controls such as fold geometry, initial irregularities such as sole markings and the like, is in general the direction of mass displacement or flow of the rock. Experiments by Riedel[2] on the squeezing of clay into an orifice afford a close resemblance (compare Figs. X−5 and 17). On either flank of the orifice the clay moves by unilateral shear on S′ or S″ as in the limbs of a fold. In the central parts, a median plane of flattening is developed, parallel with the axis of the orifice which corresponds with the direction of elongation of the folded mass. Owing to the presence of fluids and recrystallization, it may be intuitively understood that some unilateral shear may also develop in the plane of flattening, due to inhomogeneities in the plastic mass, as is seen near fold axes in slate belts.[3] Likewise, in certain circumstances, the strong unilateral shearing of fold limbs may, perhaps, have been determined rather by a preliminary flattening in the AB plane, than by a single shear direction S′ or S″. The subject is further complicated by the need to recognize secondary shearing along a primary shear direction, as in the formation of *en echelon* fissures, and by the dynamics of the processes,

[1] The structure has been described by T. G. Bonney, 'On the Geology of the South Devon Coast from Torcross to Hope Cove', *Quart. Journ. Geol. Soc. Lond.*, Vol. 40, 1884, pp. 1−27 (see p. 18), who also refers to the darkening of the shear zones between the lobes. His illustration (Fig. 9−3) is reversed, but the Plate shows lobation in the correct downward facing.

[2] W. Riedel, 'Das Aufquellen geologischer Schmelzmassen als plastischer Formänderungsvorgang', *Neues Jahrb. f. Min.*, etc., Beil. Bd. 62, Abt. B., 1929, pp. 151−70.

[3] J. H. Dieterich, 'Origin of Cleavage in Folded Rocks', *Amer. Journ. Sci.*, Vol. 267, 1969, pp. 155−65.

whereby new shears may form as a fold develops, bedding-slip takes place along the separation planes against competent beds, and other local constraints are introduced, as in the fold-cores.

Note on plane of flattening – The definition by Sander of the plane of flattening as the AB plane of a strain ellipsoid [1] includes the further statement that this ellipsoid is formed by direct compression. A similar triaxial strain ellipsoid may however be produced by unilateral shear, except that its attitude to the stress axes will in this case differ, as the AB plane of such an ellipsoid lies diagonal to the single shear direction (see Fig. V–21, p. 128). Moreover,

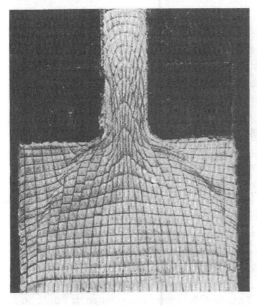

FIG. X–17. EXTRUSION OF CLAY INTO AN ORIFICE
(*After Riedel*)

the formation of such an oblique ellipsoid involves no real flattening, but only extension oblique to the shearing. The presence of two equally developed complementary shear directions or symmetrical structural detail (Fig. X–18) is good evidence for flattening in Sander's sense.

Mineral orientation and cleavage

The parallel orientation of flaky or elongated mineral grains such as mica, chlorite, graphite, and hematite, and the elongation of quartz and calcite in the cleavage dip is a notable feature of slates and cleavable sandstones. Older theories attributed the parallel orientation chiefly to rotation of randomly oriented particles as a consequence of compression and extension of the rock but this is unsatisfactory since in general there will in fact be strong orienta-

[1] B. Sander, *Einführung in die Gefügekunde der Geologischen Körper*, Teil. 1, Vienna, 1948, p. 106.

FIG. X–18. SYMMETRICAL DEVIATION OF
CLEAVAGE ABOUT QUARTZ BLEB IN BLACK SLATE
AT FOLD AXIS, BENDIGO, DEMONSTRATING
CLEAVAGE IN AB PLANE. ×30.

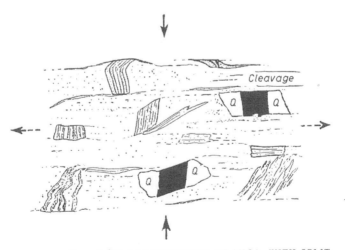

FIG. X–19. *En echelon* ARRANGEMENT OF MICA, WITH SPLIT-
TING OF SAME, IN CLEAVED LAMPROPHYRE DYKE, WOODS
POINT. ×30.

Compression normal to cleavage, with extension at right angles is
indicated (see quartz in pressure fringes and bent micas). Solid black,
arsenopyrite.

FIG. X–20. CLEAVAGE DEVELOPED ON KINK PLANES IN SLATE,
CHEWTON, VICTORIA
(*Photo: E. S. H.*)

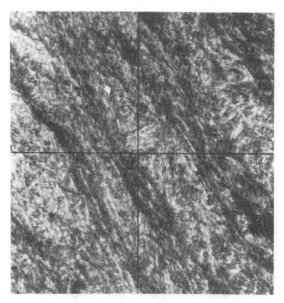

FIG. X–21. NEO-CRYSTALLIZATION OF SECONDARY
MICA ALONG SHEAR PLANES, AND RESIDUAL
SEDIMENTARY MICA FLAKES IN SLATE, BENDIGO.
×150.

tion in the bedding planes before cleavage-development. In slates the larger sedimentary mica flakes remain parallel with the original bedding and thus lie across the cleavage as *transverse micas* (Fig. X–10A, 21). In zones of strong unilateral shear some rotation of sedimentary micas is to be expected, giving

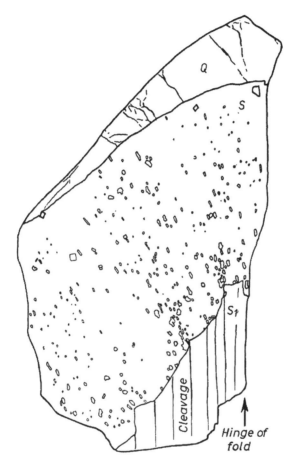

FIG. X-22. SMALL SADDLE REEF SHOWING FOLDED SANDSTONES (S), SLATE (S1) AND QUARTZ REEF (Q) WITH ARSENOPYRITE CRYSTALS ORIENTATED ALONG THE RADIAL CLEAVAGE IN SANDSTONE, BENDIGO

Natural size

rise to an *en echelon* arrangement of the flakes (Fig. X–19), and a somewhat similar effect results from the development of micro-folds due to kinking or flexing, a condition that may also be revealed by bedding (Fig. X–20); but in general the flaky particles that are parallel or nearly so with the shear planes and cleavage are due to neo-crystallization (Fig. X–21).

The principles underlying the arrangement of new minerals in relation to stress and strain are not yet fully understood.[1] Petrofabric analysis generally reveals parallel orientation of crystallographic elements such as the (001) plane of mica, along the complementary shear directions, but other statistical maxima may also be found. Minerals may also grow after deformation, along planes of low cohesion already formed or present in the rock (*mimetic crystallization*). This is important in phyllites and schists in the development of new minerals along the bedding planes, but it may also occur in slates and sandstone along the cleavage (Fig. X–22).

FIG.   X–23.   DOWNWARD   CONVERGENCE   OF
FANNING CLEAVAGE, TASMANIA
(*Photo by courtesy Emyr Williams*)

Cleavage fans and axial plane cleavage – Slaty cleavage, because of the geometrical consequences of the mechanisms by which it is developed, is often parallel with the axial plane of folds and is then referred to as *axial plane cleavage*. It is, however, generally only at and near the hinge itself that this relationship obtains, the attitude of cleavage in fold limbs being such as to form a fan-like arrangement, with either upward or downward convergence (compare Figs. X–5, 23, 24). In the minor bulges and cusps at fold hinges cleavage fans are obviously formed in relation to the local constraints and mass flowage (Fig. X–5), and a similar relationship with the plastic deformation

[1] W. D. Means and M. S. Paterson, 'Experiments on Preferred Orientation of Platy Minerals', *Contr. Mineral. and Petrol.*, Vol. 13, 1966, pp. 108–33.

and flow required by regional structure appears to apply in examples such as the South Mountain fold (Fig. VIII–17); but in other instances the fan arrangement results from actual rotation, along with the beds, in the limbs of developing folds. This accounts for the very strong upward divergence of

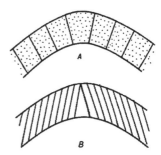

FIG. X–24. FANNING CLEAVAGE
A. Upward divergence of cleavage due to rotation in fold limbs.
B. Upward convergence of cleavage due to shearing as the limbs rotate.

fissure planes in folded sandstones (Fig. X–24A), but again in the slaty beds in which intimate and closely spaced unilateral shear goes on, it seems likely that upward convergence of cleavage may be ascribed, at least in some measure, to 'leaning over' like a pack of cards stood on edge and tilted

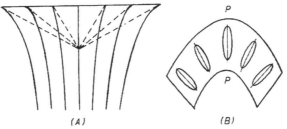

(A)   (B)

FIG. X–25. TYPES OF CLEAVAGE FANS
A. A sheaf arrangement. (After Laugel)
B. A fan due to compression in a fold with the neutral surface above the extrados. (After Lotze)

sideways (Fig. X–24B). The sliding movements involved in this may go on along axial plane cleavage already formed in the AB plane.

Regional cleavage fans – Regional variation in the dip of cleavage so as to form a 'fan' as seen in cross-section has been described in several regions.[1]

[1] This was very early referred to by Sharpe, 'On Slaty Cleavage', *Quart. Journ. Geol. Soc. Lond.*, Vol. 3, 1847, and by Laugel, 'Du Clivage des Roches', *Bull. Soc. Géol. France*, Ser. 2, Vol. 12, 1855. See also P. G. H. Boswell, *The Middle Silurian Rocks of North Wales*, London, 1949, pp. 96–108. H. Scholtz, 'Das variscische Bewegungsbild', *Fortschr. Geol. u. Pal.*, Hft. 25, Bd. 8, 1930, pp. 295–314; H. Quiring, 'Über kontravergente Transformation von Faltenzonen im Rheinischen Gebirge', *Zeitschr. Deutsch. Geol. Gesellsch.*, Bd. 91, Hft. 6, 1939, pp. 421–32. F. Lotze, 'Über Beziehungen zwischen Faltung und Schieferung', *Nachr. Ges. Wissensch. Göttingen, Math.-Phys. Kl.*, Hft. 2, 1932, pp. 113–21. J. E. Gair, 'Cleavage and the Distortion of Stratigraphic Thicknesses in Appalachian Folds', *Trans. Amer. Geophys. Union*, Vol. 30, 1949, pp. 116–18. See also E. Cloos, 'Oölite Deformation in the South Mountain Fold, Maryland', *Bull. Geol. Soc. Amer.*, Vol. 58, 1947, pp. 843–917.

FIG. X–26. CREASED SLATE

The photograph shows a cleavage plane, crossed by sub-parallel creases right to left. Note also the fine lineation. Natural size.

With local exceptions due to inhomogeneities and to the special features of individual folds, cleavage fans are related to the tectonic setting of a folded belt. In the section of the Rhineland shown in Fig. VIII–5, it is thought that the fans are due to later large-scale folding producing anticlinoria and synclinoria. According to Laugel, a cleavage fan may have the form of a sheaf or bundle, rather than a fan (Fig. X–25A), when considered in depth. Some examples (Fig. X–25B) are related to variation in the position of the plane of flattening in folds.

Secondary deformation of cleavage – Where cleavage is affected by a subsequent deformation which requires movements on new surfaces and directions, kink-bands, flexures, or shears of later development may cross the original cleavage. A series of kink-bands produces *creased or puckered slate* (Fig. X–26), but flexures or shears may be so closely spaced as themselves to constitute a crenulation cleavage, crossing the first cleavage. Such structures may imply either a separate period of later diastrophism, or a late phase of the same movements that gave rise to the original cleavage.

Cleavage and plunge – Although the attitude of cleavage relative to bedding and to the parts of a fold varies according to the geometrical relationships above described, the variation is chiefly in the dip of cleavage as seen in transverse section. Cleavage bears a much more constant relationship to fold geometry along the length of folds. With cleavage varying between S′ and S″, the intersection of the two complementary shear planes is the B axis of a

FIG. X–27. INTERSECTION OF BEDDING ON CLEAVAGE
PLANE
The stripes on the cleavage plane are thin bedding laminations.
The folded bedding is seen on the cut and polished surface. $\times\frac{1}{2}$.

strain ellipsoid, which is also the longitudinal axis of cylindrical folds formed simultaneously with the cleavage. The intersection of the two S directions may appear as a lineation on cleavage planes. The intersection of bedding and cleavage is parallel to the same axis and also appears as a lineation on cleavage planes (Fig. X–27). Statistical examination of a large number of readings of lineations of these types thus permits the recognition of the longitudinal axis of folding and its plunge [1] (see also Fig. IX–7).

[1] This statement refers to bedding-cleavage intersections within the slate. Folds in thick beds, e.g. sandstones, interbedded with slates may, because of splitting or *en echelon* folds, diverge somewhat from the overall plunge indicated by bedding-cleavage intersections.

Cleavage and facing – The use of cleavage to determine facing also depends on the near-parallelism of cleavage with the axial plane of folds in slate. The axial plane being thus determined, the further implication as to overturning of strata becomes obvious (Fig. X−28). It is also seen that where cleavage dips

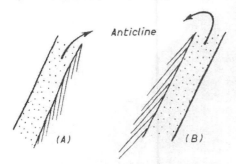

Anticline

FIG. X−28. CLEAVAGE AND FACING
A. Shows the attitude of cleavage and bedding in beds dipping normally, to the left.
B. Shows overturned beds. The arrows point towards the adjacent anticlinal crest.

(A)   (B)

less steeply than bedding, we normally have to deal with the steep limb of an asymmetrical fold.

Mullion structure and rodding – The subdivision of a bed, more particularly a sandstone interbedded with slate, into long, parallel slabs with smooth rounded surfaces is known as *mullion-structure* (Fig. X−29). *Rodding*, which

FIG. X−29. FOLD MULLIONS SEEN IN HAND SPECIMEN. ×⅓.

was formerly regarded as a synonym, is clearly distinguished by Wilson after a study of classic localities,[1] as being the structure represented by rods of quartz that has been introduced into or has segregated in the rocks, whereas mullions are formed of the rock itself. Wilson distinguished *fold-mullions* which are the exposed surfaces of small folds, generally closely spaced, and

[1] G. Wilson, 'Mullion and Rodding Structures in the Moine Series of Scotland', *Proc. Geol. Assoc. Lond.*, Vol. 64, 1953, pp. 118–51.

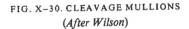

FIG. X–30. CLEAVAGE MULLIONS
(*After Wilson*)

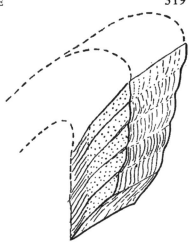

cleavage-mullions which are due to the subdivision of the rock by a coarse cleavage crossing the bedding at a large angle as in fissured sandstones, the weathered surfaces of which, especially at the sharp contact of a sandstone overlying shale (slate) in graded beds, give rise to mullions where the fissuring planes are parallel along their strike (Fig. X–30).[1]

The literature describing the microscopic features of slaty cleavage and its geometrical relationships to folds is voluminous, as the subject has been studied for over a century. Recent work has demonstrated that the essential features of cleavage are similar in comparable rocks wherever they occur.[2]

[1] In the Eifel, the strike of mullions diverges from that of fold axes, which is thought to be due to the shearing being a late effect associated with a second diastrophic episode. See A. Pilger and W. Schmidt, 'Die Mullion-Strukturen in der Nordeifel', *Abh. Hess. Landesamt. Bodenforsch. zu Wiesbaden,* 1957. Also, 'Definition des Begriffes "Mullion-Struktur" (Mullion-structure)', *Neues Jahrb. f. Geol. u. Pal.,* Monatshft. Jhg., 1957, Hft. 1, pp. 24–8.

[2] D. Sharpe, op. cit., 1847, pp. 74–105; H. C. Sorby, 'On the Origin of Slaty Cleavage', *Edinb. New Phil. Journ.,* Vol. 55, 1853, p. 137; 'On Slaty Cleavage exhibited in the Devonian Strata of Devon', *Phil. Mag.,* Ser. 4, Vols. 11, 12, 1856; M. Laugel, op. cit., 1855, p. 363; A. Harker, 'On Slaty Cleavage and Allied Rock Structures', *Rept. Brit. Ass. for 1885* (1886), p. 813; T. N. Dale, 'The Slate Belt of Eastern New York and Central Vermont', *U.S. Geol. Surv.,* 19th Ann. Rept., Pt 3, 1897–8, pp. 153–307; C. K. Leith, 'Rock Cleavage', *U.S. Geol. Surv.,* Bull. 239, 1905; F. Lotze, 'Über Beziehungen zwischen Faltung und Schieferung', *Nachr. Ges. Wissensch Göttingen, Math.-Phys. Kl.,* Hft. 2, 1932, pp. 113–21; W. J. Mead, and M. J. Mead, 'Folding, Rock Flowage, and Foliate Structures', *Journ. Geol.,* Vol. 48, 1940, pp. 1007–21; G. Wilson, 'The Relationship of Slaty Cleavage and Kindred Structures to Tectonics', *Proc. Geol. Assoc. Lond.,* Vol. 57, 1946, pp. 263–302; S. Kienow, 'Über Faltungsschieferung', *Neues Jahrb. f. Min.,* etc., Abh. B., Bd. 90, 1949, pp. 345–78; L. U. de Sitter, 'Schistosity and Shear in Micro- and Macrofolds', *Geol. en Mijnb.,* N.S. Nr. 10, Jhg. 16, 1954, pp. 429–39; P. Fourmarier, 'Remarques au sujet de la schistosité en général, etc.', ibid., N.S. Nr. 2, Jhg. 18, 1956, pp. 47–56. R. Hoeppener, 'Zum Problem der Bruchbildung, Schieferung und Faltung', *Geol. Rundsch.,* Vol. 45, 1956, pp. 247–83, gives a good bibliography. I. V. Kirillova reviews her own and other Russian works, in 'Folded Deformation in the Earth's Crust, their Types and Origin', V. V. Beloussov and A. A. Sorskii (eds), *Akad. Nauk SSSR,* 1962. Engl. translation, Jerusalem, 1965: see 'Cleavage as an Indicator of the Character of Mass Movement During the Folding Process', pp. 81–115.

CHAPTER XI

Major structures and tectonics

The association of structures of similar type in regions, zones, or belts, as in fold mountain chains, zones of nappes or of major rift faulting, slate belts, metamorphic zones, and so on, is an outstanding geological fact which leads to the consideration of the regional deployment of major structures, and their spatial relationships one to another. Taken in conjunction with other data such as the distribution of sediments, the kinds of rocks and their thickness, the location of igneous rocks and of zones of metamorphism, regional study leads to a synthesis of geological events in space and time and to an understanding, as yet far from complete, of the great units and elements in the structure of the continents and oceans. This is *tectonics*. Ed. Suess' great work *Das Antlitz der Erde* was the first major study along these lines,[1] directing attention particularly to the development of structures along their trend or regional strike as well as in cross-section. Since then geological knowledge has vastly expanded and geophysics has provided an entirely new body of data relevant to the inner structure of the globe, the former distribution of continents in the geological past, and the structure and evolution of ocean basins, which were virtually blank on the geological world map until very recently. Tectonics, from being largely an encyclopaedic exercise in the organization of knowledge, has come to be a positive element in geological philosophy, which may be used to guide regional studies and mineral exploration.

Tectonic analysis is based firstly on the recognition of regional units each of which is characterized by a certain uniformity of style as to its structure and evolution, such as geosynclinal belts of different ages, regions with little-disturbed sedimentary cover resting on a basement or platform of older origin, belts of strong folding perhaps of various dates, and the like. The recognition of these regional units is based largely on accumulated data and the correlation of results derived from all available methods of study and exploration. The reasons for linking areas together as being of one tectonic type, and of distinguishing this from other tectonic types, involve concepts as

[1] Ed. Suess, *Das Antlitz der Erde*, Prague and Leipzig, 1883–8. French edition *La Face de la Terre*, Paris, 1905–18; English edition, *The Face of the Earth*, Oxford, 1904–24.

320

to the structures and events which are concerned in the evolution of continents and ocean basins. As to the continents, these concepts include the existence and normal developmental history of geosynclines (especially eu- and mio-geosynclines): the formation of rigid *cratons* by strong folding combined with metamorphism and the bonding and buttressing due to igneous injections: the existence of various distinct types of diastrophic movements and of forces that create them, notably orogenic, epeirogenic and taphrogenic movements. While it is true that these and other concepts in tectonics are still the subject of discussion and controversy as to the totality of structures and events that may be implied in using any term (such as geosyncline) there is sufficient agreement in the broadest terms and with a minimum of associated ideas, as to the reality and usefulness of such broad concepts, and they have been applied in most tectonic maps of the continents recently published.[1] It is, however, certain that as geological and geophysical knowledge increases, tectonic concepts will be modified and added to. Already an additional system of earth movements, that involved in ocean-floor spreading and the movements of crustal plates, has been widely recognized only recently.[2] Geological and geophysical mapping and interpretation must remain as the fundamental basis for geological thought, retaining as far as possible its independence of preconceived notions. Only thus may hypotheses and concepts be tested, and unbiased information be obtained on which to base new theories.

Types of diastrophic movements

Orogenic movements – In a region of continental extent the most striking tectonic subdivision is between mountain chains on the one hand and the large areas of plateau and plain on the other. When it was found that the rocks of most mountain chains are strongly folded and thrusted, whereas corresponding formations on the plateaux and plains remain little disturbed, the movements causing mountain-building were termed *orogenic*. Since the chief structures involved are folds and thrust faults, it has subsequently become the practice to refer such structures where seen in older rocks to orogenic movements. The inference that these old structures represent in every instance a former fold mountain chain is, however, not necessarily valid, for some

---

[1] See, for example, the following tectonic maps: North America, scale 1 : 5,000,000, with discussion, *U.S. Geol. Surv.*, Prof. Pap. 628, P. B. King (ed.), 1969; Europe, scale 1 : 2,000,000, with Explanatory Note, A. A. Bogdanoff, M. V. Muratov, and N. S. Shatsky (eds), 1964; USSR, scale 1 : 2,500,000, *Ministry of Geol.*, USSR, 1968, (English legend); Australia, scale 1 : 5,000,000, *Geol. Soc. Aust.*, 1971.

[2] In the first edition of this book (1963: written in 1960) the author proposed to recognize *lineagenic movements* as a type, connected with the formation of major strike-slip faults. It now appears that such movements are for the most part closely related with ocean-floor spreading and plate tectonics, which being the broader concept supersedes the more restricted notion of lineagenic movements (see pp. 472–9).

strong folding appears to go on in troughs. Conversely, certain young, high mountain chains have a relatively simple structure, their elevation being due chiefly to vertical uplift.[1] Accordingly it is difficult to give a precise meaning to orogenesis, but strong folding with or without reverse faulting, affecting a

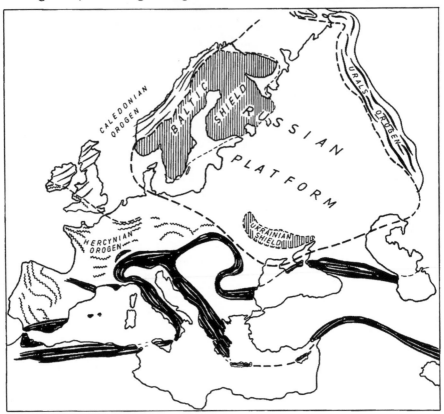

FIG. XI–1. SIMPLIFIED TECTONIC MAP OF EUROPE
Solid Black shows areas of Alpine folding.

moderately thick sedimentary succession in a long and relatively narrow zone are implied.

Epeirogenic movements – By way of contrast, epeirogenic movements affect extensive regions chiefly by up- or downwarping,[2] giving rise to depressed basins or *sags*, and raised *swells*,[3] which are broad rather than elongate features (Figs. VII–30A, XI–21).

---

[1] e.g., parts of the Andes, the Basin Ranges of the U.S.A., and eastern New Guinea: see L. Picard, 'La Structure du N.W. de l'Argentine, etc.' *Bull. Soc. Géol. France*, Ser. 5, Vol. 18, 1948, pp. 765–846. T. W. E. David, *Geology of the Commonwealth of Australia*, London, 1950.

[2] The term epeirogenic was coined by G. K. Gilbert, 'Lake Bonneville', *U.S. Geol. Surv.*, Mon. No. 1, 1890, p. 340.

[3] See R. Ruedemann, 'On the Symmetric Arrangement in the Elements of the Palaeozoic Platform of North America', *N.Y. State Mus.*, Bull. 140, 1910. In the U.S.S.R. broad

Epeirogenic movements do not give rise to individually recognizable particular structures such as a fold or a fault, and are in some measure reversible, so that a basin may become a plateau. However, the reversibility of large epeirogenic movements is doubtful, and certainly many major basins, nearly filled with sediment, have a very long history and remain depressed features over several geological periods.

As epeirogeny apparently involves chiefly vertical movements, the origination of long and narrow geosynclinal troughs is by some authors regarded as an epeirogenic event. However, ideally epeirogeny is best regarded as a regional effect, by which broad blocks of country are displaced in such a way that while some are sinking others are rising. The total displacement of rock matter in epeirogenic movements is much greater than is generally conceived, for although the vertical movements are small the volume of rock involved is much greater than that in narrow orogenic belts. Again, it is becoming known that as well as the mobile seismic belts of the earth, the cratons – shields and platforms – are also affected by young and even by present-day movements.[1]

Taphrogenic movements – The necessity to recognize a third type of movement is indicated by the tectonic importance of major belts of block faulting, notably the rift valley and graben zones, and mountain and plateau areas strongly affected by faulting with associated monoclinal warping. The term was coined for the East African rifts,[2] and is largely descriptive. In general, however, it implies tensional forces as opposed to horizontal compression for orogeny and differential vertical movements for epeirogeny.

Plate tectonics and transform faulting – The movements involved in ocean-floor spreading, the relative displacement of rigid crustal blocks (plate tectonics) and the associated phenomena of transform faulting, may, it appears, also be able to provide explanations for structures such as oceanic trenches, fold mountain belts marginal to continents, and others. It emerges from current syntheses that processes operative in the earth's mantle, including possible convection, partial melting, and phase changes are of fundamental importance; but there are many unresolved questions and it is necessary to continue the study of continental and oceanic geology along established lines in order to afford an objective basis against which hypotheses of global tectonics may be tested.[3]

---

depressions of the basement are termed synéclises (French equivalent) and broad swells antéclises. See A. Bogdanoff, 'Traits Fondamentaux de la Tectonique de l'U.R.S.S.', *Rev. Géogr. Phys. Géol. Dynam.*, Ser. 2, Vol. 1, 1957, pp. 134–65.

[1] *Map of Neotectonics of U.S.S.R.* 1 : 5,000,000. Moscow, 1960: also E. S. Hills, 'Morphotectonics and the Geomorphological Sciences with special reference to Australia': 15th William Smith Lecture, Geological Society of London, 1960.

[2] E. Krenkel, *Die Bruchzonen Ostafrikas*, Berlin, 1922.

[3] Plate tectonics and sea-floor spreading are further discussed in Chapter XIV.

Cratons, orogens, and geosynclines

The relatively rigid epeiric regions of continents are known as *cratons* and the belts of strong folding as *orogens*.[1]

Cratons, shields, and platforms – Cratons consist of old, strongly folded, and consolidated rocks, bonded and strengthened by metamorphism and by igneous intrusions. The largest and most typical have an Archaean basement, to which the folded rocks of later orogens are welded, but Palaeozoic and younger orogens still retain some mobility and generally appear as low mountain ranges, whereas the oldest Pre-Cambrian fold belts are completely immobilized and outcrop chiefly in extensive gently domed plateaux known as *shields*, of which perhaps the best known are the Baltic Shield and the Canadian Shield.

Similar Pre-Cambrian basement persist beneath much larger areas than the exposed shields, being covered with younger rocks both as relatively thin sheets and as thick accumulations in certain basins or troughs (synéclises). Such large areas with rigid cratonic basement, of the size of a sub-continent (e.g. much of European Russia) are known as *platforms* (Ger. *Tafeln*), and these typically consist of a rigid basement covered by younger platform deposits. Platforms of various ages are formed in the course of the consolidation of geosynclinal fillings by folding and igneous intrusions. In addition to the shields, large platforms may be divided into a number of smaller cratonic units in some of which small *salients* or *blocks* of consolidated basement may outcrop (see Fig. XI–2). The blocks are in places bordered by belts of thick sediments in troughs which may be folded, but do not show the intensity of folding, metamorphism, and igneous action characteristic of orogens. In the Australian continent, some of the smaller blocks into which the platform[2] is divided moved up to form plateaux or down to form basins, which tend to have polygonal outlines.[3]

In current thinking, platforms result from the consolidation of orogens, and cratonic blocks grow by the welding of these to older nuclei; but the process may be of long duration, there being intermediate stages during which a platform still retains partial mobility so that deep troughs may develop and become sediment-filled. These, however, are generally regarded today as differing from true geosynclines, and have been called *transitional basins*.

---

[1] According to Stille and others, the simatic rocks of the ocean floors act as cratons, being *low cratons* whereas the sialic cratons of continents are *high cratons*. It is now accepted that both oceanic and continental crust may behave as relatively rigid blocks or plates, despite their different constitution, but the term *craton* is used particularly for continental blocks.

[2] The Australian platform has been called the Australian Shield in earlier writings.

[3] E. S. Hills, 'The Tectonic Style of Australia', *Geotekt. Sympos. zu Ehren von A. Stille, Deutsch. Geol. Gesellsch.*, 1956, pp. 336–46. For the U.S.S.R., see N. S. Schatsky and A. A. Bogdanoff, 'Grundzüge des Tektonischen Baues der Sowjetunion', *Fortschr. d. Sowjet. Geol.*, Hft. 1, 1958 (Berlin).

Geosynclines – The concept of geosynclines arose in 1859 from the work of J. Hall on the origin of the northern part of the Appalachian chain, and the word was first used by J. D. Dana in 1873.[1] The chief characteristics of a geosyncline in the sense of Hall and Dana are the great thickness of sediments predominantly of shallow marine origin, as compared with the much smaller

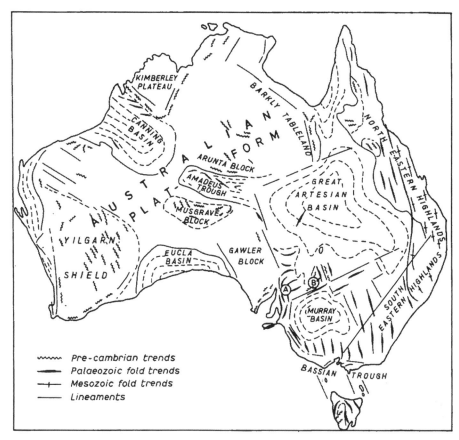

FIG. XI–2. BLOCK TECTONICS OF AUSTRALIA (SIMPLIFIED)

Blank areas in the Australian platform are mainly covered with Proterozoic and Lower Palaeozoic rocks. A = Adelaidean Geosyncline. B = Broken Hill Block.

thickness of rocks of the same age on the neighbouring cratonic block, the eventual strong folding and regional metamorphism, and the intrusion of igneous rocks, chiefly granitic. All these phenomena appear to be interrelated and, together with the elevation of the folded belt, constitute the classic notions as to the origin of fold mountain chains. Subsequently, the term

[1] J. Hall, *Natural History of New York*, Palaeontology, 1859. J. D. Dana, 'On the origin of mountains', *Amer. Journ. Sci.*, Ser. 3, Vol. 5, 1893, pp. 347–450.

geosyncline has been very widely used and extended by some geologists to include almost all types of sediment-filled basins and troughs, which according to their geological setting and tectonic history have been classified into various types of geosynclines.[1] Other authors on the contrary have restricted the use of the word to features of the type discussed by Hall and by Dana, and use other terms (e.g. troughs, basins, furrows) for sedimentary basins developed within cratons or in other ways differing from the classical geosyncline, or *orthogeosyncline* of Stille and Kay. It is undoubted that the need to use some qualifying prefix or adjective in identifying sedimentary troughs or basins that are all to be termed 'geosynclines' leads to confusion and virtually robs the word itself of any precise meaning. From this consideration alone it seems desirable to restrict usage to long and relatively narrow troughs, dominantly marine, which after sedimentation become the sites of orogeny. In the deeper parts of the geosyncline the thickness of sediments may not, in fact, be great.[2] The virtual synonym *orogen*, coined by Kober,[3] may also be used with advantage for the same group of notions, having a clear implication of the development of orogenic-type structures. Geosyncline (or orthogeosyncline) may then be used for the earlier stages of the orogen, during the time when marine sedimentation is in progress, and before the culmination of orogeny has occurred.[4] A fundamental distinction may, however, be drawn between *ensialic* geosynclines which are floored by downwarped continental crust, and *ensimatic* geosynclines in which the basement is ocean crust composed essentially of basic igneous rocks.

Orogeny and Time – The long history of the Alpine orogen, which is clearly traceable as far back as the Permian and perhaps the Carboniferous and culminated in early Tertiary folding with still later movements involved, demonstrates not only the relative permanence of an orogen, but also its tectonic significance as a persistent zone of crustal weakness.

An equally long history, during which major structures within a geosyncline slowly develop, episodic or local movements occur and various parts of the great trough are subject to different processes including rapid vertical oscillations, local folding, and igneous action, all of which

[1] See H. Stille, 'Wege und Ergebnisse der geologisch-tektonische Forschung', 25 *Jhr. Kaisser Wilhelm Gesell.*, 1936, Bd. II, pp. 77–97; *Einführung in den Bau Nordamerikas*, Berlin, 1941. Marshall Kay, 'Geosynclinal Nomenclature and the Craton', *Bull. Amer. Assoc. Petrol. Geol.*, Vol. 31, 1947, pp. 1287–93. For review articles, see M. F. Glaessner and C. Teichert, 'Geosynclines: A Fundamental Concept in Geology', *Amer. Journ. Sci.*, Vol. 245, 1947, pp. 455–82, 571–91. J. Aubouin, 'A propos d'un centenaire: les aventures de la notion de géosynclinal', *Rev. Géogr. Phys. Géol. Dynam.*, Vol. 2, Fasc. 3, 1959, pp. 135–88.

[2] R. Trümpy, 'Palaeotectonic Evolution of the Central and Western Alps', *Bull. Geol. Soc. Amer.*, Vol. 71, 1960, pp. 843–908.

[3] L. Kober, *Bau und Entstehung der Alpen*, Berlin, 1923. *Der Bau der Erde*, Berlin, 1926.

[4] E. C. Kraus, 'Das Orogen, Begriff, Bildungsweise und Erscheinungsformen', *21st Internat. Geol. Congr. Norden*, 1960, Rep. 18, pp. 236–47.

affect also the sedimentation, may be demonstrated for all well-studied orogens. For the Appalachians, the period covered ranges from Cambrian to Permian.

After the major folding has culminated, orogens are commonly rejuvenated by later block uplifts, giving mountain systems of secondary elevation as compared with young folded zones of alpine type. Rejuvenated orogens are

FIG. XI–3A

Model showing the Southern Flinders and Mt Lofty Ranges, South Australia, a rejuvenated Palaeozoic orogenic belt.

well known in the Appalachians, the Urals, and the Flinders and Mount Lofty Ranges in South Australia (Fig. IX – 3).

In thinking about the structures within orogens, although we are aware of their long period of development, we often erroneously imagine the final array of structures to be the result of one great movement. This fallacy is doubtless influenced by model experiments on folding, but, as Gilluly has pointed out, present day movements of the ground even in non-orogenic areas are such that, if continued in the same sense, they would

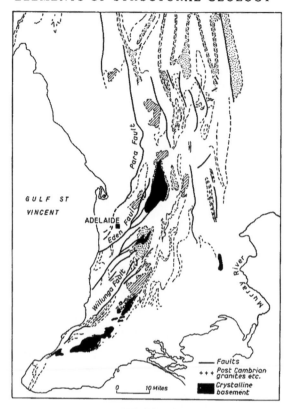

FIG. XI–3B

Geological Map of the area shown in Fig. 3A. Fold pattern Pre-Cambrian and lower Palaeozoic. Topography
results from Cainozoic movements on faults. (*By
courtesy, Geological Society of Australia*)

give rise to high mountains in a span of time much less than a geological
period.[1]

Structure of orogens

It is a general feature of folded orogenic belts, well exemplified in the
European Alps and the Appalachians, that a sequence of tectonic zones is
represented, each zone being generally parallel with the trend of the belt and
characterized by a certain similarity in structures and rock-types, both sedimentary and metamorphic (Figs. XI–4, 5, 8). Igneous action and related ore
deposits may also, where present, exhibit a zonal arrangement.

To some extent, the zoning is a reflection of the original sedimentary facies

[1] J. Gilluly, 'Distribution of mountain-building in geological time', *Bull. Geol. Soc. Amer.*,
Vol. 60, 1949, pp. 561–90.

in the sedimentary trough, which are governed by the geological setting of the trough and by events such as vulcanicity, by the onset of orogenic movements, and the presence or absence of a ridge or geanticline within the main geosyncline. At an early stage, a marine trough, not necessarily deep, exists into which little terrigeneous sediment is spilled because the neighbouring craton is low. Basic submarine lavas (spilites) are extruded and mingle with the thin pelagic deposits (radiolarian cherts and red clays, pelagic limestone, shales with manganese nodules) on the floor of the trough. Along the margins and on the ridges (or geanticlines) within the trough, neritic deposits are laid down. Sedimentation of terrigenous derivation originates from growing mountains formed by the first orogenic phases. Sometimes spoken of as 'rapid', this sedimentation of flysch type with greywackes, microbreccias and the like deposited by turbidity currents, and shales laid down from matter in

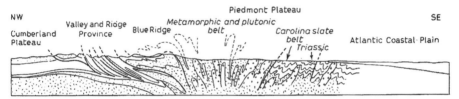

FIG. XI-4. THE STRUCTURE OF THE APPALACHIAN OROGEN, SHOWING THE CHIEF ZONES AND PROVINCES

Length of section 360 miles. (*After P. B. King*, Bull. Amer. Assoc. Petrol Geol., *Vol. 34, 1950, Fig. 2*)

suspension, is in reality strongly episodic, with long periods of non-deposition between the 'floods' of sediment.

The original arrangement of sedimentary facies may be changed by the great horizontal translations of nappes, and, in fact, it was from the maladjustment of facies as they now lie in the Western Alps that the concept of such great translations originated. The zoning is, however, also a reflection of the transverse variation in type and intensity of deformation within an orogen. In this it is usual to find a transition from the strongest deformation where slaty cleavage or schistosity is developed (or again only very strong folding) to close asymmetrical folding and thrusting, and finally to gentle or widely spaced folds on the margin of the neighbouring craton.

The majority of deformed belts are asymmetrical, the folds being overturned towards a neighbouring craton which is called the foreland, and the intensity of folding and metamorphism increasing with distance from the foreland (Figs. XI–4, 5).

There can, indeed, be no question but that the structure of orogens marginal to continents, and especially of the great circum-Pacific mobile belt with island arcs and foredeeps flanking the Asiatic land-mass, is fundamentally asymmetrical. The totality of topographic and geological features indicative

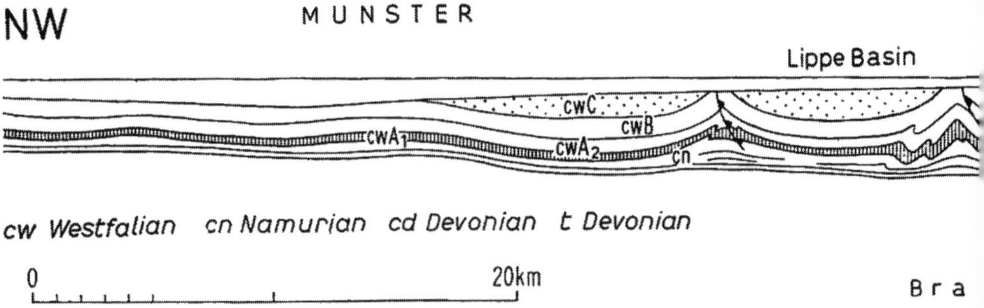

NW   M U N S T E R

Lippe Basin

cw Westfalian   cn Namurian   cd Devonian   t Devonian

0   20km
|_,_,_,_,_,_,_____|_____| B r a

FIG. XI–5. THE STRUCTURE OF PART OF
The section approximately corresponds with the northwestern half of the Appalachian

of this is matched by seismic evidence for the existence of an inclined shear zone, dipping beneath the continent and penetrating to depths of up to 700 km[1] (Fig. XI–6). It is, however, already clear that there is wide variability in the structure of orogens, and much further study will be required to find an underlying and unifying concept. According to Kober a symmetrical arrangement is normal, with inwardly inclined axial planes and thrusts on either side of a central, relatively undisturbed region of *betwixt-mountains*, termed the median mass[2] (Fig. XI–15), which is disrupted by faulting. In Peach's view the Scottish Highlands also show bilateral symmetry,[3] and Aubouin suggests that varying combinations of a fundamental 'bicouple' consisting of a non-magmatic *miogeosyncline* adjacent to the foreland craton and a magmatic *eugeosyncline*, separated by a ridge, may result in either asymmetrical or symmetrical orogens (Fig. XI–7).[4] It has, however, proved to be difficult if not impossible to substantiate any general constant spatial relationship between so-called mio- and eugeosynclines in different continents.

Folding of orogens

Orogenic folding is generally thought to result from horizontal compression of the geosynclinal sediments between cratonic blocks by a vice-like mechanism. This concept derives from the classic experiments of Hall on folding,[5] and has led some to formulate purely hypothetical high cratons on the oceanic side of marginal geosynclines, notably for the Appalachians where especially in earlier writings a continental 'Appalachia' was postulated to the east. The

[1] B. Gutenberg and C. F. Richter, *Seismicity of the Earth*, Princeton, 1949. M. Ewing and F. Press, 'Structure of the Earth's Crust', *Encyclopaedia of Physics*, Vol. XLVII, Berlin, pp. 246–57.

[2] L. Kober, *Der Bau der Erde*, Berlin, 1928, p. 173.

[3] B. N. Peach and J. Horne, *Chapters on the Geology of Scotland*, 1930, Fig. 27.

[4] J. Aubouin, 'A propos d'un centenaire: les aventures de la notion de géosynclinal', *Rev. Géogr. Phys. Géol. Dynam.*, Vol. 2, Fasc. 3, 1959, pp. 135–88.

[5] J. Hall, 'On the Vertical Position and the Convolutions of certain strata, and their relation with Granite', *Trans. Roy. Soc. Edinburgh*, Vol. 7, 1815, p. 79.

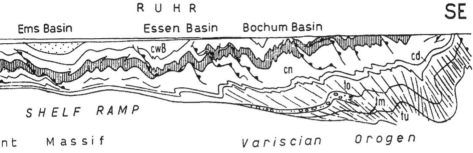

R U H R   **SE**

Ems Basin   Essen Basin   Bochum Basin

S H E L F   R A M P

nt   M a s s i f   V a r i s c i a n   O r o g e n

THE VARISCAN OROGEN OF THE RHUR

section of Fig. 4. (*After M. and R. Teichmuller, Geol. en Mijnb., N.S. Jhg. 20, 1958*)

evidence for the purely simatic nature of the northern Pacific Ocean, however, makes such an idea untenable for the Japanese and adjoining orogens.

Equally intriguing problems are posed by non-orogenic marginal shelves to cratons, where very thick sedimentation has gone on, in the Westralian shelf possibly up to 40,000 ft (approximately the figure quoted by Hall for the Appalachians and greater than the total thickness involved in the Alps), or

FIG. XI–6. EASTERN MARGIN
            OF ASIA

(*a*) Topographic section.

(*b*) Location of earthquake foci.

(*c*) Diagrammatic cross-section showing zones of shearing and deep crustal structure. (*After Gutenberg and Richter, 1949*)

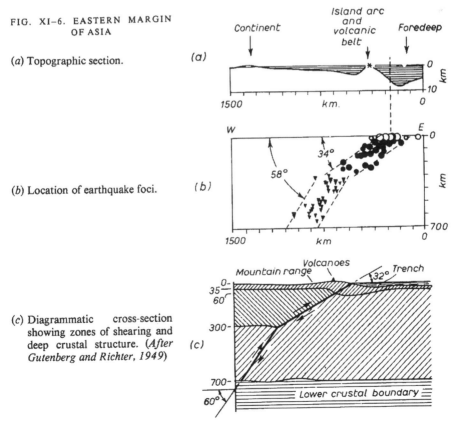

again in the so-called Gulf Coast 'Geosyncline' of the U.S.A. Both exhibit only minor folding, lack igneous phenomena, and are nevertheless major tectonic features of considerable duration (the Westralian shelf basins probably originated in the Lower Palaeozoic or even Upper Proterozoic). By whatever name we choose to call such major features, whether geosynclines, shelves, or marginal troughs, it is true that we are profoundly ignorant of the causes underlying their formation and development.

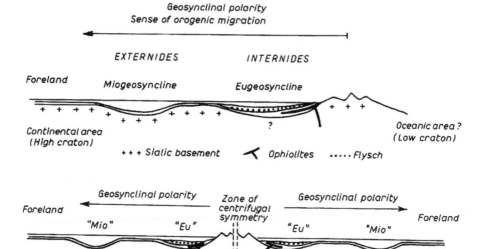

FIG. XI-7. GEOSYNCLINAL BICOUPLES

A. A single bicouple (othogeosyncline), consisting of adjoining eugeosyncline and miogeosyncline.

B. Adjacent bicouples giving rise to a geosynclinal zone with symmetrical structure. (*After Aubouin, 1959*)

Fold trends – The trend of the internal tectonic zones in an orogen, and of its major folds, is generally parallel with the edge of the neighbouring craton, although modified locally, as in the Alps and the Juras, by pre-existing massifs, troughs, or ridges on the craton, to which the folds appear to be moulded[1] (Fig. XI–8A). This suggests translation of the rock-masses in a direction normal to the trend of the belt. On the other hand folding may result from horizontal shearing movements between the resistant blocks bordering the geosyncline, combined with some lateral compression. Experimental studies have shown that folding can result from such movements,[2] but the arrangement of the folds within the deformed zone is characteristically differ-

[1] W. H. Bucher, 'Deformation in Orogenic Belts', in *Crust of the Earth, Geol. Soc. Amer.,* Spec. Pap. No. 62, 1955, pp. 343–68.

[2] W. J. Mead, 'Notes on the Mechanics of Geologic Structures', *Journ. Geol.,* Vol. 28, 1920, pp. 505–23. C. K. Leith, *Structural Geology,* New York (revised edn), 1923, p. 192.

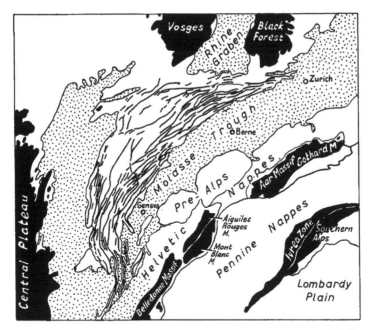

FIG. XI–8A. MAJOR TECTONIC ZONES OF THE WEST ALPINE
REGION

Solid black, resistant massifs; stipple, sedimentary troughs; lines north of
Molasse Trough, Jura folds. (*After Bucher, 1955*)

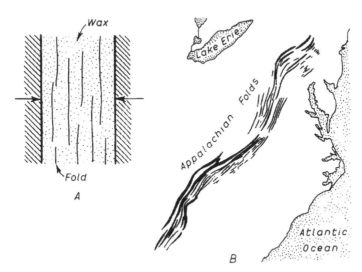

FIG. XI–8B. THE NOTION OF FOLDING BY DIRECT COMPRESSION
A. Experimental folds in wax. (*After Mead, 1920*)
B. The folds of the Appalachian chain (*after Willis* [1]) (generalized).

[1] B. Willis, 'Mechanics of Appalachian Structure', *13th Ann. Rept. U.S. Geol. Surv.*, Pt 2, 1893, pp. 217–81.

ent from that in a zone of soft rocks deformed by direct compression. In folding by direct compression (Fig. XI—8B), the direction of greatest relief is upwards, the least strain axis is in the direction of the active forces, and the mean strain axis is horizontal and at right angles to this direction. The folds are long and parallel with each other and also with the margins of the trough. Folds produced by shearing movements are arranged *en echelon* within the deformed zone, with the fold axes at approximately 45° to the direction of the shearing couple [1] (Fig. XI—9A).

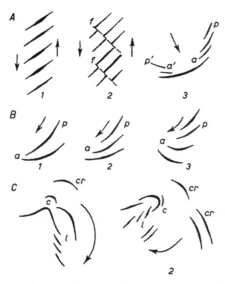

FIG. XI–9. TYPES OF FOLD PATTERN
ACCORDING TO LEE

The arrows indicate the principal directions of movement.

A. Forms of xi (Greek ξ) structure.
1. Parallel xi type; 2. Parallel xi type with tear faults f, f; 3. Convergent xi type: a and a′, anterior ends; p and p′, posterior ends.
B. Forms of nu (Greek ν) structure.
1. Simple nu; 2. Intermediate variety; 3. Spiral type.
C. Forms of eta (Greek η) structure.
1. Elongate form; 2. Broad form; c, cranium; cr, crown; l, link.

Horizontal shearing movements have been postulated for the mountain chains of the western Pacific region and China, where the arrangement of the orogenic arcs is explained by combinations of direct compression, horizontal shear, and torsion about a vertical axis.[2] The various patterns of folds observed in eastern Asia have been reproduced by Lee in experiments with soft wet tissue paper manipulated in various ways on a polished board.[3] The patterns represented are shown in Fig. XI—9. Lee does not discuss the relationship of these folds to a cratonic basement, but it may be that some are secondary effects of strike-slip faulting in the basement, affecting a cover of younger rocks.

[1] M. K. Hubbert, 'The Direction of the Stresses producing given Geologic Strains', *Journ. Geol.*, Vol. 36, 1928, pp. 75–84; R. T. Chamberlin, 'The Strain Ellipsoid and Appalachian Structures', ibid., pp. 85–90.

[2] S. Tokuda, 'On the Echelon Structure of the Japanese Archipelagoes', *Jap. Journ. Geol. Geog.*, Vol. 5, 1926–7, pp. 41–76.

[3] J. S. Lee, 'Some Characteristic Structural Types in Eastern Asia and their bearing on the Problem of Continental Movements', *Geol. Mag.*, Vol. 66, 1929, *passim. Idem. 'Distortion of Continental Asia'*, The Palaeobotanist, Vol. 1, 1952, pp. 298–315. Biq Ching-chang, 'The Structural Pattern of Taiwan compared with the *eta* type of shear form', *Proc. Geol. Soc. China*, Vol. 2, 1959, pp. 33–46.

A major difficulty that confronts the appeal to any vice-like mechanism between cratons is to explain close folding in a belt scores or in places hundreds of miles in width, since the soft rocks of a deforming geosyncline could not transmit stress over such great distances. Argand[1] postulated that the basement itself yields by *plis de fond*, but there is rarely evidence for

FIG. XI-10. THE DÉCOLLEMENT OF THE FOLDED JURAS
(*After Heim, from* Kober, Der Bau der Erde)

folding of the basement other than that of large radius of curvature. Where the covering rocks can slide over the crystalline basement, as a cloth slides or wrinkles on a table-top, the mechanical difficulties are less. Argand refers to folds of this type as *plis de couverture*. The Juras afford a classic example, where it is established on geological and geophysical grounds that the Mesozoic sediments exhibiting the typical Jura-type folds were folded independently of the basement, which is a complex of old rocks planed off along

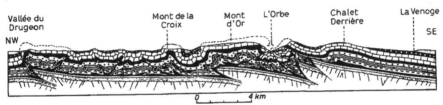

FIG. XI-11. POSSIBLE RELATION OF JURA FOLDS TO BASEMENT FAULTS
(*After Aubert, 1949*)

an unconformity on which the sediments rest. The basal Mesozoic rocks are the Anhydrite Formation of the Middle Muschelkalk, a highly incompetent unit which has acted as a lubricating layer, permitting the sliding of the beds above. This is the so-called *décollement* of the folded Juras[2] (Fig. XI-10). The folding is most intense in the south, nearer the Alps, the more northerly sectors exhibiting extensive flat-lying areas in the Tablelands (*Tafel-Jura*), with intervening narrow folded zones. Aubert[3] has argued that these latter originate as a result of the push of upthrust faults in the basement[4] (Fig. XI-11), but the origin of the Jura folds chiefly as the result of a décollement in

[1] E. Argand, 'La Tectonique de l'Asie', *Congr. Géol. Internat.*, Session 13, Belgium, 1922 (1924).

[2] L. W. Collet, *The Structure of the Alps*, London, 1927. A. Buxtorf quoted in A. Heim, *Geologie der Schweiz*, Vol. 1, Leipzig, 1919.

[3] D. Aubert, 'Le Jura', *Geol. Rundsch.*, Vol. 37, 1949, pp. 2–17.

[4] According to some, the main Jura folding took place on the foreland block fronting the Alps, by forces originating from the Alpine orogen, but an alternative view invoking gravitational sliding has been put forward – see p. 350.

which Middle Triassic and younger rocks were deformed independently of the rocks on which they now rest, is currently accepted.[1]

The origination of folds from buried fault blocks pushing into an overlying cover is however recognized in the autochthonous nappes of the Alps, and is demonstrated in cliff sections in the Isle of Anglesey, North Wales.[2] The concept is widely applicable.

Evidence for the shortening of the section of basement rocks beneath a sedimentary cover during folding is available especially in the foreland of the geosynclinal trough, where thinner sediments covering the cratonic margin permit the exposure and study of basement structures. Lees[3] adduces examples from many regions where the shortening occurs by reverse faulting (Fig. XI–12), and Migliorini discusses the geometry of basement wedges with

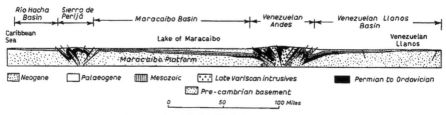

FIG. XI–12. FORELAND FOLDING DUE TO BASEMENT FAULTING
(*After G. M. Lees, 1952*)

special regard to the generation of slump-masses from the affected cover rocks.[4] Folds formed over basement faults have been termed *plis de revêtement*,[5] but the corresponding English term *drape-folds* is normally applied for open folds of the Plains type, rather than for close or overturned folds. Folds produced by strike-slip faulting in the basement are also *plis de couverture*, but we are here discussing the deformation of thick geosynclinal successions lying marginal to or between cratons, where more radical disturbance of the basement is required to produce the complexities of orogenic folding.

Argand's notions as to *plis de fond* clearly do not apply on a regional scale, as there is ample evidence for the near-permanent rigidity of crystalline basement rocks in the cratons, but the notion that deeply buried basement rocks may soften and yield plastically is implicit in many writings on folded

[1] W. G. Pierce, 'Jura Tectonics as a Décollement', *Bull. Geol. Soc. Amer.*, Vol. 77, 1966, pp. 1265–76.

[2] R. M. Shackleton, 'The Structural Evolution of North Wales', *Liv. and Manch. Geol. Journ.*, Vol. 1, Pt 3, 1953, pp. 261–97.

[3] G. M. Lees, 'Foreland Folding', Anniversary Address, *Quart. Journ. Geol. Soc. Lond.*, Vol. 108, 1952, pp. 1–34; idem, 'The Evolution of a Shrinking Earth', ibid., Vol. 109, 1953, pp. 217–57.

[4] C. I. Migliorini, 'Composite Wedges and Orogenic Landslips in the Apennines', *Rept. 18th Internat. Geol. Congr.*, Pt 13, 1952, pp. 186–98.

[5] M. Casteras, 'Recherches sur la structure du versant nord des Pyrénées centrales et ornietales', *Bull. Serv. Carte Géol. de France*, Vol. 37, 1933.

geosynclines and in the notions of syntexis and anatexis, rheomorphism and granitization under ultraplutonic conditions, which are not always related to this context. The concept of vertical zones characterized by different folding processes is even older (zone of flowage, zone of fracture), and has recently

FIG. XI-13. VERTICAL TECTONIC ZONING IN GREENLAND

1. Structures of the Basement. 2. Structure in the zone of differential movements between the basement and the cover. 3. Structures of the cover. The complete array of structures is shown below. (*From J. Haller,* Geol. Rundsch., *Vol. 45, 1946–7, p. 166*)

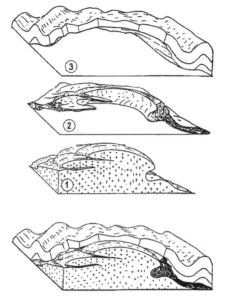

been more specifically enunciated by Haller for folds in Greenland[1] (Fig. XI–13). A sharp division between such zones appears, however, to be unlikely, unless aided by lithological differences.

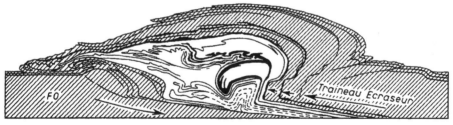

FIG. XI–14. THE NAPPE STRUCTURES OF THE WESTERN ALPS, SHOWING THE TRAINEAU ÉCRASEUR PUSHING AGAINST THE ALPINE EDIFICE AND OVER-RIDING THE FORELAND OF EUROPE (FO)

(*After Argand, from Collet*)

Shortening of the basement – In Argand's synthesis of the western Alpine structure, the required shortening of the crystalline basement is achieved by major overthrusting of the African block over the European block, so forming a great crushing sledge (*traineau écraseur* – Fig. XI–14). The *traineau*

[1] J. Haller, 'Probleme der Tiefentektonik, Bauformen im Migmatit-stockwerk der ostgrönlandischen Kaledoniden', *Geol. Rundsch.*, Vol. 45, 1956, pp. 159–67.

écraseur is one craton (the *hinterland* or African block) thrusting over the other (the *foreland* or European block).

Other theories again propose the underthrusting of the active block. This was postulated by Lawson[1] and is also inherent in Chamberlin's Wedge Theory of diastrophism, whereby the settlement of the heavy simatic rocks of the ocean basins causes them to press against the neighbouring continent, and

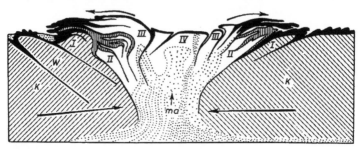

FIG. XI–15. THE STRUCTURE OF THE WESTERN ALPS SHOWING UNDERTHRUSTING OF THE GEOSYNCLINAL SEDIMENTS BY BASEMENT BLOCKS

K, crystalline basement; *W*, crystalline wedge; *I*, autochthonous nappes; *II*, parautochthonous nappes; *III*, exotic nappes; *IV*, median massif. *ma*, granitic magma. (*After Kober*)

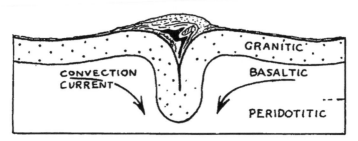

FIG. XI–16. THE TECTOGENE OR CRUSTAL DOWNBUCKLE

The west Alpine nappes are shown to scale on the buckled granitic layer. (*After Griggs, 1939*)

thrust under them.[2] Somewhat similar ideas on deep underthrusting were expressed by Hobbs.[3]

The movements postulated by Kober for the Alps are, however, of a different kind, picturing only the underthrusting of the geosynclinal sediments by both the flanking cratons as they move together[4] (Fig. XI–15). Such a

[1] A. C. Lawson, 'Insular Arcs, Foredeeps, and Geosynclinal Seas of the Asiatic Coast', *Bull. Geol. Soc. Amer.*, Vol. 43, 1932, pp. 353–81.

[2] R. T. Chamberlin, 'The Wedge Theory of Diastrophism', *Journ. Geol.*, Vol. 33, 1925, pp. 755–92.

[3] W. H. Hobbs, 'Mechanics of Formation of Arcuate Mountains', *Journ. Geol.*, Vol. 22, 1914, pp. 71–90, 166–88, 193–208.

[4] L. Kober, *Bau und Entstehung der Alpen*, Vienna, 1955.

notion is bound up with many current theories, especially those that involve the drag of convection currents in the mantle, on the base of the sialic crust. In these theories the shortening of the basement section results from the downward turning of the granitic layer [1] or its actual folding to form a *crustal down-buckle* [2] which is the basic feature of a *tectogene* (Fig. XI–16). As the down-buckle develops the sedimentary cover is skimmed off and develops complex structures independently of the simple downfold. Experiments by Griggs support the hypothesis (p. 138).

In the above theories, relative displacements of crustal blocks amounting to many miles are postulated, and similar displacements are now believed to be

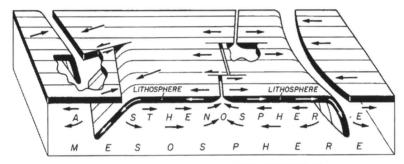

FIG. XI–17. A VERSION OF GLOBAL TECTONICS

Arrows on lithosphere indicate relative movements of adjoining blocks. Arrows in asthenosphere represent possible compensating flow in response to downward movement of segments of lithosphere. One arc-to-arc transform fault appears at left between oppositely facing zones of convergence (island arcs), two ridge-to-ridge transform faults along ocean ridge at centre, simple arc structure at right. (*From Isacks et al., 1968*)

connected with spreading ocean floors (see pp. 471–9). Where moving blocks impinge against each other at less than six centimetres per year, crustal down-buckling as in a tectogene can result; with higher relative speeds one plate plunges beneath the other and is partly destroyed in the hot asthenosphere, part of the upper mantle. In this latter case a trench, with associated vulcanicity, results [3] (Fig. XI–17).

Folding due to strike-slip faulting – The above brief outline of folding in orogens indicates that while the cratons are relatively rigid they are shattered at their edges during orogeny (crystalline wedges) and in depth beneath the former geosyncline they probably undergo much more drastic deformation

[1] A. Holmes, *Principles of Physical Geology*, London, 1965.

[2] F. A. Vening Meinesz, 'Gravity Expeditions at Sea', *Netherlands Geodetic Comm.*, Vol. 2, 1934, Delft. D. Griggs, 'A Theory of Mountain-Building', *Amer. Journ. Sci.*, Vol. 237, 1939, pp. 611–50.

[3] B. Isacks, J. Oliver, and L. R. Sykes, 'Seismology and the New Global Tectonics', *Journ. Geophys. Research*, Vol. 73, 1968, pp. 5855–99. H. W. Menard, 'The Deep-Ocean Floor', *Sci. Amer.*, Vol. 221, 1969, pp. 126–42.

and transformation. It is particularly in this deeper zone that we have evidence for synchronous granitic intrusions and the formation of schists and gneisses by metamorphism or transfusion of the deep sediments and of the basement, producing metamorphic belts that appear to be the roots of orogens. The folding of the geosynclinal rocks is, however, widely distributed throughout the thick succession of sediments. By way of contrast, *plis de*

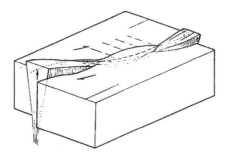

FIG. XI–18. SHATTERING ALONG A
STRIKE-SLIP FAULT

The wedges would form anticlines and synclines in cover-rocks overlying a crystalline basement. (*From Kingma*)

couverture are found in a soft cover lying over a more rigid basement. Strike-slip faulting in the basement gives rise to narrow zones of folding and secondary faulting with relatively flat-lying beds of wide extent flanking the folded zone on either side.

The secondary folds generated in cover-rocks by strike-slip faults in the basement must be distinguished from primary folds that may be formed by the same stresses that give rise to the faults. The geometrical relationships of these to the faults is referred to in an earlier chapter.

FIG. XI–19. FOLDING NORMAL TO A
STRIKE-SLIP FAULT
(*From Cotton*)

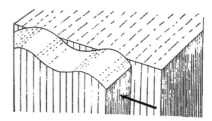

Secondary folds parallel to a strike-slip fault form where a fault zone rather than a single clean shear plane is present. Slivers of basement are squeezed upwards (Fig. XI–18) producing anticlines in the cover, or again highly plastic clays or salt-beds may be mobilized and squeezed upwards, as in parts of the San Andreas fault zone.[1] Slices of the cover-rocks caught up between the walls of a strike-slip fault zone are, broadly regarded, synclinal structures.

Folds normal to a strike-slip fault arise where one flank is compressed more than the other, as shown by Cotton[2] (Fig. XI–19).

[1] See R. E. Wallace, 'Structure of a Portion of the San Andreas Rift in Southern California', *Bull. Geol. Soc. Amer.*, Vol. 60, 1949, pp. 781–806.
[2] C. A. Cotton, 'Geomechanics of New Zealand Mountain-Building', *N.Z. Journ. Sci. & Technol.*, Vol. 38, 1956, pp. 187–200.

The third and probably the most common type is where drape-folds are arranged *en echelon* in a zone above the strike-slip fault. Whether right- or left-handed echelons will be formed will depend on the physical properties of the rocks affected, and examples where the sense of movement on the basement fault can be determined with certainty are not available. The possibility that the basement itself may fracture by Riedel shears or tension gashes also exists, and these would also yield *en echelon* drape-folds in the cover (Fig. XI–20).[1]

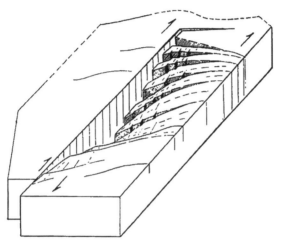

FIG. XI–20. *En echelon* FAULTING ALONG A STRIKE-SLIP FAULT
Drape-folds would form in cover-rocks over the wedges. (*From Kingma*)

Resurgent tectonics – It is well known that faults of great antiquity may be rejuvenated under later stress, and in fact any large structural element in the crust may influence structures of subsequent origin. Thus the application of idealized concepts such as the strain ellipsoid must be modified in dealing with extensive areas by inhomogeneities of the crust, which is very far from being isotropic in relation to stress. Ruedemann,[2] from a study of Pre-Cambrian trend lines, first showed that the fundamental framework of continental structure is outlined by the trends of ancient foliation, folds, and shear zones, which have, presumably, influenced certain of the younger structural lines. The rejuvenation of old orogens and the renewal of movements on major faults have recently received considerable attention, especially in the great shields of Africa and Australia. Many of the faults of the East African Rifts are as old as Mesozoic, and there is good evidence that some are even Pre-Cambrian strike-slip faults, transformed during later time into normal faults.

[1] J. T. Kingma, 'Possible Origin of Piercement Structures, Local Unconformities, and Secondary Basins in the Eastern Geosyncline, New Zealand', *N.Z. Journ. Geol. Geophys.*, Vol. 1, 1958, pp. 269–74. D. R. Shawe, 'Strike-Slip Control of Basin-Range Structure Indicated by Historical Faults in Western Nevada', *Bull. Geol. Soc. Amer.*, Vol. 76, 1965, pp. 1361–78.
[2] R. Ruedemann, 'The Existence and Configuration of Pre-Cambrian Continents', *N.Y. State Mus.*, Bull. 239–40, 1920, pp. 67–152.

The evidence that in shields and cratons, old structures influence younger, is, as would indeed be suspected on *a priori* grounds, very strong. However, where the deformation of cratons by later movements is more profound, especially if it involves rock-flowage in depth, new directions may obliterate old, or again, in an intermediate condition, competent massifs may persist and influence later trends, while less competent zones yield in a new direction.[1]

The commonly held idea that orogeny welds geosynclinal rocks to neighbouring cratons and thus immobilizes them is only partially correct. The orogen may be later rejuvenated, and again, cratonic basement rocks underlie many great geosynclines, so that the notion of concentric growth of all continental masses about cratonic nuclei is fallacious. Indeed, many geosynclinal zones appear to have arisen from the fracturing or fragmentation of larger Pre-Cambrian platforms.[2] This concept, so well applicable to many Lower and Upper Proterozoic geosynclines and troughs, may not apply for the margins of the Pacific Ocean, or even for the Mediterranean orogens, each of which may in fact be unique as to its fundamental geotectonics.

Block tectonics

The subdivision or fragmentation of a large platform by mobile zones, whether geosynclines *sensu stricto*, or troughs, furrows or synéclises, has been recognized in many continents. Indeed, the obvious fact that great faults or zones of *Bruchfalten* divide cratons into smaller blocks which retain a certain individuality as tectonic units for very long periods of time implies that within large cratons an array of smaller cratonic blocks may be recognized. These latter, sub-rectangular, lozenge-shaped, or polygonal, may be depressed to form sedimentary basins or elevated as outcropping coigns of old rocks. Their margins, and the zones between the blocks, are tectonically active in various degrees, and may exhibit faulting, folding, downwarping to form hinge-lines at the edges of basins or geosynclines, or upwarping to form plateau margins[3] (Fig. XI–21).

Subdivision in this way has been demonstrated by Krenkel for Africa,[4] by

[1] W. F. Brace, 'Interaction of Basement and Mantle during Folding near Rutland, Vermont', *Amer. Journ. Sci.*, Vol. 256, 1958, pp. 241–56. See also U.S.–Canada Basement Symposium, *Amer. Assoc. Petrol. Geol.*, Vol. 49, No. 7, 1965, pp. 887–1041.

[2] E. S. Hills, 'Some Aspects of the Tectonics of Australia', *Journ. & Proc. Roy. Soc. N.S. Wales*, Vol. 79, 1946, pp. 67–91. A. V. Peyoe and V. M. Sinitzin. 'Certains problèmes fundamentaux de la doctrine des géosynclinaux', *Izv. Akad. Nauk.*, Sér. Géol. No. 4, 1950, pp. 28–52. H. Stille, 'Recent deformations of the earth's crust in the light of those of earlier epochs', *Geol. Soc. Amer.*, Spec. Pap. 62, 1955.

[3] J. H. Rattigan, 'Fold and Fracture Patterns Resulting from Basement Wrenching in the Fitzroy Depression, Western Australia', *Proc. Aust. Instit. Mining & Met.*, No. 223, 1967, pp. 17–22; J. G. Smith, 'Tectonics of the Fitzroy Wrench Trough, Western Australia', *Amer. Jour. Sci.*, Vol. 266, 1968, pp. 766–76.

[4] E. Krenkel, *Geologie Afrikas*, Berlin, Vol. 1, 1925, Fig. 4.

Cloos for Europe[1] and by the author and others for Australia.[2] Among the effects are the formation of polygonal basins 'framed' by outcropping basement rocks but underlain by depressed blocks, and a notable tendency for the sedimentary filling of basins to increase very rapidly in thickness at the

FIG. XI–21A. THE MURRAY BASIN, SOUTHERN AUSTRALIA
Photograph of relief model.

margin of a block rather than to overlap as a gradually thickening sedimentary wedge. Again, the weak zones becomes belts of sedimentation and subsequent folding, varying from minor disturbance to intense orogeny. According to H. Cloos basement block-structure determines the complex pattern in plan of the folded zones of Europe (Fig. XI–1, 22), and van Bemmelen classes the effects of block movements in Europe as meso-tectonic Cainozoic

[1] H. Cloos, 'The Ancient European Basement Blocks – Preliminary Note', *Trans. Amer. Geophys. Union*, Vol. 29, 1948, pp. 99–103.
[2] E. S. Hills, 'Some Aspects of the Tectonics of Australia', *Journ. & Proc. Roy. Soc. N.S. Wales*, Vol. 79, 1946, pp. 67–91.

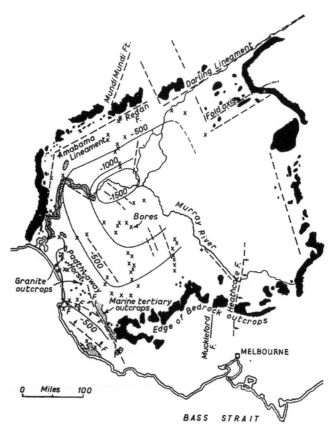

FIG. XI-21B. THE MURRAY BASIN, SOUTHERN AUSTRALIA

Sketch map to show some major faults that frame the basin, and approximate structure contours drawn near the base of the Tertiary rocks in the basin.

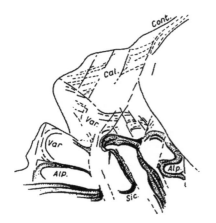

FIG. XI-22. EUROPEAN BASEMENT BLOCKS

Showing the main subdivisions of the basement and their relationship to Caledonian, Variscan and Alpine fold belts. (*After H. Cloos*)

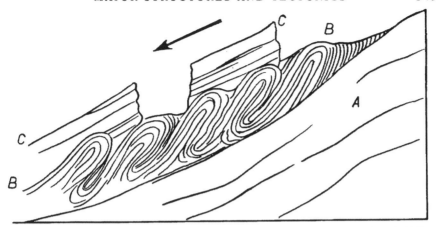

FIG. XI–23. GRAVITATIONAL TECTONICS ACCORDING TO NAUMANN (1850)
(*Lehrbuch der Geognosie* Leipzig, 1819–1850)

deformations distinct from megatectonic effects such as continental drift and ocean floor spreading.[1]

Questions as to the antiquity and permanence of crustal blocks are not readily answered, for although they may be recognized as individual units for periods of time of the order of 100–150 million years, it is clear that at times they may be broken and new unit blocks formed. This is demonstrated by the sedimentary troughs of Lower Proterozoic age in northern Australia, which in places cross at right angles. Such cross-fracturing generally occurs along the same pronounced trend directions as determine the older block outlines.[2] The subject is linked closely with lineament tectonics (see Chapter XIV).

Gravitational tectonics

The term gravitational tectonics (*tectonique d'écoulement*; *Fliesstektonik*) implies the gliding of very large rock-masses down gentle slopes which originated chiefly through the action of hypogene forces. The gliding is accompanied by folding and faulting, ranging from rather simple structures to those of orogenic, and particularly of Alpine style. It is the formation of important geological structures during sliding, which implies a general creep of the rocks rather than disruptive slipping, that distinguishes gravitational tectonics from related phenomena such as slumping and major landsliding. The idea is a very old one in geological philosophy, having been evoked by Gillet-Lamont in 1799, by Kühn in 1836, and by Naumann (1850)[3] (Fig. XI–23),

---

[1] R. W. van Bemmelen, 'The Alpine Loop of the Tethys Zone', *Tectonophysics*, Vol. 8, 1969, pp. 107–13.

[2] E. S. Hills, 'The Tectonic Style of Australia', *Geotekt. Sympos. zu Erhren v. Hans Stille*, Deutsch. Geol. Ges., Stuttgart, 1956, pp. 336–46.

[3] F. P. W. Gillet-Lamont, 'Observations géologiques sur le gisement et la forme des replis successifs, etc.', *Journal des Mines*, No. 59, Ann. 7, 1799, pp. 449–54. K. A. Kühn, *Handbuch der Geognosie*, Freiberg, 1836.

but until the 1930s it was completely subordinated by the adherence of most geologists to the contraction theory, with its attendant stress on folding by lateral compression between rigid earth blocks – the 'jaws of a vice' mechanism. The recognition of nappe structures in the western Alps, and the drawing of geological cross-sections that in themselves virtually pose the proposition of flowage for the vast yet relatively thin recumbent folds, led Schardt to propose gravitational sliding for the Prealps,[1] but in later years it was chiefly from physicists that the notion received any support.[2] In 1930 Haarmann published his Oscillation Theory, according to which structures of

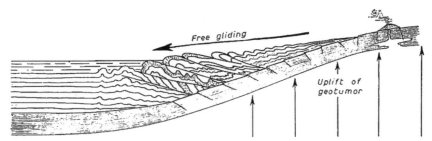

FIG. XI–24. SECONDARY (GRAVITATIONAL) TECTONICS
(*After Haarmann*)

orogenic type may be formed by *free-gliding* under gravity, down the slopes of an uplifted geotumor, or by gravitational sliding in a sedimentary basin (*full-trough gliding*), or by gliding of plastic rocks enclosed between resistant beds as in drag-folding (*Expressionsgleitung*)[3] (Fig. XI – 24).

Haarmann adduced examples of free-gliding from the nappes, but it was the adherence of the alpine field-geologists Schneegans and Lugeon to the theory in relation to the nappes Ubaye–Embrunais in 1938–40[4] that gave a guarantee of respectability to the notion of gravitational tectonics and stimulated its wide application in the Alps, the Rif of Morocco, Algeria, the Apennines, and even for the folding of the Jura.[5]

[1] H. Schardt, 'Sur l'origine des Préalpes romandes (Zone du Chablais et du Stockhorn)', *Eclog. geol. Helvet.*, Vol. 4, 1893, p. 129. E. B. Bailey, *Tectonic Essays Mainly Alpine*, Oxford, 1935, gives a good regional description.

[2] R. M. Deeley, 'Mountain Building', *Geol. Mag.*, Dec. 6, Vol. 5, 1918, pp. 111–20; H. Jeffreys, 'On the Mechanics of Mountains', ibid., Vol. 68, 1931, pp. 435–42.

[3] H. Haarmann, *Die Oszillationstheorie Eine Erklärung der Krustenbewegungen von Erde und Mond*, Stuttgart, 1930.

[4] D. Schneegans, 'La Géologie des nappes de l'Ubaye–Embrunais entre la Durance et l'Ubaye', *Mém. Serv. Carte. Géol. France*, 1938. M. Lugeon and D. Schneegans, 'Sur le diastrophisme alpin', *C.R. Acad. Sci.*, No. 210, 1940, pp. 87–90.

[5] An excellent review is given by E. Gagnebin, 'Quelques problèmes de la tectonique d'écoulement en Suisse orientale', *Bull. Lab. Géol. Univ. Lausanne*, No. 80, 1945. See also J. Goguel, *Traité de Tectonique*, Paris, 1952, pp. 207–14. M. Gignoux, 'La Tectonique d'Écoulement par Gravité et la Structure des Alpes', *Bull. Soc. Géol. France*, Ser. 5, Vol. 18, 1948, pp. 739–64; idem, 'Sur les nouvelles Théories de la Tectonique d'Écoulement', *C.R. Congr. Géol. Internat.*, 19th Session, Algiers, 1952 (1953), Sect. III, Fasc. III, pp. 193–6. Jan R. van de Fliert, 'Tectonique d'Écoulement et Trias Diapir au Chetthaabas, Algérie', ibid., pp. 71–88. L.

Van Bemmelen's Undation Theory, which was propounded shortly after Haarmann's Oscillation Theory and resembles it in the emphasis placed on primary vertical movements in the crust followed by secondary tectogenesis due chiefly to gravitational sliding, is potentially of general applicability, but its fundamental tenets relate largely to sub-crustal phenomena rather than to structural geology.[1]

Gravitational tectonics affords an acceptable mechanism to explain, in particular, certain thin nappe sheets that have been translated several tens of

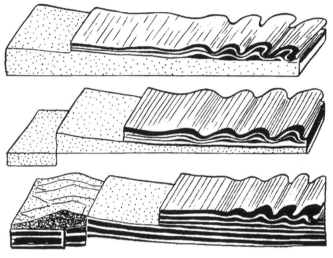

FIG. XI–25. GRAVITATIONAL TECTONICS ACCORDING TO REYER (1888)

miles from their roots, and which, because of the predominance of soft rocks in the sheet, cannot be assumed to have been thrust along by end-on push. Gravitational forces are body forces, and differ fundamentally in their action from applied forces in that they are not subject to the necessity of being transmitted from point to point through a body, but act, of their own nature, on every particle. The necessity for transmitting forces for scores of miles through soft sediments by end-on push introduces a difficulty which may be overcome if recourse is had to a resolute of gravity acting on rocks resting on an inclined plane. This was proposed by Reyer in 1888 (Fig. XI–25).[2] The

---

U. de Sitter, 'Gravitational Gliding Tectonics: An Essay in Comparative Structural Geology', *Amer. Journ. Sci.*, Vol. 252, 1954, pp. 321–44. The reported major slides of the northern Apennines in Italy are briefly discussed in English by de Sitter, *Amer. Journ. Sci.*, Vol. 252, 1954, pp. 321–44. See also H. Korn and H. Martin, 'Gravity tectonics in the Naukluft Mountains of South West Africa', *Bull. Geol. Soc. Amer.*, Vol. 70, 1959, pp. 1047–78.

[1] R. W. van Bemmelen, 'On the Geophysical Foundations of the Undation Theory', *Kongl. Akad. Weten., Amsterdam*, Vol. 36, 1933, pp. 337–43; 'Die Anwendung der Undationstheorie auf das alpine System in Europe', ibid., pp. 686–93.

[2] E. Reyer, *Theoretische Geologie*, Stuttgart, 1888.

inclined plane may be an original geological surface such as an unconformity or even a bedding-plane, especially if lubricated by soft rocks such as clay, shale, gypsum, or salt, or again, a surface of rupture, that is, a low-angle fault comparable with the basal slip-surface of a landslide. In order to produce complex folding in the sliding mass it is, however, necessary that it should deform freely by hydro-plasticity, crystal plasticity, or both. Thus only soft rocks such as plastic clay, gypseous shale, or rock-salt can be expected to

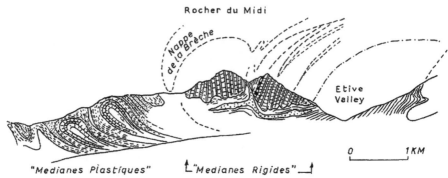

FIG. XI–26. THE *Medianes Rigides* OF THE PREALPS, SHOWING DISRUPTION OF MASSIVE LIMESTONES DURING GRAVITATIONAL SLIDING
(*After Lugeon and Gagnebin*)

flow under low confining pressures, and the flow of stronger rocks should leave its mark in the crystalline structure, as in schists. The reaction of limestones remains somewhat obscure, in that in some instances a thin limestone will be disrupted; in other instances (probably the more common) thin bedded limestones yield plastically by folding and stretching while more massive limestones in the same environment act as relatively rigid rocks (Fig. XI–26). Since it is known from experiments that dolomite is less readily deformable than calcite,[1] it is most likely that much depends on the composition of the limestone, and again on the presence of agents capable of inducing rehealing, or even, on the presence or absence of planes of mechanical weakness on which small movements may go on, such as joints and bedding (*movements sur joints*).

Although slopes down which gliding may have occurred are demonstrable in many Alpine nappes, it has been necessary also to postulate for certain localities where the slope is now reversed, that this may have been brought about by movements subsequent to the gliding. However, with masses some thousands of feet thick a slope of only a few degrees would appear to be necessary in the case of the sliding of soft sediments. On the other hand, the shearing and slicing of the crystalline basement that is seen in the roots of the

[1] F. J. Turner *et al.*, 'Plastic Deformation of Dolomite Rock at 380°C', *Amer. Journ. Sci.*, Vol. 252, 1954, pp. 477–88.

autochthonous and parautochthonous nappes is surely an indication of the
action of applied forces of hypogene origin, for which the great elevation of
the alpine chain itself is a witness, and which afforded the *primary tectogenesis* to which the gravitational effects are secondary. Indeed, as the elevation
of the Alps is still very considerable, it is perhaps a little strange that
·gravitational flowage is not more in evidence at the present time, despite the
unloading due to erosion, and Goguel has given a timely warning that in order
to justifiably hypothecate gravitational flowage, a thickness of several thousand feet is required to give the necessary plasticity to rocks such as massive
limestones, so that a thicker cover than exists today must be assumed for
many nappes. In the case where massive limestones are enclosed in shale,
however, the limestones may be broken into slices or blocks and carried along
with the yielding shale as in the *Medianes rigides* of the Prealps[1] (Fig. XI–
26).

The frontal region of many nappes shows overfolds in which the reverse
limb is but little if at all thinned, so that the total thickness of the section has
been nearly doubled (Fig. XI–27). This indicates a forward movement of the

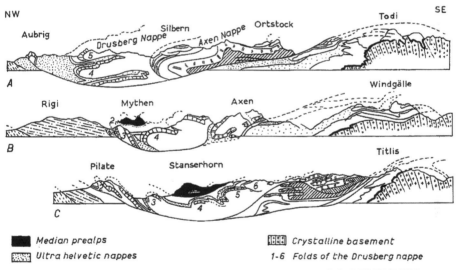

Median prealps   Crystalline basement
Ultra helvetic nappes   1-6 Folds of the Drusberg nappe

FIG. XI–27. THE PREALPS – AN EXAMPLE OF GRAVITATIONAL TECTONICS
(*After Lugeon and Gagnebin*)

normal limb, which advances over the reverse limb and would therefore imply
either drag by a still higher advancing thrust sheet, or gravitational flow. The
increase in thickness, however, would be resisted by the weight of an over-
riding thrust nappe. Thus neither the notion of end-on push to produce the
nappes, nor the notion of drag (*entrainement*) beneath an over-riding thrust

[1] M. Lugeon and E. Gagnebin, 'Observations et vues nouvelles sur la géologie des Préalpes
romandes', *Bull. Lab. Géol., Univ. Lausanne*, No. 72, 1941.

mass (the mechanics of which itself would involve equal difficulties) offers a real alternative to gravitational flow. Certainly the notion of the 'crushing sledge' (*traineau écraseur*), whereby the Alpine edifice was thought to have been produced by a southern block over-riding Europe,[1] was both grandiose and unsatisfactory, and the newer concepts are more in accord both with the flowing lines of Alpine tectonics and with the lobal and arcuate plan of the termination of nappes, whether they plunge into troughs filled with sediments, or mould themselves to the outline of resistant massifs in their path. The extension of the gliding hypothesis to the Juras[2] had to contend with the difficulty that the postulated gliding towards the north was opposed by the tilt of the basement, which today is towards the south, and a reversal of tilt was therefore proposed. This hypothesis has, however, been criticized because of the existence of various definite phases in the Jura folding over a long period of time, and because there is evidence of an influence of basement fracturing on the folding of the sedimentary cover.[3]

Gravity collapse structures. — Among the simplest and still perhaps the most convincing examples of limited sliding tectonics in mountainous regions is that described from Iran by Harrison and Falcon as *gravity collapse structures*.[4] These occur in a series of interbedded limestones and marls which are folded into regular anticlines and synclines of simple form, approximating to sine curves, but of great amplitude (12,000–15,000 ft). A thick Pre-Albian limestone forms the core of the folds and is exposed in simple dome-shaped anticlinal mountains, but in the marls, sandstones, and silts above, ending with the Miocene Lower Fars series which contains anhydrite (with gypsum near the surface), a variety of structures is seen that is explained by sliding of these beds down the flanks of the great anticlines into the synclinal valleys. The structures, which are illustrated in Fig. XI–28, show a sequence of increasing amount of flowage, some with folding, others with faulting (*slip-sheets*), and finally, in the extreme case, complete overturning of limestone beds to form a *flap*.

In these structures it will be noted that the folds ascribed to collapse have an attitude that is the reverse of the normal drag-fold relationship to the anticline, and point to movement of the upper beds towards the syncline. In this as in other examples referred to gravitational tectonics, there remains a

[1] L. W. Collet, *The Structure of the Alps*, London, 1927.

[2] M. Lugeon, 'Une hypothèse sur l'origine du Jura', *Bull. Lab. Géol., Univ. Lausanne*, No. 73, 1941.

[3] D. Aubert, 'Le Jura et la tectonique d'écoulement', *Bull. Lab. Géol., Univ. Lausanne*, No. 83, 1945. H. Suter, 'Tektonische Juraprobleme, ein historischer Rückblick', *Eclog. Geol. Helvet.*, Vol. 49, 1956, pp. 363–72.

[4] J. V. Harrison and N. L. Falcon, 'Collapse Structures', *Geol. Mag.*, Vol. 71, 1934, pp. 529–39; idem, 'Gravity Collapse Structures and Mountain Ranges, as exemplified in Southwestern Persia', *Quart. Journ. Geol. Soc. Lond.*, Vol. 92, 1936, pp. 91–102. G. M. Lees, 'The Geology of the Oilfield Belt of Iran and Iraq', in *The Science of Petroleum*, Oxford, 1938, pp. 140–8.

possibility that the movements went on under a cover of superincumbent rocks that have now been eroded away, rather than with the topography now existing. Thus the highly mobile Lower Fars beds at the top of the succession, which were originally about 10,000 ft thick, would almost certainly have flowed towards the synclines and dragged the Asmari Limestone and lower

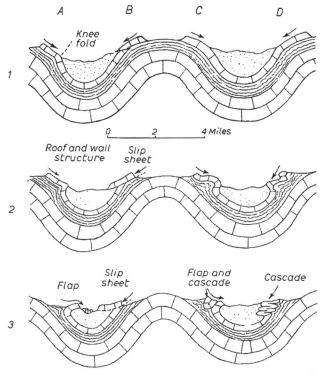

FIG. XI-28. GRAVITY-COLLAPSE STRUCTURES

The diagrams represent successive stages in the development of a flap (sequence below A), a slip-sheet (below B), a flap and cascade (below C) and a cascade (below D). Natural scale. (*After Harrison and Falcon, 1934*)

beds with them. Nevertheless the movement pattern of the Persian collapse structures is so clearly down-flank that a primary tectonic origin is ruled out and their origin by gravitational sliding seems incontestable.

Sliding and vulcanicity — Sliding that may have been induced by the weight of volcanic rocks piled on an elevated dome has been suggested by Reeves[1] and by van Bemmelen,[2] but in each example effects such as magmatic or regional

[1] F. Reeves, 'Origin and Mechanics of the Thrust Faults adjacent to the Bearpaw Mountains, Montana', *Bull. Geol. Soc. Amer.*, Vol. 57, 1946, pp. 1033–48.

[2] R. W. van Bemmelen, 'Ein Beispiel für Sekundärtektogenese auf Java', *Geol. Rundsch.*, Vol. 25, 1934, pp. 175–94; *Mountain Building*, The Hague, 1954, p. 89.

forces may need to be considered. The Javanese example is illustrated in Fig. XI–29, which is self-explanatory, and that of the Bearpaw Mountains of Montana in Fig. XI–30. In the latter, a series of superficial upthrusts, each associated with an anticline, is limited to the beds lying above a bentonite horizon that afforded a gliding surface extending for 20–30 miles from the volcanic centre from which the gliding developed. Between the faults the beds remain quite flat, and the sliding must have occurred on a slope of only 3°,

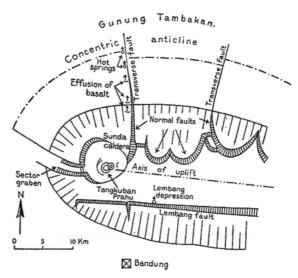

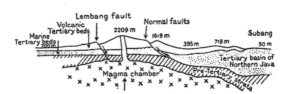

FIG. XI–29. GRAVITATIONAL TECTONICS IN JAVA
(*After van Bemmelen*)

aided by the accumulation of some 5000 ft (max.) of volcanic rocks at the Bearpaw Mountains, and the disturbing effects of several hundred small earthquake shocks per day.

Gravitational flowage in sedimentary basins

Quite apart from the slumping of superficial deposits, it is conceived that sediments already deposited to a considerable thickness in a basin may, as a result of further sinking of the axis of the rough, of tilting of the flanks, or a succession of earthquake shocks, creep down-slope with the accompaniment of folding and low-angle thrusting. Haarmann termed the process

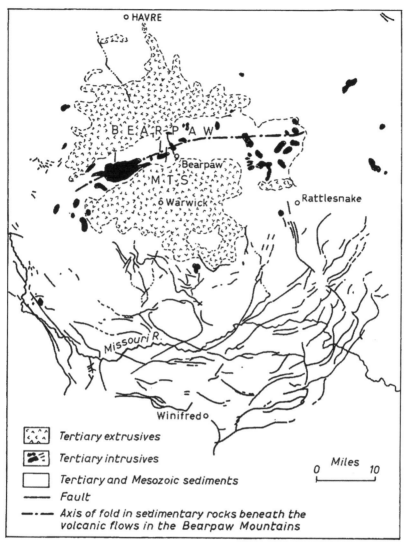

○ HAVRE

B E A R P A W

Bearpaw

M T S

○ Warwick

○ Rattlesnake

Missouri R.

Winifred ○

```
⟨⟨⟨⟩⟩⟩  Tertiary extrusives

◆◆◆     Tertiary intrusives

☐       Tertiary and Mesozoic sediments

——      Fault

—·—·—   Axis of fold in sedimentary rocks beneath the
        volcanic flows in the Bearpaw Mountains
```

Miles
0   10

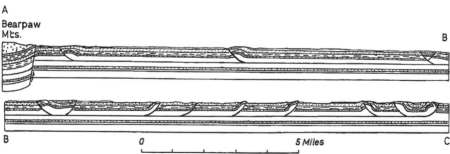

A
Bearpaw
Mts.

B

B

C

0   5 Miles

FIG. XI–30. THE BEARPAW MOUNTAINS

An example of sliding tectonics in a region of relatively simple structure. (*After Reeves*)

Volltroggleitung (full-trough gliding), and suggested that the increasing intensity of folding in depth in the Ruhr coal-basin might be so explained. Fourmarier[1] has also explained the common increase of fold intensity in depth in geosynclines as due to a 'slow descent of material' towards the centre of the trough. G. W. Bain has discussed this in dealing with the marble belt of Vermont.[2] Like Reyer and Haarmann he points to strongly asymmetrical folds inclined away from the flanks of broad anticlinal arches, where the beds are thinned to one-fifth or one-tenth of their original thickness; fold-intensity varies very little across a fold-train, whereas folds due to lateral thrust are

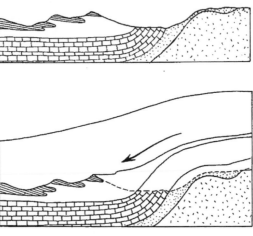

FIG. XI–31. FLOWAGE FOLDING IN VERMONT
(*After G. W. Bain*)

most intense near the moving mass that applied the thrust, and the folding dies away very rapidly at the distal end; great variations occur in the thickness of the incompetent beds (Fig. XI–31).

Many of the structures ascribed to gravitational flowage are strong orogenic-type folds and faults, but increasing attention is being given to the idea that much simpler structures such as the small, narrow anticlines of the Gulf Coast oilfield of U.S.A. may be of gravitational origin. In the Gulf Coast geosyncline a thick sedimentary succession was deposited on a somewhat irregular, but regionally sloping basement. The thinning of incompetent formations over anticlines and their thickening in troughs, and the parallelism of anticlinal crests with the depositional strike of the sediments, point to slow creep down the slope, which is inclined at 2°–3° or more. Shale and salt are the mobile beds involved, and their lateral flow is thought to be induced by the

[1] P. Fourmarier, 'Efforts tangentiels et efforts verticaux dans la tectogenèse?', *Bull. Soc. Géol. Belg.*, Vol. 69, 1946, No. 4, pp. 87–182.
[2] G. W. Bain, 'Flowage Folding', *Amer. Journ. Sci.*, Vol. 22, 1931, pp. 503–30. See also H. de Terra, 'Structural Features in Gliding Strata', ibid., Vol. 21, 1931, pp. 204–13.

great weight of super-incumbent deposits, and the regional slope. The corollary to folding, that is, tension-faulting in the up-slope part of a mass subject to gravitational creep, is demonstrable in the Gulf Coast region.[1] Ramberg has made model studies simulating gravity tectonics by centrifuging. The results suggest that not only the superficial rocks are involved but also the rise of gneiss-filled anticlines and the sinking of major synclines, as found in the Scandinavian Caledonides.[2]

[1] M. Bornhauser, 'Gulf Coast Tectonics', *Bull. Amer. Assoc. Petrol. Geol.*, Vol. 42, 1958, pp. 339–70; E. Cloos, 'Experimental Analysis of Gulf Coast Fracture Patterns', *Amer. Assoc. Petrol. Geol.*, Vol. 52, 1968, pp. 420–44.

[2] H. Ramberg, 'Experimental Study of Gravity Tectonics by Means of Centrifuged Models', *Bull. Geol. Instit. Univ. Uppsala*, Vol. 42, 1963, pp. 1–97; idem, 'Experimental Models of Fold Mountains', *Geol. Foren. Stockholm Förhandl.*, Vol. 87, 1966, pp. 484–91; idem, 'The Scandinavian Caledonides as Studied by Centrifuged Models', *Bull. Geol. Instit. Univ. Uppsala*, Vol. 43, 1967, pp. 1–72.

CHAPTER XII

Igneous rocks

The structural mapping of igneous massifs received a great stimulus when Hans Cloos suggested ways in which flow structures, joints, and faults might be systematically related to the shape of intrusions and their mode of emplacement.[1] According to Cloos, intrusions may be considered firstly as fluids which move under pressure in chambers within the crust, which chambers may also be enlarged and moulded by the intrusive processes. During this stage *flow structures* are formed. In the second stage when the intrusion has cooled somewhat, a two-phase system is produced consisting of crystals forming a mesh-work with residual liquid between the grains. This is capable of sustaining shearing stress and may yield by faulting or by the opening of tension gashes. At this stage, during which forces are still operating to enlarge the intrusion and to force the walls apart, structures due to *fracture* are first formed, and they continue to form while the rock solidifies. Finally, after complete solidification, further fracturing due to crustal stresses and also to cooling stresses may be superimposed on the earlier, but in Cloos' synthesis these post-solidification structures, which previously had been thought to be chiefly responsible for the jointing of granites, are of minor importance. These notions find their most fruitful application in the study of intrusions of plutonic type (plutons), but they also, afford a basis for the study of all igneous rocks and, with some modification, to zones of granitization or transfusion.

STRUCTURES OF THE FLUID AND PLASTIC STAGES

During the fluid stage of igneous rocks some structures are formed due to the crystallization processes, and others due to flow in the liquid. Apart from

[1] H. Cloos, *Der Mechanismus tiefvulkanischer Vorgänge*, Braunschweig, 1921, Sammlung Vieweg; idem, 'Das Batholithenproblem', *Fortschr. Geol. u. Pal.*, Hft. I, 1923, p. 80; idem, 'Einführung in die tektonische Behandlung magmatischer Erscheinungen: Pt I, Das Riesengebirge', Berlin, 1925 (with Bibliography to 1925); idem, 'Einige Versuche zur Granittektonik', *Neues Jahrb. f. Min.*, etc., Beil. Bd. 64, Abt. A, 1931, pp. 829–36. The most complete account of the subject in English is given by R. Balk, 'Structural Behaviour of Igneous Rocks', *Geol. Soc. Amer.*, Mem. No. 5, 1937.

micro-structures and textures, with which we are not particularly concerned for the moment, most of these structures are represented by some form of banding or layering, or in sub-aqueous lava flows by pillow structure.

Banding due to crystallization-processes – Banding of basic rocks is often strikingly shown by a sharp contrast between white felspar-rich layers and dark-green or black ferromagnesian-rich layers seen for instance in banded gabbros and norites (Fig. XII–1), and strong banding in nepheline syenites,

FIG. XII–I. LAYERING IN THE SKAERGAARD INTRUSION, EAST
GREENLAND
(*By courtesy L. R. Wager*)

involving mineral grading, has been studied in Greenland and the Kola Peninsula.[1] The structure is believed to be caused by successive crops of crystals segregating themselves according to their densities as they sink through the liquid. This notion is strongly supported by the gradation seen in the crystal distribution, resembling graded bedding in sediments, and by the concentration of heavy minerals such as chromite and ilmenite among the ferromagnesian silicates of the dark bands.[2] There is, moreover, much evidence to suggest that such banding is normally developed horizontally, and

[1] H. Sörensen, 'Rhythmic Igneous Layering in Peralkaline Intrusions', *Lithos*, Vol. 2, 1969, pp. 261–83.
[2] L. R. Wager and W. A. Deer, 'The Petrology of the Skaergaard Intrusion, Kangerdlugssuaq, East Greenland', *Med. om Grönld.*, Vol. 105, 1939; B. V. Lombaard, 'On the Differentiation and Relationships of the Rocks of the Bushveld Complex', *Trans. Geol. Soc. S. Africa*, Vol. 37, 1935, pp. 5–52; H. H. Hess, 'Primary Banding in Norite and Gabbro', *Trans. Amer. Geophys. Union*, 19th Ann. Meet., Vol. 19, 1938, pp. 264–8.

A

B

FIG. XII–2. PILLOW LAVAS, MOTUTARA, NEW ZEALAND

A. Flattening at the base where the flow rests on tuffs.
B. Downward-pointing projections between rounded tops of pillows.
(*Photo: F. A. Singleton*)

accordingly it has been widely used in mapping, notably in the Bushveld Complex and the Great Dyke of Southern Rhodesia (Fig. XII – 29).[1]

Top and bottom criteria and pillow-structures – The recognition of means of determining the top or bottom of ancient lava-flows has been of the greatest value in deciphering the structure of Pre-Cambrian mineral fields notably in the greenstones of Canada and Western Australia. This depends largely on the characteristic shapes of individual 'pillows' in *pillow lavas*. At the base of a flow these are bun-shaped with a flat underside and rounded top, but the pillows resting on these, and occurring within a flow have rounded tops, and 'tails' that fit into the V-shaped hollows between lower pillows, the 'tails' often connecting with breaks in the outer skin of other pillows (Fig. XII – 2).[2] Pillow structure appears to result from the chilling of the outer skin of a sub-aqueous flow, while the inside remains liquid. Liquid escaping from cracks in

FIG. XII–3. TOP AND BOTTOM
CRITERIA IN A LAVA FLOW

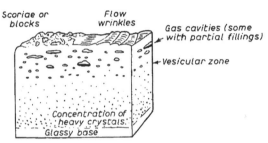

the skin forms new pillows, linked to the parents by 'tails'. The whole mass resembles a pile of liquid-filled balloons.

In massive flows, the features that may assist in recognizing top and bottom surfaces are a chilled glassy and rather uniform base lacking scoriaceous or brecciated lava, and a much more irregular, highly vesicular scoriaceous or brecciated top, sometimes with flow wrinkles (Fig. XII–3). A gradation in mineral composition may also occur which from petrological study may afford guidance as to facing.[3]

Flow structures

Flow structures in igneous rocks are formed in the fluid stage, but plastic flow occurs in softened rocks adjoining some intrusions, and rheomorphic effects occur in plutons originating by transfusion. Both fluid and plastic flow have therefore to be considered under flow structures.

[1] H. H. Hess, 'Vertical Mineral Variation in the Great Dyke', *Trans. Geol. Soc. S. Africa*, Vol. 53, 1951, pp. 159–68; L. R. Wager and G. M. Brown, 'Funnel-shaped Layer Intrusion', *Bull. Geol. Soc. Amer.*, Vol. 68, 1957, pp. 1071–4.

[2] M. E. Wilson, 'Structural Features of the Keewatin Volcanic Rocks of Western Quebec', *Bull. Geol. Soc. Amer.*, Vol. 53, 1942, pp. 54–69 (with Bibliography). See also R. R. Shrock, *Sequence in Layered Rocks*, New York, 1948, for illustrations.

[3] B. S. Butler and W. S. Burbank, 'The Copper Deposits of Michigan', *U.S. Geol. Surv.*, Prof. Pap. No. 144, 1929, discuss the features of flow tops in ancient lavas.

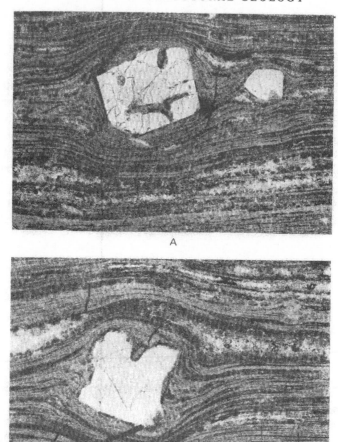

FIG. XII–4. PHOTOMICROGRAPHS OF FLOW STRUCTURE IN
RHYOLITE, ROCKLANDS, VICTORIA

A. Vertical compression above quartz phenocrysts.
B. Slight rotation (right-handed) after formation of curved flow layers.

Fluid flow – The record of fluid flow is best seen in rocks as *fluidal texture*,[1] and the macroscopic flow structures of lavas and dyke rocks.

Textures and structures due to flow are revealed by the parallel orientation of tabular or platy inclusions such as phenocrysts or xenoliths, which indicate *flow layers*, or by the parallel alignment of elongated, needle-shaped crystals giving *linear flow structure*. In rhyolites, differences in the colour or texture of parallel laminae constitute flow banding or layering, and in granitic rocks

---

[1] In German *Fluidaltextur* is still used more strictly for textures that in English would be termed flow structures, as in trachytes.

schlieren, which are layers containing the same minerals as the average of the rock but in different proportions, may in some instances represent flow layers.

Flow layering is normally parallel to the contacts of igneous masses. In lavas the layers are parallel with the base of the flow, in dykes with the walls, and in larger magma chambers with the walls and roof. Exceptions are, however, to be expected locally, especially in lava flows.[1]

The flow layers in rhyolites, and the aligned felspar laths that are characteristic of trachytes, mould themselves around phenocrysts and foreign inclusions in smooth curves (Fig. XII−4) which are generally compared with stream-lines. The double curvature at both ends of inclusions suggests, however, that the movement involves compression normal to the layers, and it may be imagined that this develops during the lateral spreading of a thick flow

FIG. XII−5. LINEATION OF FLOW LAYERS DUE TO CRUMPLING AT RIGHT ANGLES TO THE FLOW DIRECTION

Rhyolite, Rocklands, Victoria. $\times\frac{1}{2}$.

under its own weight, the flow becoming thinner at the same time after the manner of treacle. This notion also explains why rhyolite flow planes are well developed throughout a flow, rather than only near the base where viscous drag is greatest, but it does not exclude differential lateral flow combined with the vertical compression. Strong lineation of the flow layers is caused by phenocrysts, but may also be an effect of crumpling at a viscous stage of cooling (Fig. XII−5) the crumples lying, like the flow wrinkles described below, across the general direction of flow in the lava.

[1] G. T. Benson and L. R. Kittleman, 'Geometry of Flow Layering in Silicic Lavas,' *Amer. Journ. Sci.*, Vol. 266, 1968, pp. 265–76.

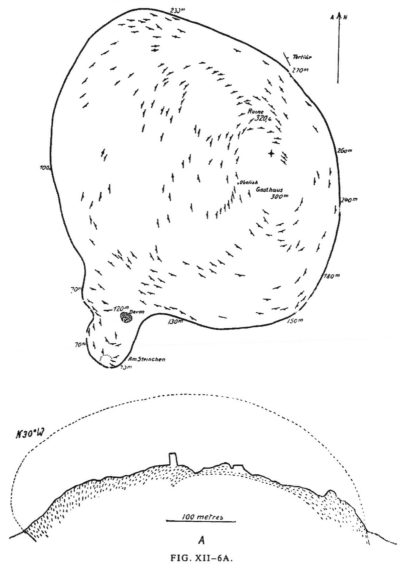

FIG. XII–6A.

Map and section of dome of platy flow structures, shown by the parallel orientation of tabular felspar phenocrysts in the Drachenfels trachyte. (*After H. and E. Cloos, 1927*)

The notion that there is compression normal to flow layering, with extension parallel to the layers, enables comparison to be made with boudinage, with the 'stretching' of slates and schists about crystals (*Streckungshöfe*) and with the effects of compression about hard inclusions in rocks such as coal.[1]

A somewhat similar mechanism is involved in the development of flow

[1] A comparison should be made between Figures XII–4 and III–4.

structure in knob-like volcanic cupolas such as that of the Drachenfels[1] (Fig. XII – 6). There the well-known trachyte containing large phenocrysts of sanidine exhibits fluidal texture parallel with the walls of the cupola and thus

FIG. XII-6B. RECONSTRUCTION OF THE DRACHENFELS

The reconstructed form of the trachytic body of the Drachenfels, which is a swelling at the upper end of a volcanic neck, beneath a cover of tuffs. (*After H. and E. Cloos, 1927*)

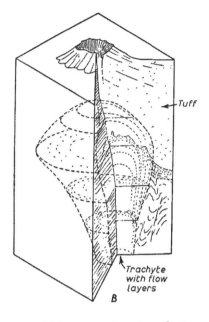

Tuff

Trachyte with flow layers

B

normal to the pressure of the rising magma which pressed outwards to enlarge the dome, squeezing the magma against the walls as it did so.

Two somewhat different mechanisms may thus be involved in producing parallel orientation of inclusions in magma. The first is flattening combined

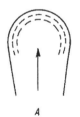

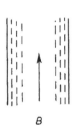

A

B

FIG. XII-7. FLOW LAYERS IN INTRUSIONS

A. Normal to pressure at a surface blocking flow.
B. Parallel to the walls in laminar flow.

with lateral spread, generally under the action of a force directed normal or nearly so to the plane of spreading (Fig. XII–7A). This type is illustrated in many common analogues.[2] The second is by laminar flow in a channel-way, in dykes, and in confined lava flows (Fig. XII–7B). The analogue commonly

---

[1] H. and E. Cloos, 'Die Quellkuppe des Drachenfels am Rhein', *Zeitschr. f. Vulkanologie*, Bd. II, 1927, pp. 33–40.
[2] It is shown in bread baked in a container, where the pressure of gases gives a laminar texture parallel with the sides.

drawn is with clay or metal extruded through an orifice.[1] The arrangement of logs floating in a river illustrates both types. Where freely moving they lie in the stream-lines parallel with the banks, but at an obstruction they tend to lie across the stream, forming an arc convex downstream due to the push of the stream and drag at the banks. Any line that at first was directly transverse to the stream is stretched along its length to accommodate the arching, an effect that is important in the interpretation of the fractures in igneous intrusions. There is a close analogy with lava flows that are confined to a channel, the

FIG. XI–8. ALIKA BASALT FLOW, MAUNA LOA, JUNE, 1919, SHOWING FLOW LINES
PARALLEL WITH THE BANKS
(*By courtesy T. A. Jaggar*)

flow lines being parallel with the sides (Fig. XII–8) and the flow planes with the bottom of the flow, while the surface skin is thrown into transverse wrinkles as it becomes more viscous (Fig. XII–9). These latter are *pressure ridges*, the comparable smaller ridges an inch or so in height formed on pahoehoe lava tongues being known as *flow wrinkles*. There are also close analogies with glacial features, which are discussed below.

Laminar flow structures, whether developed by flattening or by differential laminar flow, are similar in one important regard, that there is essentially only one 'flow plane' involved in the movements. In welded ash-flow tuffs

[1] W. O. Williamson, 'Ceramic Products: Their Geological Interest and Analogies', *Amer. Journ. Sci.*, Vol. 247, 1949, pp. 715–49; idem, 'The Effects of Extrusion and Some Other Processes on the Microstructure of Clay', ibid., Vol. 251, 1953, pp. 89–108; idem, 'Lineations in Three Artificial Tectonites', *Geol. Mag.*, Vol. 92, 1955, pp. 53–62. See also the experiments of Riedel, p. 310.

entrapped gases greatly facilitate flowage even when the particles have been deposited from the ash flow, and chemical factors are also involved.[1] Linear flow structures may also form during plastic flow, and their general properties are further discussed below.

Plastic flow – Under plastic flow are included here those structures and textures which, being distributed throughout the rock, affect the majority of mineral grains. Sharply defined joints and faults are accordingly dealt with as

FIG. XII–9. BASALT FLOW AT KILAUEA, SHOWING PRESSURE RIDGES TRANSVERSE TO THE FLOW

(By courtesy T. A. Jaggar)

fractures. Effects due to plastic flow may be expected in the intrusion after a crystal mesh has formed, and in the wall-rocks if these suffer rheomorphism and deformation during the emplacement of the pluton. The development of plasticity in the invaded rocks under deep-seated plutonism is admirably illustrated by Sederholm's studies of anatexis and syntexis,[2] but few detailed studies of the structural pattern so developed have been made. Osborn, however, describes complementary shear planes intersecting in the linear flow structure (which is parallel to the dip of the flow planes) in the Val Verde tonalite,[3]

---

[1] H. U. Schmincke and D. A. Swanson, 'Laminar Viscous Flowage Structures in Ash-Flow Tuffs from Gran Canaria, Canary Islands,' *Journ. Geol.*, Vol. 75, 1967, pp. 641–64; D. C. Noble, Discussion, ibid., Vol. 76, 1968, pp. 721–3.

[2] J. J. Sederholm, 'On Migmatites and Associated Pre-Cambrian Rocks of South-western Finland', *Bull. Comm. Géol. Finl.*, Vol. 10, No. 58, 1923.

[3] E. F. Osborn, 'Structural Petrology of the Val Verde Tonalite, Southern California', *Bull. Geol. Soc. Amer.*, Vol. 50, 1939, pp. 921–50.

and den Tex has illustrated them in the Kosciusko pluton in southeastern Australia.[1]

Plastic deformation of wall-rocks is exhibited around the Bald Rock batholith, California (Fig. XII–10), in which it is seen that a broadly similar attitude to that reported by Osborn, with vertical lineation at steeply dipping contacts, and shear planes intersecting in this lineation, obtains.[2] Structural patterns such as these indicate circumferential stretching of the batholithic walls, combined with compression and flattening normal to the walls.

In plastic flow the most obvious linear structure is generally the line of intersection of the complementary slip planes, which in the examples quoted above is vertical. Needle-shaped crystals, especially such as form at a late

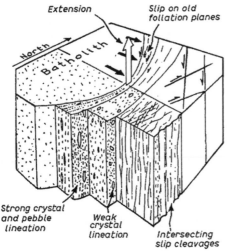

FIG. XII–10. STRUCTURES AT THE VERTICAL MARGIN OF THE BALD ROCK BATHOLITH, BIDWELL BAR, CALIFORNIA

(*After Compton, 1955*)

stage, may grow in this direction, and accordingly will be mapped as a linear structure. While this may correspond with the general direction of upward flow of the magma, it clearly does not represent flow lines analogous with those in stream-line flow.

Recalling also the lineation of rhyolite flow planes due to wrinkling across the direction of flow, it would seem preferable not to designate linear structures in igneous rocks as linear flow lines but simply as lineations, the origin of which must be carefully investigated in each area of study.

Structure of plutons

Flow structures have found their greatest use as an aid in revealing the form of individual plutons and the internal structure of complex batholiths. Thick

[1] E. den Tex, 'Complex and Imitation Tectonites from the Kosciusko Pluton, Snowy Mountains, New South Wales', *Journ. Geol. Soc. Aust.*, Vol. 3, 1956, pp. 33–54; see also Chapter XIII, pp. 427–9.

[2] R. R. Compton, 'Trondhjemitic Batholith near Bidwell Bar, California', *Bull. Geol. Soc. Amer.*, Vol. 66, 1955, pp. 9–44.

lava complexes, especially where interbedded sediments from which dips may
be judged are rare or absent, may also be mapped using flow structures in the
lavas (Fig. XII–11).

In deciphering the form of plutons, the parallelism of platy flow structure
with the walls enabled H. Cloos to demonstrate that an intrusion in Bavaria is

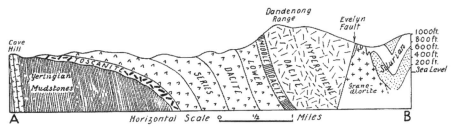

FIG. XII–11. STRUCTURE OF THE DANDENONG RANGES VOLCANIC COMPLEX,
VICTORIA

Based on the mapping of flow layers in the lavas.

not a batholith but a *harpolith*, with sub-parallel and rather low-dipping
contacts, giving a sickle-shape to the intrusion (Fig. XII–12).

In complex batholiths, the presence of several intrusive centres or units
within the large massif may be indicated by local domes of platy flow structure, sometimes with narrow and impersistent *screens* of country-rock separating the several sub-batholiths (Fig. XII–13).

The type of pluton to which the notions of granite-tectonics are best
applied is one in which platy flow structure gives a moderately strong

FIG. XII–12. AN 'APPARENT-PLUTON' IN BAVARIA, SHOWING THE STRUCTURE OF A
HARPOLITH

(*After H. Cloos, Das Batholithenproblem, Fortschr. Geol. u. Pal., Hft. 1, 1923*)

banding, both in the intrusion and in the adjacent wall-rocks. Accordingly
they hold for intrusions in an intermediate depth zone of the crust, below the
superficial zone of fracture where the magma may break through to yield
thick lava flows and the contacts are themselves determined largely by fracturing. Indeed the lack of prominent flow structures in such superficial intrusions, and the joint patterns commonly found, do not permit the wide application of Cloos' synthesis to them. Complexes of this type are separately
discussed on pp. 399–401.

In the interpretation of flow-layering and similar planar structures in granitic masses, due consideration must be given to granitization and transfusion [1] and to the possible presence of relic structures in a granitized terrain.

According to Cloos, the space occupied by a granite mass is due largely to forcible pushing apart of the walls and roof of an original magma chamber, a

FIG. XII–13. STRUCTURE OF A COMPLEX BATHOLITH
Rock Creek Salient, California. (*After Mayo, 1941*)

concept that does not fully overcome what is called the 'space-problem' for batholithic masses unless the original chamber is in some way accounted for. On the other hand, there is much evidence to indicate that, even in schistose terrains where transfusion is commonly postulated to account for granitic

[1] H. H. Read, Presidential Addresses to the Geological Society of London for 1948 and 1949: 'A Commentary on Place in Plutonism', *Quart. Journ. Geol. Soc. Lond.*, Vol. 104, 1948–9, pp. 155–206, and 'A Contemplation of Time in Plutonism', ibid., Vol. 105, 1949, pp. 101–56. These papers afford an excellent introduction to the subject and a bibliography of earlier works. See also 'Meditations on Granite, Parts 1 and 2', *Proc. Geol. Assoc. Lond.*, Vol. 54, 1943, pp. 64–85; Vol. 55, 1944, pp. 45–93.

massifs, the space problem may, in fact, be resolved largely by forceful shouldering apart and upward displacement of the invaded rocks, combined with stretching over the roof of an intrusion. This is indicated by Mayo[1] and emphasized by Noble[2] for the great batholiths of the Sierra Nevada, California, and also for the Bald Rock batholith, California[3] (Fig. XII–10).

Wegmann regards many batholiths as diapiric injections having structural relationships to the wall rocks comparable with salt domes,[4] an idea which is

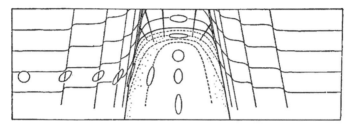

FIG. XII–14. GENERALIZED CROSS-SECTION OF A GRANITE DIAPIR ACCORDING TO DE WAARD (1949)

also implicit in Grout's analogies drawn with experimental studies on salt domes, and which is supported by the field studies of de Waard[5] (Fig. XII–14).

The geometry of structures due to flow and fracture will indeed be very similar if, instead of a clearly defined chamber filled with magma, a zone of softening due to transfusion, granitization and local magmatic injection be substituted, combined with strong upward pressure. These processes may be conceived to act more radically in the centre of the active zone, where the 'granite' will correspondingly be freer from regular banding and from 'inclusions', thus giving rise to the gradation outwards to more strongly gneissic 'flow structures' that is also indicated in Cloos' synthesis. A structural pattern of the appropriate type may, therefore, afford evidence either for the intrusion of magma or for rheomorphic effects in a massif originating by granitization. But where relic structures can be traced with little disturbance from the country-rock into the granite, the evidence would indicate that the

[1] E. B. Mayo, 'Deformation in the Interval Mt Lyell–Mt Whitney, California', *Bull. Geol. Soc. Amer.*, Vol. 52, 1941, pp. 1001–84.

[2] J. A. Noble, 'Evaluation of Criteria for the Forcible Intrusion of Magma', *Journ. Geol.*, Vol. 60, 1952, pp. 34–57.

[3] R. R. Compton, 'Trondhjemite Batholith near Bidwell Bar, California', *Bull. Geol. Soc. Amer.*, Vol. 66, 1955, pp. 9–44.

[4] C. E. Wegmann, 'Über Diapirismus (Besonders im Grundgebirge)', *Bull. Comm. Géol. Finl.*, Vol. 15, No. 92, 1930, 58–76. M. K. Akaad, 'The Ardava granitic diapir of County Donegal, Ireland', *Quart. Journ. Geol. Soc. Lond.*, Vol. 112, 1956, pp. 263–90.

[5] F. F. Grout, 'Scale Models of Structures Related to Batholiths', *Amer. Journ. Sci.*, Vol. 243A (Daly Volume), 1945, pp. 260–84. D. de Waard, 'Diapiric Structures', *Proc. Kon. Ned. Akad. Wetensch.*, Vol. 52, 1949, pp. 1–14.

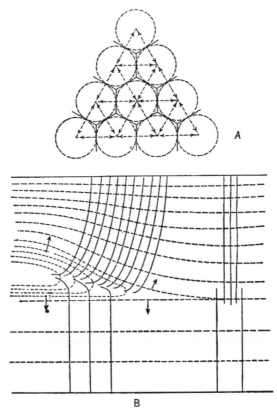

FIG. XII–15. DIAGRAMS SHOWING COLUMNAR
JOINTING IN A LAVA FLOW

(*After J. P. Iddings*, Igneous Rocks, *Vol. I, Lond.,
1909, Figs. 14 and 16*)

A. Cracks forming at right angles to the lines joining
equidistant points form a hexagonal pattern.
B. Development of jointing in a lava flow. Broken lines
are isothermal lines, full lines, columnar jointing.

transformation was quiet, and lacking in synchronous pressures that might
cause movement in the granitized rocks.[1]

STRUCTURES DUE TO FRACTURE

Jointing in lava flows, dykes, and sills

Columnar jointing – Columnar jointing is best seen in lava flows and sills; but
it is also developed in tuffs and even in rocks heated by igneous intrusions

[1] A most informative review, 'Granite emplacement with special reference to North
America' has been made by A. F. Buddington, *Bull. Geol. Soc. Amer.*, Vol. 70, 1959, pp. 671–
747.

(Fig. V−6). The rock is divided into columns that are ideally hexagonal in cross-section, but which in fact vary from three- to eight-sided, and which range from less than an inch to many feet in cross-section. The long axes of the columns are generally at right angles to cooling surfaces, and thus to the top and bottom of flows and sills, and to the walls of dykes, in which latter they are accordingly approximately horizontal.

FIG. XII−16. APPEARANCE OF TWO FLOWS DUE TO DEVELOPMENT OF COLONNADE
AND ENTABLATURE IN TERTIARY BASALT, BACCHUS MARSH, VICTORIA

The sides of the columns commonly show horizontal markings called 'chisel marks' which appear to represent stages in the growth of the cracks, with a slight deviation in the direction of cracking at each stage. Where a column breaks off, one fracture surface is concave and the other convex, giving *cup and ball jointing*. Columns that develop in the lower part of a lava flow form the *colonnade* and are in general larger and more regular than those that formed above (the *entablature*). The two sets meet within the flow, giving a false appearance of the existence of two flows (Fig. XII−16), and in some lavas the top layer has distinctive structures, giving rise to a three-fold division of the flow.[1] In sills such as the dolerites of Tasmania or the Karroo sills of South Africa, which attain a thickness of more than 1500 ft, extremely regular, straight columns are characteristic, but in many basalt flows the columns are locally somewhat irregular. They may be radially arranged apparently about cooling centres, and where two sets meet, the axes of the columns tend to bend towards each other forming a cusp. Columnar jointing

FIG. XII–17. DEVIL'S WATCHTOWER, WYOMING
(*Photo: U.S.G.S.*)

in large volcanic necks such as the Devil's Watchtower, Wyoming, is commonly parallel with the vertical axis of the neck towards the top, but curves outwards towards the walls in the lower parts (Fig. XII–17).

The geometrical features of columnar jointing may all be accounted for by fracturing by tensional stresses developed during the cooling of the solidified but still hot rock down to normal temperatures.[2] Equidistant points in a plane

[1] A. Spry, 'The Origin of Columnar Jointing, Particularly in Basalt Flows', *Journ. Geol. Soc. Aust.*, Vol. 8, Pt 2, 1962, pp. 191–216.

[2] An excellent account is given by J. P. Iddings, 'The columnar structure in the igneous rock on Orange Mountain, New Jersey', *Amer. Journ. Sci.*, Vol. 31, Ser. 3, 1886, pp. 321–31; idem, *Igneous Rocks*, Vol. 1, pp. 320–7, New York, 1909. C. B. Hunt, 'A Suggested Explanation of the Curvature of Columnar Joints in Volcanic Necks', *Amer. Journ. Sci.*, Vol. 36, Ser. 5, 1938, pp. 142–9. A. V. G. James, 'Factors producing columnar structure in Lavas, and its occurrence near Melbourne, Victoria', *Journ. Geol.*, Vol. 28, 1920, pp. 458–69.

parallel with a cooling surface lie at the apices of equilateral triangles, and if the length of the side is taken as that in which shrinkage will produce a crack, it is seen that the cracks will be arranged hexagonally. A close but random distribution of cooling centres produces a crack configuration more closely resembling that observed in real basalt flows.[1] Cup-and-ball jointing and spheroidal cracking in sectors along the length of a column may be explained by lengthwise contraction after the columns form.[2]

FIG XII-18. PLATY JOINTING IN LOWER TERTIARY BASALT, HARKAWAY, VICTORIA
(*Photo: E. S. H.*)

The possibility that an alternative mechanism involving the generation of convection cells in the liquid may be involved in columnar jointing has also been mooted, chiefly because of evidence for slight differentiation within each column. Some columns have a central core that differs somewhat from the outer margins,[3] but this might perhaps be due to late-stage alteration. Although it is possible that the positioning of columns may be determined by convection cells, it is quite certain that columnar jointing is due to fracture in the solid stage, since the joints cut across large individual crystals.

[1] I. J. Smalley, 'Contraction Crack Networks in Basalt Flows', *Geol. Mag.*, Vol. 103, 1966, pp. 110–23.

[2] S. I. Tomkeieff, *Bull. Vulcanol.*, Vol. 6, 1940, p. 89. See review by A. Holmes in *Nature*, Vol. 156, 1945, p. 425.

[3] R. B. Sosman, 'Types of Prismatic Structure in Igneous Rocks', *Journ. Geol.*, Vol. 24, 1916, pp. 215–34, gives an excellent discussion. See also D. Lafeber, 'Columnar Jointing and Intra-columnar Differentiation in Basaltic Rocks', Gedenkboek H. A. Brouwer, *Verh. Kon. Ned. Geol. Mijnb. Gen.*, Dl. 16, 1956, pp. 1–11.

In plutonic intrusions, cooling joints are not recognized with certainty according to current ideas, although the characteristic two vertical and one sub-horizontal sets of joints in granites were formerly attributed to cooling.[1] *Platy jointing* – In addition to columnar jointing, sub-horizontal *platy jointing* is well-developed in many lava flows, producing slabs a few inches thick. Both types may be present in the one flow (Fig. XII–18). The origin of platy jointing of this type is not known.

Fractures in plutons

Although we are concerned here chiefly with structures that are related to events accompanying the emplacement of igneous massifs, it should be mentioned that faults and joints of entirely subsequent origin may be so prominent and numerous as to render the recognition of primary structures a matter of great difficulty. Furthermore, younger deformation affecting an already jointed rock tends to cause displacements on the existing joints, which may thus become slickensided. Indeed, many fractures in granites are termed 'joints' largely because evidence for fault displacements on them is not obtainable. Faulted xenoliths or dykes from which displacements may be gauged[2] are more numerous in the border zones than in the central parts of granitic masses. There is, further, a transition between flow and fracture as a mass cools. Earlier-formed structures may at any stage be displaced or flexed by later movements, and indeed such displacements can yield valuable information on the physical state of the rock at the time.

Structures formed at the stage when fracture is possible in plutons include joints, faults, and dykes related to the intrusion, especially aplites. Dating of the structures is, indeed, facilitated by late stage effects such as the injection of aplites, the formation of pegmatites, and the development of minerals along joints and faults, since these permit the recognition of structures formed during the intrusive period (Fig. XII–19).

Regional fracture pattern – The fracture pattern of plutons is less clearly related to the shape of individual batholiths or stocks than is the pattern of flow structures. Steeply dipping joints and faults in particular often conform with a regional pattern that may be over-printed on a series of small igneous massifs, virtually irrespective of their shape, and which is also prominent in the country rock. This is well shown, for instance, in the tin lodes of Cornwall, which follow the E.N.E. trend of the mineralized belt, within which several granitic stocks occur at the surface (Fig. XII–20). Again, the Tertiary

---

[1] J. Geikie, *Structural and Field Geology*, Edinburgh, 1905, p. 149. It is possible that *S*-joints in granites (see p. 377) may be caused by contraction on cooling: see H. Cloos, 'Über Ausbau und Anwendung der Granittektonischen Methoden', *Abh. Preuss. Geol. Landesanst.*, N.F., Vol. 89, 1922, p. 5.

[2] Always bearing in mind the fact that random exposures exhibiting one plane only may give a false impression of the actual displacement.

ring-complexes of western Scotland clearly show that a regional fracture pattern, as well as that locally developed at each igneous centre, may indeed be of considerable tectonic significance and interest (Fig. XII–35).

In these quoted examples the regional pattern and that locally found at each plutonic centre are clearly related, but in other instances a pattern that is connected with active or recently active faulting may be prominent. There is, however, in certain of the few examples that have been studied, a strong indication that the young pattern may follow much older lines, even those

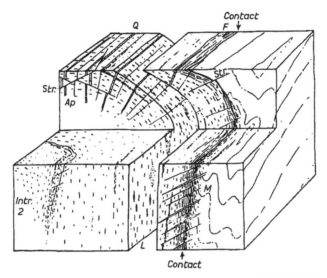

FIG. XII–19. BLOCK DIAGRAM OF PART OF A BATHOLITH, WITH A SUBSIDIARY INTRUSION, DISSECTED ALONG A FOLIATION SURFACE

M, Marginal thrusts, some with injected aplite; *F*, flow layers, and foliation; *L*, linear flow structure; *Q*, cross joints, some with injected aplite; *Str.*, planes of stretching. Schistosity parallel to the granite contact is developed in the wall rocks, and the axes of folds are tilted away from the intrusion.

connected with the original folding and fracturing associated with an intrusion or group of intrusions.[1]

Accordingly there would seem to be a wide range of possibilities with regard to the genesis of fractures in plutons, and the regional as well as the local pattern must be given due consideration. This calls for the integration of studies in the invaded rocks with those on the plutons themselves, which has rarely been attempted.

Local fracture patterns – In the domed roof of a magma chamber the fracture pattern is fairly clearly related to the lateral stretching and doming of the

[1] E. den Tex, 'Geology of the Grey Mare Range in the Snowy Mountains of New South Wales', *Proc. Roy. Soc. Vict.*, Vol. 71, Pt 1, 1959, pp. 1–24.

mass under continued upward pressure from the magma below. The joints represent tension cracks formed normal to the doming, and accordingly they radiate outwards from the centre like the ribs of a fan in section. Being tensional, they do not exhibit slickensided surfaces, but are often filled with aplite dykes (Fig. XII−21).

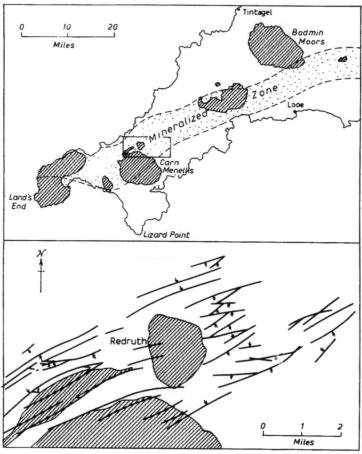

FIG. XII–20. MAP OF THE GRANITIC STOCKS OF S.W. ENGLAND, AND ENLARGEMENT OF INSET, SHOWING CORNISH TIN LODES
(*After* Geol. Surv. Great Britain)

In general these are to be termed *tension joints*, but where they are also normal to the dome of flow layers they may be termed *cross-joints* (*Q*-joints or *Quaderklüfte*).[1] Kuznetsov suggests that as cooling proceeds from above, a

[1] H. Cloos, 'Tektonik und Magma, Bd. 1', *Abh. Preuss. Geol. Landesanst.*, N.F., Vol. 89, 1922. H. Cloos, 'Der Sierra-Nevada Pluton in Californien', *Neues Jahrb. f. Min.*, etc., Beil. Bd. 76, Abt. B, 1936, p. 392. R. Balk, 'Structural Behaviour of Igneous Rocks', *Geol. Soc. Amer.*, Mem. No. 5, 1937, p. 33.

succession of tension joints may be formed progressively in depth,[1] but an alternative view is that along the steeply dipping contracts of the sides of a stock, the tension joints are formed *en echelon* in the zone of differential upward movement[2] (Fig. XII–19), in which case they cannot be true cross-joints.

Two sets of steeply dipping primary joints approximately at right angles to each other are generally prominent in granites. The second set in some massifs strikes parallel with the trend of linear flow structures projected on to a horizontal plane, and in such cases are known as *S*-joints (*Längenklüfte*).

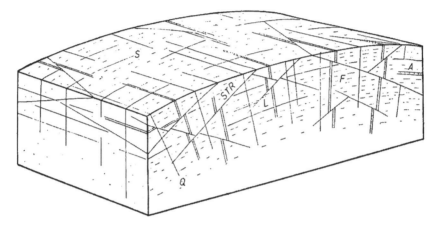

FIG. XII–21. CLOOS' DIAGRAM SHOWING THE CHIEF TYPES OF JOINTS IN A BATHOLITH

Q, cross-joints; *S*, longitudinal joints; *L*, flat-lying joints; *STR*, planes of stretching; *F*, linear flow structures; *A*, aplite dykes.

They are best developed where the linear flow structures approach the horizontal, that is, near the roof of a large intrusion.

Flat-lying or gently dipping joints (*Lagerklüfte*) in granite include primary joints which are often filled with aplite or pegmatite, and other secondary joints due to a number of causes. Those of primary origin may lie in the flow planes especially near the roof of a cupola. They commonly show an increase in dip as the margins of the pluton are approached, where in fact they are no longer flat-lying. This illustrates the difficulty of recognizing and categorizing the various joints in granite, especially where flow structures are not recognizable.

Sheeting – Flat-lying joints of secondary origin are often so closely spaced as to divide the rock into relatively thin slabs, giving *sheeting* that commonly

[1] V. I. Kuznetsov, 'The Place of Pegmatites in the Process of Forming a Granite Massif', *L'Vovskoe Geol. Obsch., Mineralogy Coll.*, No. 5, 1951, pp. 99–112.
[2] E. Cloos, 'Feather joints as Indicators of the Direction of Movements on Faults, Thrusts and Magmatic Contacts', *Proc. Nat. Acad. Sci. Wash.*, Vol. 18, 1932, pp. 387–95. R. Balk, 'Structural Behaviour of Igneous Rocks', *Geol. Soc. Amer.*, Mem. No. 5, 1937.

conforms with the surface of the granite on hillsides (Fig. XII–22). The fractures in sheeted granite are gently curved partings rather than parallel planar joints, and indeed it is in those granites that are relatively free from primary joints that sheeting is usually best developed – doubtless because the stresses causing sheeting may be taken up by movements on already existing joints. Although sheeting becomes more widely spaced in depth, it remains prominent in the New England granite down to more than 300 ft and the spontaneous and sometimes explosive splitting of these granites after quarrying indicates that they are in a state of strain.[1] The most favoured notion to explain sheeting is that it is due to elastic expansion upwards on relief of load due to the removal by erosion of the former cover.[2] However, the *exfoliation*

FIG. XII–22. SHEETING IN GRANITE

The quarry-face shows sheeting sub-parallel with the ground surface, wider-spaced in depth, and in part lenticular. (*Based on Dale, 1923*)

of granite into thin sheets concentric with the surface of large boulders and cliffs appears to be related to some types of sheeting, and is clearly due chiefly to weathering (Fig. III–22), whereby the outer layers of the rock are expanded by the formation of hydrated minerals having a larger volume than the primary minerals of the rock.[3] Again, sheeting is particularly common in glaciated regions, and it may be that alternate freezing and thawing of water in the rocks or the effects of temperature changes during glaciation and deglaciation, have led to the development of the structure in such regions.[4]

Rift, grain, and hardway – These terms are used by quarrymen for the directions along which the rock may be most conveniently split into rectan-

[1] It is normal for large quarried blocks of nearly all rocks to split along curving cracks known as 'drys'. Specimens obtained from deep bore holes may also exhibit strain, but the most marked effects are perhaps those in deep mining known as 'rock-bursts'.

[2] R. H. Jahns, 'Sheet Structure in Granites: its origin and use as a measure of glacial erosion in New England', *Journ. Geol.*, Vol. 51, 1943, pp. 71–98, with many references; C. A. Chapman and R. L. Rioux, 'Statistical Study of Topography, Sheeting, and Jointing in Granite, Acadia National Park, Maine', *Amer. Journ. Sci.*, Vol. 256, 1958, pp. 111–27; W. V. Lewis, 'Pressure Release and glacial erosion', *Journ. Glaciol.*, Vol. 2, 1954, pp. 417–22 (with good bibliography).

[3] E. Blackwelder, 'Exfoliation as a Phase of Rock Weathering', *Journ. Geol.*, Vol. 33, 1925, pp. 793–806.

[4] W. B. Harland, 'Exfoliation Joints and Ice Action', *Journ. Glaciol.*, Vol. 3, 1957, pp. 8–10; R. Wolters, 'Zur Ursache der Entstehung oberflächenparalleler Klüfte', *Rock Mechanics*, Vol. 1, 1969, pp. 53–70.

gular blocks. They do not necessarily represent directions of jointing in the quarry. Because of the lack of precision in our knowledge of the genesis of joints and directional structures in granite and because the quarryman's terms refer chiefly to the ease of splitting, interpretation on a structural basis is rarely possible. The *rift* is the direction of easiest parting, the *grain* is another, approximately at right angles to the rift, and the *hardway* is the third direction, along which it is necessary to split the rock in order to obtain a rectangular block.[1]

The rift is very often a sub-horizontal direction, like that of sheeting, and in many cases it may be caused by expansion of the surface layers on weathering. It is, however, a primary direction in some granites, in which it is parallel

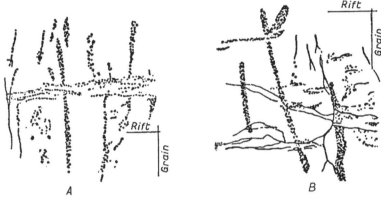

FIG. XII–23. LINES OF MINUTE BUBBLES PARALLEL TO THE RIFT AND
GRAIN IN GRANITE
(*After Dale*)

with the direction of minute parallel cracks and strings of bubbles in quartz crystals.[2] In the granites of New England, the grain is also parallel with a second set of micro-fractures and bubble strings (Fig. XII–23).

Faults – Thrust faults known as *marginal thrusts* occur in the steeply dipping flanks of plutons. Although each is of small displacement, their overall effect is considerable, and as the movement is outwards from the intrusion they represent an enlargement of the upper zones of the magma chamber. E. Cloos explains them as pinnate shear planes (see p. 375) due to the rise of the intrusion relative to the wall rocks, as if the contact were itself a fault zone.[1] Minor sets of pinnate shear joints may develop along the marginal thrusts themselves.

In the roof of a cupola, the lateral stretching results in the formation of gently dipping normal faults (*Streckfläche*) in two complementary sets, the

[1] T. N. Dale, 'The Commercial Granites of New England', *U.S. Geol. Surv.*, Bull. No. 738, 1923; idem, 'The Granites of Vermont', ibid., No. 404, 1909. J. A. Howe, *The Geology of Building Stones*, London, 1910, pp. 54–8.
[2] T. N. Dale, *U.S. Geol. Surv.*, Bull. No. 738, 1923.
[3] E. Cloos, 'Feather Joints as Indicators of the Direction of Movements on Faults, Thrusts and Magmatic Contacts', *Proc. Nat. Acad. Sci. Wash.*, Vol. 18, 1932, pp. 387–95.

acute bisectrix of which lies in the direction of stretching, that is, in the flow layers. It may be recalled that it is often only by the slickensiding of their surfaces that faults can be recognized in granites, and it is common to call those that do not reveal themselves as faults by displacement of dykes or other objects, *shear-joints*.

The full measure of stretching in a pluton is given not only by the displacements on faults but also by the total width of dykes filling tension cracks.

SILLS, LACCOLITHS, AND DYKES

Stress-relationships

The attributes of sills and dykes in which we are particularly interested in structural geology relate to the forces that are connected with their injection, their regional pattern and mutual arrangement, and their effects on the structure and attitude of the invaded rocks. Since open gashes comparable with the spaces occupied by dykes and sills are unknown in the earth it is clear that in all instances magmatic action and stress are very closely connected in the localization and emplacement of hypabyssal rocks.

The origin of the channelways occupied by dykes and sills would intuitively appear to be as tension gashes or cracks normal to a principal axis of tensional stress. While this may be so for dykes injected in near-surface rocks, it would appear that sill channels can very rarely be conceived to have formed under regional tensions, because the vertical stress axis is one of compression due to the load of superincumbent rocks. Likewise, at depths such that this load exceeds the crushing strength of the rocks, there can be no principal stress which is tensional under steady stress and the failure of rocks will be by shear fracture or flow. From geological evidence it seems clear, however, that dykes occupying cracks or gashes do, in fact, originate at depths greater than the 2–5 miles at which the crushing strength of granite and other crystalline rocks is exceeded, and accordingly some factor other than the regional field of stress is likely to be involved. This appears to be the wedging effect of fluids penetrating cracks, although the possible influence of shock waves rather than steadily applied forces may, in some instances, require to be considered especially in the initiation of cracks.

Anderson[1] shows that at the wedge-like tip of a fluid injection, a large tensional stress is generated by the fluid pressure, at right angles to the tip, which is in general sufficient to extend the fracture. According to Griffith's theory of brittle failure, such cracks are most likely to be formed in the direction of the maximum stress axis and normal to the minimum stress, whether this be compressive or tensile.[2] This has been experimentally demonstrated by

---

[1] E. M. Anderson, *The Dynamics of Faulting and Dyke Formation*, Edinburgh, 1942.
[2] A. A. Griffith, 'The Phenomena of Rupture and Flow in Solids', *Phil. Trans.*, Ser. A, Vol. 221, 1921, pp. 163–98.

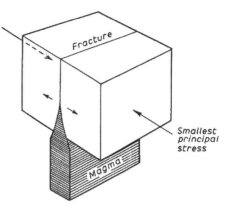

FIG. XII–24. STRESS-RELATIONS FOR
THE INJECTION OF A DYKE UNDER HIGH
CONFINING PRESSURE
(*After Anderson and Hubbert*)

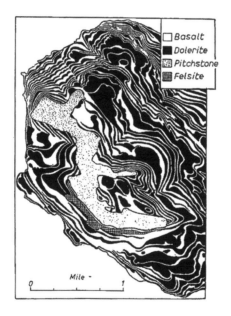

FIG. XII–25. A SILL COMPLEX, SGURR OF EIGG (SKYE)
(*After Harker*, Mem. Geol. Surv. (Skye), *1904*)

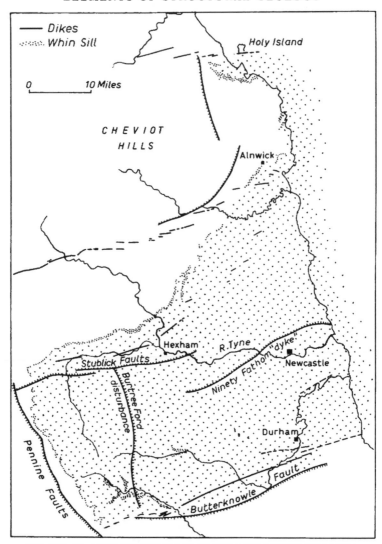

FIG. XII–26. AREA KNOWN TO BE UNDERLAIN BY THE WHIN SILL
(*After A. Holmes and H. F. Harwood, 1928*)

Hubbert, and theories of fluid fracture have been more fully developed in connection with the penetration of rocks by drilling muds[1] (Fig. XII–24). While many aspects remain to be clarified, the theory is adequate to account for nearly vertical dykes injected at great depths, in which case the overburden pressure need not be exceeded by the fluid magmatic pressure, and the dykes form normal to the minimum stress which, although compressional, is re-

[1] M. K. Hubbert and D. G. Willis, 'Mechanics of Hydraulic Fracturing', *Amer. Inst. Min. Met. Eng., Petrol. Branch*, 31st Ann. Meeting, Paper 686–G, 1956. H. K. van Poollen, 'Theories of Hydraulic Fracturing', *Col. School Mines Quart.*, Vol. 52, No. 3, 1957, pp. 112–31.

duced by regional extension. It also suggests that sills require magmatic pressure sufficient to lift the overburden, and hence that they are more appropriate to shallow crustal depths. The formation of sills would also be aided by lateral compression, as well as by the presence of pre-existing sub-horizontal planes of weakness afforded by bedding planes and nonconformities.

Sills are injections parallel or nearly parallel with the stratification of flat-lying or gently dipping beds (Fig. XII−25). Where an intrusion of this type follows an unconformity it is distinguished as an *interformational sill* or *sheet*. Sills may subsequently be folded along with the invaded rocks, in which case they assume steep dips, but their original injection into relatively flat beds is fundamental.[1] Sills are often of considerable lateral extent. The type example from which the word sill was taken is the Whin Sill in the north of England, which although only from 20 ft to 240 ft thick (averaging 80 ft) has an area of some 1500 sq. miles (Fig. XII−26.)[2]

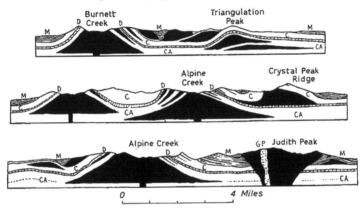

FIG. XII–27. LACCOLITHS OF THE JUDITH MOUNTAINS, MONTANA
(After Weed and Pirsson, from R. A. Daly, Igneous Rocks and their Origin, *N.Y., 1914)*

Individual sills among the Jurassic dolerites of Tasmania and the Karroo dolerites exceed 1500 ft in thickness, but especially in the Karroo examples there are also hundreds of much thinner sills, linked in various ways with inclined sheets and dykes, forming a sill-swarm. The volumes involved are enormous − for the Karroo, some 50,000–100,000 cubic miles of dolerite magma.[3]

Laccoliths are similar to sills in that they are formed by injection of magma

[1] It is true that there are many injections into already folded rocks, which locally follow the bedding but are initially steeply dipping. These are not sills. Likewise where a sill transgresses the bedding at an acute angle, it is not a dyke.

[2] A. Holmes and H. F. Harwood: 'The Age and Composition of the Whin Sill and the related dikes of the north of England', *Min. Mag.*, Vol. 21, 1928, pp. 493–542.

[3] A. L. du Toit, 'The Karroo Dolerites of South Africa: A Study in Hypabyssal Injection', *Trans. Geol. Soc. S. Africa*, Vol. 23 (for 1920) 1921, pp. 1–42. A. B. Edwards, 'Differentiation in the Dolerites of Tasmania', *Journ. Geol.*, Vol. 50, 1942, pp. 451–610.

along the bedding in flat-lying beds, but they are much thicker compared with their lateral extent, and are correspondingly mushroom-shaped. The uplift of the beds over laccoliths produces domes (Fig. XII–27). It is usual to assume that, whereas sills are fed by dykes, laccoliths originate from pipes resembling volcanic necks. This is so because the laccoliths are rounded or sub-elliptical in plan, but many bodies believed to be laccoliths because of the doming of the rocks above them may in fact be stocks, and thus lack a definite floor.[1]

FIG. XII–28. DYKES CUTTING STEEPLY DIPPING ROCKS
Note the matching walls of the left-hand dyke. Studley Park, Victoria. (*Photo: E. S. H.*)

Dykes – The openings that are occupied by dykes rarely show evidence of displacement such as would indicate that they were active faults at the time the dykes were intruded. A notable exception to this statement is however afforded by ring dykes, along which fault displacements of some thousands of feet have been measured. Some dykes locally follow older faults but even this is rare, and indeed dyke channels originate at the time of intrusion of the dykes, and represent cracks formed by deformation accompanying magmatic injection as discussed above (Fig. XII–28).

The size of dykes varies from less than an inch to some miles in width. The

[1] R. A. Daly, *Igneous Rocks and the Depths of the Earth*, New York, 1933, gives an excellent account of laccoliths.

Great Dyke of Southern Rhodesia is the largest known, being from 2 to 8 miles wide at outcrop, and 337 miles long. It is composed of basic and ultrabasic rocks that exhibit strong banding, and appears to be funnel-shaped in cross-section (Fig. XII – 29). The layers of different composition are nearly horizontal along the whole length of the dyke. Parallel to the Great Dyke are a few smaller basic dykes of normal tabular form.[1]

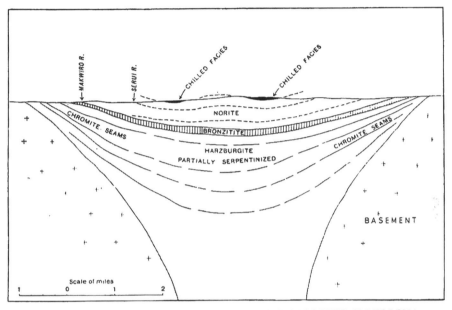

FIG. XII–29. SECTION ACROSS THE GREAT DYKE OF SOUTHERN RHODESIA
(*After Lightfoot, 1940*)

Even quite thin dykes a foot to a few feet wide may persist for considerable distances. At Bendigo, monchiquite dykes 2–3 ft wide have been followed vertically downwards in mines for over 4000 ft without notable change, and they must extend very much farther. The same applies to lamprophyre dykes at Woods Point (Fig. XII – 30) and Walhalla (Victoria).[2] It is evident that such thin dykes must have been injected with extreme rapidity to avoid chilling, so that they are sometimes termed *flash injections*.

Dyke swarms

Dykes rarely occur alone, but are commonly grouped in *dyke swarms* consisting of hundreds or thousands of individual dykes, and, in regional swarms,

[1] B. Lightfoot, 'The Great Dyke of Southern Rhodesia', *Proc. Geol. Soc. S. Africa*, Vol. 43, 1940, pp. xxvii–xliii. B. G. Worst, 'The Differentiation and Structure of the Great Dyke of Southern Rhodesia', *Trans. Geol. Soc. S. Africa*, Vol. 61, 1958, pp. 283–354.
[2] E. S. Hills, 'The Woods Point Dyke-Swarm, Victoria', *Sir Douglas Mawson Anniversary Volume*, Univ. of Adelaide, 1952, pp. 87–100.

extending for tens or hundreds of miles. Some swarms are clearly related to local effects, as around volcanic centres. Regional swarms are related to major crustal structures and stresses. Others again require for their explanation the combined effects of regional and local forces.

Local dyke swarms – Dyke swarms localized about volcanic centres or in the neighbourhood of stock-like intrusions are, apart from ring dykes and cone sheets (discussed below) commonly radial in pattern, and centred on the

FIG. XII–30. THE DIORITE DYKE AT THE MORNING STAR MINE, WOODS POINT, VICTORIA

Gold-bearing quartz reefs are developed as 'quartz floors' on many of the faults transverse to the dyke.

volcanic neck or stock (Fig. XII–31). This is due probably to up-doming by magmatic pressure giving rise to tension cracks, rather than to central collapse due to the weight of volcanic products or withdrawal of magma pressure, since active injection-pressure appears to have operated to form the dykes.

Where the local geology is complex the radial pattern may be modified. These relationships are best illustrated by Odé's analysis of the swarm at Spanish Peaks, Colorado[1] (Fig. XII–32A), where a swarm of 500 dykes is

[1] H. Odé, 'Mechanical Analysis of the Dike Pattern of the Spanish Peaks Area, Colorado', *Bull. Geol. Soc. Amer.*, Vol. 68, 1957, pp. 567–76.

centred on small stocks of granite and diorite. Closely adjacent on the west is the front of the Culebra Range which is not penetrated by the dykes, and which in the stress–analysis is regarded as a rigid mass. The dyke swarm has a median axis that is normal to this mountain front and which accordingly is taken as the direction of maximum compressive stress. Assuming also that the magma rose under pressure through a cylindrical hole in the crust beneath West Spanish Peak, and that the mountain front acted as a rigid boundary, the principal stress trajectories in a horizontal plane have been calculated and plotted, and it is seen (Fig. XII–32B) that there is excellent agreement with

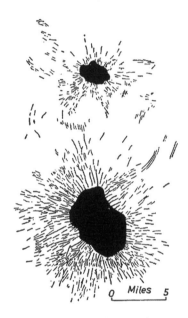

FIG. XII–31. RADIAL DYKE SWARMS, CRAZY
MOUNTAINS, MONTANA
(*After W. H. Weed*, U.S. Geol. Survey)

the actual pattern, with the dykes following the trajectory of the maximum stress.[1]

Regional dyke swarms – Wager and Deer[2] have described the dyke swarm closely following a major crustal flexure along the coast of East Greenland (Fig. XII–33, 34). The flexure is demonstrated in the Tertiary basaltic series, which attains seaward dips of 60°. The dykes are present in the zone where the basalt dips from 30° to 60°, but are absent on the plateau to the west. The swarm changes direction about Kangerdlugssuak in a curve having a radius of about 10 miles, within which region there are vertical dykes radial to this curve. The marginal dyke swarm occurs only along the flexure, its intensity

[1] In Odé's calculations the effect of the rigid boundary was introduced by assuming an 'image' of the circular hole across the boundary. He gives two solutions according to whether the sign of the stresses deriving from the hole was positive or negative, but the two solutions are very similar and only one is figured here.

[2] L. R. Wager and W. A. Deer, 'A Dyke Swarm and Crustal Flexure in East Greenland', *Geol. Mag.*, Vol. 75, 1938, pp. 39–46.

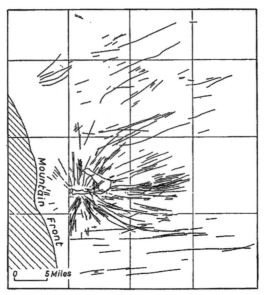

FIG. XII–32A. DYKE SWARM AT SPANISH PEAKS,
COLORADO
(After Knopf, from Odé, 1957)

FIG. XII–32B. STRESS, TRAJECTORIES CAL-
CULATED BY ODÉ, FOR COMPARISON WITH 32A

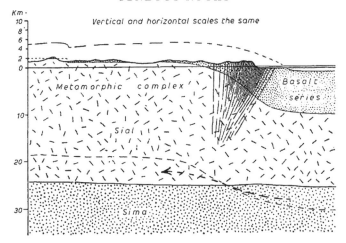

FIG. XII–33. DYKE SWARM, EAST GREENLAND, IN SECTION
(*After Wager and Deer, 1938*)

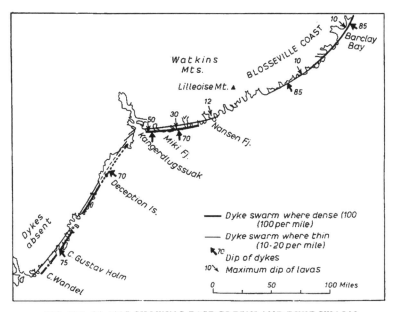

FIG. XII–34. MAP SHOWING EAST GREENLAND DYKE SWARM
(*After Wager and Deer, 1938*)

(number of dykes in unit distance) is proportional to the intensity of the flexure, and the occurrence of the dykes is accordingly determined chiefly by the stretching associated with the movement.[1] The concentration of the dykes on the short, steeply dipping part rather than on the crest of the flexure

---

[1] For the Nyasaland dyke swarm in relation to rift-valley tectonics, see F. Dixey, 'The East African Rift System', *Colonial Geol. and Min. Res., Supp. Series*, Bull. Suppt. No. 1, 1956, p. 8.

suggests, however, that the tension cracks were not formed so much by stretching at the extrados of an arch, as from the secondary stresses resulting from the fault-like movement of two major earth masses, as shown in Fig. IV–18. The more steeply dipping dykes may represent those that have been

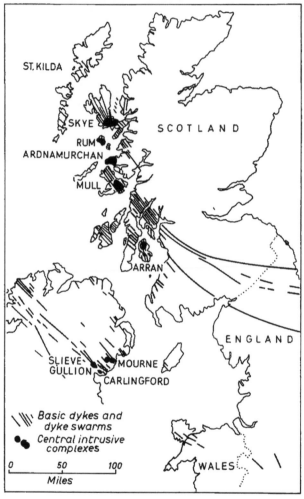

FIG. XII–35. MAP OF TERTIARY DYKES AND VOLCANIC
CENTRES, SCOTLAND
(*After Richey*)

least rotated and which have formed last. The great Lebombo Flexure in Eastern Transvaal and Mozambique has a similar but less well-known dyke swarm along it.[1]

The Tertiary tholeiite dyke swarm of western Scotland and northern

[1] H. Cloos, 'Hebung-Spaltung-Vulkanismus', *Geol. Rundsch.*, Vol. 30, Zwischen-Hft. 4A, 1939, see p. 504.

England is notable for the strongly marked concentration of the dykes as they pass through the central complexes of Mull, Ardnamurchan, Rum, and Skye, which are aligned on the N.N.W. trend of the dykes in that area (Fig. XII–35). Passing to the S.E., the dykes curve towards the east and become wider spaced.[1]

The Middle Devonian Woods Point dyke swarm in Victoria, Australia, is notable for the curving outline of the edge of the swarm, but more so for the

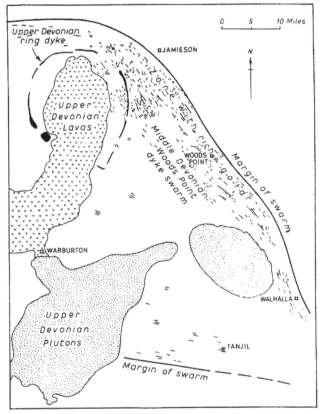

FIG. XII-36. THE WOODS POINT DYKE SWARM, VICTORIA, AND THE CERBEREAN CAULDRON WITHIN THE UPPER DEVONIAN RING-DYKE

rich gold deposits associated with the dykes, which range from peridotite to granophyre but are predominantly mica lamprophyres and quartz-diorites[2] (Fig. XII–36).

Although regional dyke swarms are known in many areas their integrated study is much neglected despite the useful data that it may afford.

[1] J. E. Richey, 'Scotland – The Tertiary Volcanic Districts', *British Regional Geology*, Edinburgh, 1948.
[2] E. S. Hills, 'The Woods Point Dyke-Swarm, Victoria', *Sir Douglas Mawson Anniversary Volume*, Univ. of Adelaide, 1952, pp. 87–100.

Cone sheets and ring dykes

First recognized in Skye by Harker[1] who called them *inclined sheets*, the geometry of cone sheet swarms has been more fully elucidated in Mull and Ardnamurchan. As generalized from the myriads of small dykes, each group of cone sheets constitutes a family of cones dipping inwards towards a common centre or *focus* which lies some $2\frac{1}{2}$–3 miles below the present ground-level. The average dip of cone sheets is 45°, but those nearer the centre dip at much steeper angles, up to 70°, and those farther out at lesser inclinations down to 30°.[2] The central area, however, is free of cone sheets and there is also an outer limit to them (Fig. XII–37A). In the Scottish centres the cone sheets represent an upward extension of the crust of 3000–4000 ft. Owing to

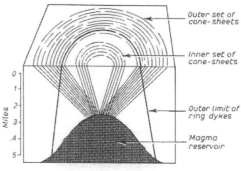

FIG. XII–37A. BLOCK DIAGRAM OF CONE SHEETS

the absence of cone sheets from the core area, it is seen that their injection most probably resulted in the elevation of a rim surrounding a lower central plateau, giving a marked resemblance to many of the craters of the moon.

There can be no doubt that cone sheets occupy fractures generated by the upward pressure of magma in a chamber below the centre of the sheets. According to Anderson, the stress-distribution in the rocks above a magma-chamber of cupola-like shape is virtually identical with that which would be derived from a point source of upward pressure, and the planes normal to the axis of minimum stress are those along which cone sheets would be introduced. This does not account well either for the central area lacking cone sheets, or for the presence of a well-defined outer limit.[3] The resemblance of these conical fractures to those produced in glass or flint by percussion or by pressure at a point has been noted by Jeffreys,[4] and this analogue appears to

[1] A. Harker, 'The Tertiary Igneous Rocks of Skye', *Mem. Geol. Surv.*, 1904.
[2] The set of centre 1 at Ardnamurchan dips at 10°–20°.
[3] E. M. Anderson, 'The Dynamics of the Formation of Cone-sheets, Ring-dykes, and Cauldon-subsidences', *Proc. Roy. Soc. Edinburgh*, Vol. 56, 1935–6, pp. 128–57.
[4] H. Jeffreys, 'Note on Fracture', ibid., pp. 158–63.

be more likely, as it accounts well both for the central core and the outer limit of the sheets. Experiments show that pressure cones in glass are spiral cracks which grow laterally, rather than concentric cones that develop one by one[1] (Fig. XII–38). This suggests the interesting possibility that some cone-sheets may also be spiral in plan, and that in their origination a fracture travels laterally around the spiral as well as upwards from the magma chamber.[2]

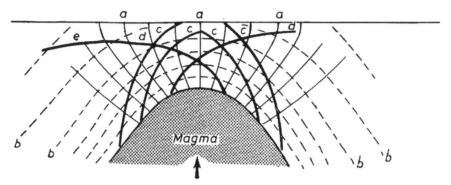

FIG. XII–37B. STRESSES IN RELATION TO CONE SHEETS AND RING DYKES

The stresses are such as would be generated by pressure at the point of the arrow. The thin lines (*a*) and broken thin lines (*b*) are principal stress trajectories. *a* represents maximum (compressive) stress and is the direction of tension cracks which are occupied by cone sheets. A magma chamber of cupola shape, as shown, produces comparable stresses to a point of pressure. (*After Anderson*)

Thick lines represent trajectories of maximum shear-stress and may be represented by ring dykes (*c*) on reduction of upward pressure (subsidence into the magma-chamber). Shear fractures of type *d* (dips low, or even reversing in direction (*e*)) are thought to be represented as dykes on Chilwa Island (Garson).

On the other hand nearly vertical 'cone sheets' occur in central complexes associated with the Lebombo Flexure in south-east Africa, and on Chilwa Island (Nyasaland) cone sheets with very low dips appear to occupy shear-fractures[3] (Fig. XII–37B).

[1] S. Tolansky and V. R. Howes, 'Optical Studies of Ring Cracks in Glass', *Proc. Phys. Soc. Lond.*, Vol. 67, Sect. B, 1954, pp. 467–72. See also J. W. French, 'The Fracture of Homogeneous Media', *Trans. Geol. Soc. Glasgow*, Vol. 17, Pt 1, 1922, pp. 50–68.

[2] This could account for the splitting-off of dyke offshoots in an *en echelon* plan, and also for the apparently great number of ring dykes, which are not readily explained as concentric tensional cracks since the first crack to be filled with magma would free the remaining block from tension, or alternatively would radically alter the stress-distribution and hence the pattern of dykes. For description of cone sheets see J. E. Richey, H. H. Thomas *et al.*, 'The Geology of Ardnamurchan, North-west Mull, and Coll', *Mem. Geol. Surv., Scotland*, 1930. J. E. Richey, 'Tertiary Ring Structures in Britain', *Trans. Geol. Soc. Glasgow*, Vol. 19, 1932, pp. 42–140. J. E. Richey, 'Scotland – The Tertiary Volcanic Districts', *British Regional Geology*, Edinburgh, 1948.

[3] Reports on the S.E. African complexes by the Research Institute of African Geology, University of Leeds (unpublished). For Chilwa Island, see M. S. Garson and W. Campbell Smith, 'The Geology of Chilwa Island', *Geol. Surv. Nyasaland*, Mem. No. 1, 1958.

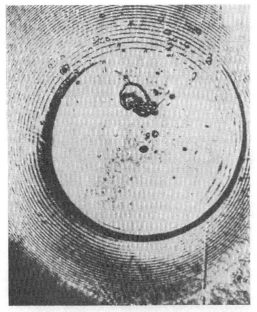

A

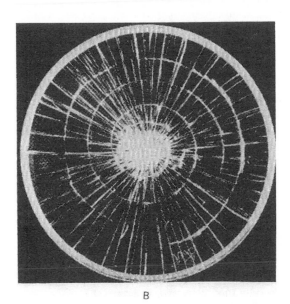

B

FIG. XII–38. SPIRAL CRACKING IN GLASS

A. 'Cone of percussion' in glass, demonstrating the
spiral nature of the crack, and the central 'core'.
(*After Tolansky and Howes, 1954*)

B. Spiral and radial cracking in laminated glass under
concentrated load. (*From Nadai*)

Ring dykes – Ideally a ring dyke forms a closed ring as seen in its outcrop after erosion, but many arcuate dykes that do not encompass a complete ring are also to be regarded as ring dykes where they form part of a concentric series of intrusions in a *central complex*. Most ring dykes are elliptical rather than circular in plan, and some allied dykes exhibit polygonal outlines, but a circular shape is very probably that which may be expected in homogeneous rocks which are not, at the same time, subject to regional stresses (Fig. XII–39).

Ring dykes are generally several hundreds of feet wide, and are composed of plutonic or coarsely porphyritic rocks, but chilled margins are common

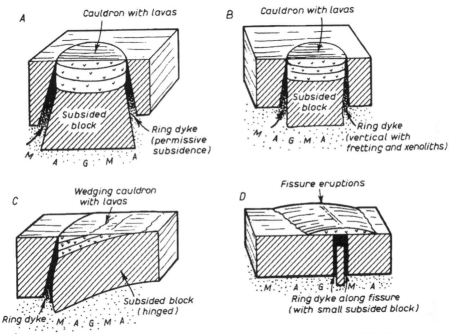

FIG. XII–39. TYPES OF RING DYKES AND RELATED STRUCTURES

and some ring dykes have broken through to the surface feeding thick lava sequences, so that they are clearly characteristic of the zone of fracture rather than the deeper crustal layers.

In some cases only one ring dyke is present, but in others there is a concentrically arranged series of them, often with a more or less complete *screen* of country rock separating the individual dykes; but again screens may be lacking so that the successive dykes make contact with each other. This is perhaps best seen in the Tertiary central complexes of western Scotland in Mull and Ardnamurchan, where also intersecting series of ring dykes and associated cone sheets result from a shift of the centre for successive families of dykes.

The association of ring dykes with lava flows that occupy a central area of subsidence bounded by the dyke itself was first demonstrated at Glencoe,[1] and is now recognized in many other complexes notably in New Hampshire,[2] Africa,[3] and Australia[4] (Figs. XII–36, 40, 43).

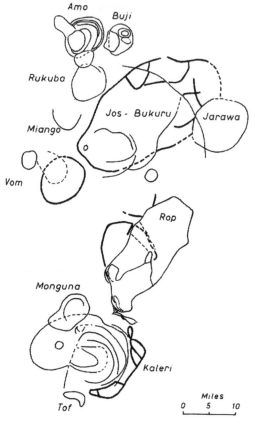

FIG. XII–40. RING-FRACTURES ON THE JOS PLATEAU, NORTHERN NIGERIA

These have controlled the emplacement of the Younger Granites. Earliest fractures in thick lines. (*After Jacobson, Macleod, and Black*)

The central area of subsidence, filled with lavas, is a *volcanic cauldron subsidence*. Beneath the lavas, the sunken bedrock block generally sags and cracks, so that the measured throw on the ring fault occupied by the dyke is

[1] E. B. Bailey, H. B. Maufe *et al.*, 'Geology of Ben Nevis and Glencoe', *Mem. Geol. Surv. Scotland*, 1916; C. T. Clough, H. B. Maufe, and E. B. Bailey, 'The Cauldron-Subsidence of Glencoe, and the Associated Igneous Phenomena', *Quart. Journ. Geol. Soc. Lond.*, Vol. 65, 1909, pp. 611–78.

[2] M. P. Billings, 'Mechanics of Igneous Intrusion in New Hampshire', *Amer. Journ. Sci.*, Vol. 243A (Daly Volume), 1945, pp. 40–68.

[3] R. R. E. Jacobson, W. N. MacLeod, and R. Black, 'Ring Complexes in the Younger Granite Province of Northern Nigeria', *Geol. Soc. London*, Mem. No. 1, 1957. For other African ring structures see discussion in *Proc. Geol. Soc. Lond.*, No. 1545, Session 1956–7, pp. 22–8, and A. L. du Toit, 'The Karroo Dolerites of South Africa', *Trans. Geol. Soc. S. Africa*, Vol. 23 (for 1920), 1921, pp. 1–42.

[4] For Australia, see E. S. Hills, 'Cauldron Subsidences, Granitic Rocks, and Crustal Fracturing in S.E. Australia', *Geol. Rundsch.*, Vol. 47, 1959, pp. 543–61.

less than the total subsidence in the centre. Thicknesses of 5000–10,000 ft of lavas may accumulate in the cauldron, generally in very thick flows that welled out of the dyke-channel as the cauldron floor sank, which must have been a relatively rapid process. There is, however, evidence at some centres for a preceding period of vulcanicity during which a succession of thin flows and pyroclastics accumulated, and the loss of magmatic and gaseous pressure that permitted the final subsidence may in such cases be connected with these earlier events.

The Cerberean cauldron subsidence, one unit of a larger complex of Devonian age in Victoria, Australia, illustrates many features typical of volcanic cauldrons, although it lacks the cone sheets that are associated with the Scottish complexes (Fig. XII–36). The cauldron is truly circular in outline in the north, with a diameter of 15 miles, but tapers to the south where it also becomes involved in the fracture systems of a neighbouring cauldron. Stratigraphical mapping permits an estimate of the downthrow in the immediate vicinity of the ring fracture, of some 2000–3000 ft,[1] but as the lavas dip at angles of up to 50° around their edges, the central parts must have subsided farther, and in fact they contain at least 5000–6000 ft of acid lavas. That the thick upper flows of rhyolite and dacite were extruded from the ring fracture is clear both from the field relationships and from the petrological similarities of the chilled edges and the centre of the dyke with the upper flows. The ring dyke dips outwards at 75°–80° where the dyke is 600 ft wide, which agrees with the stratigraphically determined down-throw of 2000–3000 ft.

High outward dips have been recorded for many ring dykes although Billings reports that the majority of New Hampshire ring dykes are vertical.[2] With an outward dip, the opening for a ring dyke may be produced by 'permissive' subsidence of the central block into an underlying magma chamber, the dyke-channel opening as the central block subsides. This mechanism also explains the presence of ring dykes in parallel series, by successive subsidences. It accounts for the smooth contacts shown by many ring dykes, and also for the presence of large rafts of country rock in them, as some spalling off along the ring fractures would be inevitable. Vertical ring fractures, however, cannot yield an opening simply by movement of the central block, and accordingly it must be assumed that ring dykes occupying such fractures made their way in by piecemeal stoping or by forceful pushing apart of the country rocks. Stoping will leave its mark in irregular contacts, xenoliths, and contamination. Evidence for forceful injection is rare, although the concentric folding of the rocks adjoining the southeast caldera of Central Mull affords a possible example.[3]

[1] S. E. Thomas, 'Geology of the Eildon Dam Project', *Geol. Surv. Vict.*, Mem. No. 16, 1947.

[2] M. P. Billings, 'Mechanics of Igneous Intrusion in New Hampshire', *Amer. Journ. Sci.*, Vol. 243A (Daly Volume), 1945, p. 52.

[3] H. H. Thomas, in 'Geology of Ardnamurchan, North-west Mull, and Coll', *Mem. Geol. Surv. Scotland*, 1930, p. 55.

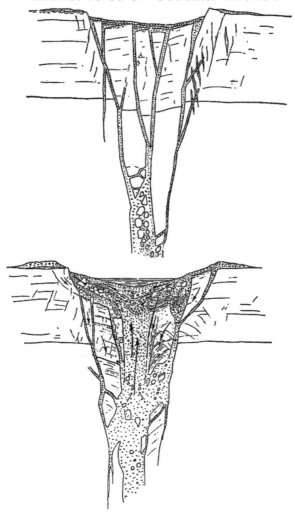

FIG. XII-41. EXPLOSION AND COLLAPSE WITH
FORMATION OF SUB-CIRCULAR DEPRESSIONS,
TYPIFIED BY SWABIAN VOLCANO-EMBRYOS
(After H. Cloos, Geol. Rundsch., Vol. 32, 1941, Figs. 34, 35)

Taubeneck[1] records that the main boundary fault of the Glen Coe cauldron, while in places nearly vertical, elsewhere dips inward at an average of
70°. The subsided block thus tapers downwards so that upward pressure
must have been involved in the opening up of the fracture. Upward doming is
exhibited in several known ring intrusions,[2] and also indicates upward pressure rather than permissive subsidence.

[1] W. H. Taubeneck, 'Notes on the Glen Coe Cauldron Subsidence, Argyllshire, Scotland',
Bull. Geol. Soc. Amer., Vol. 78, 1967, pp. 1295–316.
[2] R. L. Smith, R. A. Bailey, and C. S. Ross, 'Structural Evolution of the Valles Caldera, New
Mexico, and its Bearing on the Emplacement of Ring Dikes,' *U.S. Geol. Surv., Prof. Pap.* 424–
D, 1961, pp. 145–9.

The majority of known ring dykes occur without cone sheets, but where these latter are present the time-relationships show an overlap, in places with earlier cone sheets cut by ring dykes, elsewhere the reverse.

Alkaline complexes, with many of which carbonatites are associated, afford many excellent examples of ring intrusions including ring dykes and cone sheets, although some concentric ring structures associated with them are of metasomatic origin.[1]

Among additional circular fractures related to vulcanism may be mentioned circular calderas bounded by normal faults caused by collapse into a magma chamber on partial withdrawal of the magma[2] (Fig. XII–41), and large-scale circular fractures on which individual vents are themselves situated, as in the Azores.[3]

Many related effects are also seen in *cryptovolcanic disturbances* which may be topographically expressed either as hills (due to up-doming) or as basins due to differential erosion. The origin of many such structures is, however, doubtful, and some may be meteorite craters rather than volcanic forms.[4] A group of dissected domes over a mile in diameter in otherwise flat-lying Karroo rocks described by Spence from East Tanganyika, may perhaps be cryptovolcanic and related to sub-surface carbonatite plugs.[5] More clearly of volcanic origin are the circular explosion vents (*maars*) of the Eifel and other regions (Fig. XII–41).

Ring dykes, batholiths, and stocks

In regions where volcanic cauldron subsidences are known, small batholiths and stocks of plutonic rock, subcircular, elliptical, or polygonal in outline, are generally also present, and appear to have been emplaced by some mechanism similar to that which gave rise to the cauldrons, but without the escape of magma to form lavas. This is seen for example in the Devonian massifs of Lorne, Scotland, that are related to the Glencoe cauldron, also in New Hampshire, Nigeria, and Australia.[6] Concentric structure in the intrusions, which are generally acid or intermediate rather than basic in character, is

[1] B. C. King and D. S. Sutherland, 'Alkaline Rocks of Eastern and Southern Africa – Part 1. Distribution, Ages and Structures', *Sci. Progr.*, Vol. 48, No. 190, 1960, pp. 298–321, give a useful review.

[2] H. Williams, 'The Geology of Crater Lake National Park, Oregon', *Carnegie Inst. Wash.*, Pub. 540, 1942, 157 pp.; H. Kuno, 'Formation of Calderas and Magmatic Evolution', *Trans. Amer. Geophys. Union*, Vol. 34, 1953, pp. 267–80.

[3] F. Machado, 'The Fracture Pattern of the Azorean Volcanoes', *Bull. Vulcanol.* Vol. 17, 1955, pp. 119–25.

[4] W. H. Bucher, 'Cryptovolcanic Structures in the United States', *16th Internat. Geol. Congr.*, 1936, pp. 1055–84. See also J. H. Freeberg 'Terrestrial Impact Structures – A Bibliography' *Geol. Surv. Bull.* 1220, 1966.

[5] J. Spence, 'The Geology of Part of the Eastern Province of Tanganyika', *Geol. Surv. Tanganyika*, Bull. No. 28, 1957.

[6] See references under Ring Dykes.

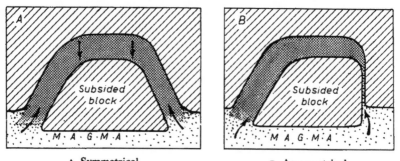

A. Symmetrical   B. Asymmetrical

FIG. XII–42. RING-FRACTURE STOPING AND PLUTONIC INTRUSION

commonly demonstrated by petrological differences between the peripheral and central parts, and partial screens may be present between these. Some such intrusions show evidence of an arched roof, indicating that the magma was confined to a sub-surface chamber. This may be attributed to the sinking of a truncated cone-shaped block bounded by a ring fracture, which Daly has termed *ring-fracture stoping*, as opposed to the piecemeal stoping of small

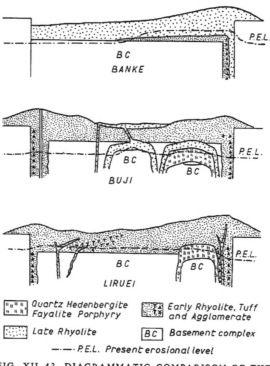

FIG. XII–43. DIAGRAMMATIC COMPARISON OF THE
VOLCANIC STRUCTURES OF THE BANKE, BUJI AND
LIRUEI COMPLEXES, NORTHERN NIGERIA

(*After Jacobson, Macleod, and Black*)

blocks[1] (Fig. XII–42). The mechanism was proposed by Richey for the Mourne granites,[2] and is similar to that postulated for the successive ring dyke intrusions of the Tertiary ring complexes of western Scotland. A second possibility, that the roof of the intrusion was bodily forced upwards, along fractures, has also to be considered and may be necessary to account for several Victorian examples where the lavas of associated cauldrons, after solidification, were intruded by granodiorite which rises far above the base of the lavas and makes relatively straight, vertical contacts with them.[3]

It is a notable fact that the several regional suites of volcanic and subsurface ring and polygonal complexes already known demonstrate the presence, in each region, of a parent magma having clearly defined petrological characteristics, which, with its differentiates, gives rise to a petrographic province. That this parent magma may have migrated laterally to the region as a sheet at some appropriate horizon a few miles below the surface seems highly probable, but it is in any case clear that the ring complexes as well as linear dyke swarms are intimately related to major phenomena of magma genesis, magmatic movements, and major crustal structures and processes.

[1] R. A. Daly, *Igneous Rocks and the Depths of the Earth*, New York, 1933, p. 269.

[2] J. E. Richey, 'The Structural Relations of the Mourne Granites (Northern Ireland)', *Quart. Journ. Geol. Soc. Lond.*, Vol. 83, 1928, pp. 653–88.

[3] E. S. Hills, op. cit. supra, 1959. Other references to Australian ring-complexes are: N. C. Stevens, 'Ring-structures of the Mt Alford District, South-east Queensland', *Journ. Geol. Soc. Aust.*, Vol. 6, 1959, pp. 37–50. P. J. Stephenson, 'The Mt Barney Central Complex, S.E. Queensland', *Geol. Mag.*, Vol. 96, 1959, pp. 125–36. C. D. Branch, 'Volcanic Cauldrons, Ring Complexes, and Associated Granites of the Georgetown Inlier, Queensland', *Bur. Min. Res.*, Bull. 76, 1966.

CHAPTER XIII

Structural Petrology

By E. DEN TEX

There are two major aspects of structural petrology. The first relates to the recognition and description of the geometrical properties of rock fabrics, and is known as *petrofabric analysis*. The second involves the attempt to establish relationships between these geometrical properties and the movements or strains involved in the generation of the fabric, and even to infer possible causative forces or stresses. This is known as petrotectonics. These two aspects of the subject are further discussed below.

PETROFABRIC ANALYSIS

In 1911 Sander[1] redefined a term previously of little consequence in petrology. His *fabric* (*Gefüge*) comprises all textures and structures of a rock and is meant to describe all the spatial data it contains, irrespective of its exterior shape or boundaries.

Petrofabric analysis (*Gefüge-analyse*) is a somewhat misleading name for the study of rock fabrics because it often neglects the megascopic structures and completely ignores the non-directional textures – crystallinity, grain-size, shape, and porosity – which constitute the *scalar fabric* of the rock. Moreover, the directional structures and textures forming the *vectorial fabric* lose their kinematic and dynamic significance when distributed in a random manner so as to render the fabric *isotropic*. Only the preferred orientations of directional structures and textures of an *anisotropic fabric* provide a geometry which can be interpreted in terms of movement, strain, or stress.

Fabric elements and components

The essential notions in petrofabric analysis relate to the definition of fabric in terms of fabric elements and fabric components and to the orientation of the fabric in space.

Fabric elements – If the exterior shape of rock-masses is to be of no signifi-

[1] B. Sander, 'Ueber Zusammenhänge zwischen Teilbewegung und Gefüge in Gesteinen', *Min. Pet. Mitt.*, Vol. 30, 1911, pp. 281–314.

cance for the vectorial fabric, the latter must be defined by directional units within the rock, which are commonly known as the *fabric elements*. Fabric elements include both the shape of grains or of homogeneous aggregates, and crystallographic planes or lines within them. The former may be defined by crystals of planar, linear, or triaxial habit such as mica, tourmaline, and hornblende, or by veins, rods, lenses, and other non-equant aggregates of minerals. The preferred orientation of these is shown in their external form which is therefore a *preferred dimensional orientation* (Fig. XIII–1). On the

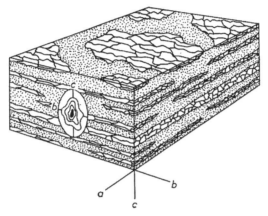

FIG. XIII–1. BLOCK DIAGRAM OF A ROCK CONSISTING OF THIN LAYERS AND LENTICULES OF COARSE-GRAINED QUARTZ SET IN A MATRIX OF FINE-GRAINED QUARTZ, FELDSPAR, GARNET, AND PYROXENE

To a first approximation in the field, the fabric appears to be simply planar parallel to the lamination (*ab*). Inspection of thin sections, cut at right angles to the *ab* plane, reveals that the coarse quartz grains have a triaxial habit, the short diameter of which is preferentially oriented parallel to the *c* fabric axis, whereas the long diameter is oriented to form a lineation corresponding with the *b* axis of the fabric (dimensional orientation). The circular inset of the front face is a contoured equal-area plot of orientations of optic axes from coarse quartz crystals. It shows the presence of another linear structure due to preferred orientation of the quartz lattice parallel to the *a* fabric axis. The habit of the coarse-grained quartz is not a common crystal habit of this mineral but a fabric habit. The symmetry of the fabric is orthorhombic if allowance is made for the different nature of the *a* and *b* lineations.

other hand, planes and directions in crystal lattices, even of cubic minerals, may serve as fabric elements to form a *preferred lattice orientation* (Fig. XIII–1). These fabric elements are crystallographic and optic axes, twin planes, cleavages, and translation glide planes and glide lines.

Fabric components – Considered geometrically, the fabric may be resolved into *planar* and *linear* components. Workers in macro-structures recognize, for instance, stratification and schistosity, fold axes, and slip-striae, occurring in various combinations with different emphasis in one tectonic unit. In micro-structures a planar component is defined by a more or less rigorous planar parallelism of fabric elements such as the basal cleavage of mica crystals or the optic axis of calcite crystals. All planar fabric components are known as *s-planes*, in which *s* is a mnemonic for stratification, schistosity, slip, shear, etc. A linear component is defined by linear parallelism of fabric elements such as the *c*-axis of hornblende or the *b*-pinacoid of feldspar crystals. Linear fabric components visible with the unaided eye are known as *lineations*. The independence of the geometry of a structure from the predominant habit of its fabric elements may not be clearly realized, especially when a structure of a weak statistical nature is concerned. For instance, in natural exposures a planar fabric component, consisting of a weak parallelism of tourmaline crystals, appears nearly always as a lineation whether a linear component is present or not.[1] Statistical treatment is indeed essential to reveal the geometry of structural data of the kind used in a fabric analysis. Even in the presence of a well-defined schistosity or lineation the structure may consist of more components than casual observations in the field would suggest (Fig. XIII–1).

Fabric axes and indices

Like the geometry of a crystal the planar and linear components of the vectorial fabric are conveniently referred in space to three co-ordinates or fabric axes. Sander[2] has argued that geometrical crystallography presents the structural petrologist with a familiar system of notation. As fabric axes he adopts the three mutually perpendicular axes *a*, *b*, and *c*. Any joint system, cleavage, bedding, or other *s* plane in the fabric may thus be described by the Miller indices of the corresponding crystal forms such as {*h k l*} for planes not parallel to any axis, {*o k l*} for planes parallel to the *a* axis only, {*h o l*} for planes parallel to the *b* axis only, and {*h k o*} for those parallel to the *c* axis only. *S* planes parallel to more than one fabric axis are commonly designated as *ab*, *ac*, and *bc* rather than by the corresponding Miller indices {*001*}, {*010*}, and {*100*}. Similarly a fold axis, mineral stretching, common intersection of *s* planes or other lineation in the fabric may be described by the zone symbols of the corresponding crystal edges or zone axes such as [*u v w*] for a lineation within {*h k l*}, i.e. a direction intersecting the *a*, *b*, and *c* axes; and so on. Only lineations parallel to a fabric axis are commonly designated as *a*, *b*, or *c* lineations rather than by the corresponding zone symbols [*100*], [*010*], and [*100*] (Figs. XIII–1, 10). As with crystallographic axes the fabric axes should be selected purely for convenience of description. Thus the most

[1] K. E. Lowe, 'A graphic solution for certain problems of linear structure', *Amer. Miner.*, Vol. 31, 1946, pp. 425–34. R. H. Clark and D. B. McIntyre, 'A macroscopic method of fabric analysis', *Amer. Journ. Sci.*, Vol. 249, 1951, pp. 755–68. E. den Tex, 'Stereographic distinction of linear and planar structures from apparent lineations in random exposure planes', *Journ. Geol. Soc. Aust.*, Vol. 1 (for 1953), 1954, pp. 55–66.
[2] B. Sander and O. Schmidegg, 'Zur petrographisch-tektonischen Analyse III', *Jahrb. Geol. Bundesanst., Vienna*, Vol. 76, 1926, pp. 323–404.

prominent and persistent s plane, whether it be bedding, cleavage, or foliation is referred to as the ab plane with the c fabric axis oriented at right angles to it. The b axis is further specified as the normal to the most obvious symmetry plane of the fabric or – if that plane cannot be recognized – it is tentatively assumed to be parallel to the most conspicuous and persistent lineation lying within the ab plane. Failing a lineation in ab, b is fixed arbitrarily within that plane and a of necessity at right angles to b in the same plane. Failing an s plane, the b axis is taken to be normal to the most obvious symmetry plane or parallel to the most prominent lineation, while a is selected at right angles to b and parallel to a second lineation or to the plane containing the first and second lineations if more than one lineation is available, and in the proper angular relationships.

In the absence of any macroscopic s planes and lineations the geographical co-ordinates (N., E., and zenith) should be chosen as fabric axes since they have to be shown anyway in most fabric studies at some stage.

In a fold lacking persistent cleavage the obvious choice for ab is the bedding while the fold axis should be selected for b as it is either normal to the principal symmetry plane or the most pronounced lineation or both. In this case the a and c axes do not have a fixed relation to the geographical co-ordinates and it is then imperative that the latter be shown in addition to the fabric axes in all diagrams made from different locations in the fold. In folds that are not entirely of the flexural type there is often a prominent cleavage which has a more uniform orientation than the bedding, and is therefore to be preferred as ab. The fold axis or the intersection of bedding and cleavage is then assumed to indicate b. Should the geographical orientation of the cleavage be constant throughout the fold, a uniform relation exists between all fabric axes and geographical co-ordinates so that it suffices to show the latter only once.

Sander's system of fabric axes and reference symbols is the one most widely adopted by structural petrologists but different notations have been proposed from time to time.[1] The tendency to attach kinematic or even dynamic significance to the fabric axes and their common planes stems from Sander's early work.[2] It has unfortunately led to much confusion of thought especially so since the axes of the strain ellipsoid are usually labelled A, B, and C (see p. 87).[3] Inasmuch as the fabric axes happen to coincide with the deformation axes of the strained body of rock, they may inherit some of their kinematic significance but it is unwarranted to assume, for instance, that the

[1] A useful review of the various systems of references and symbols used in structural geology may be found in 'Tectonic facies, orientation, sequence, style and date', by W. B. Harland, Geol. Mag., Vol. 93, 1956, pp. 111–20.

[2] B. Sander, Gefügekunde der Gesteine, Vienna, 1930, pp. 56–60, 108.

[3] This confusing analogy is maintained by denoting the fabric axes as x, y, and z, and the axes of the deformation ellipsoid as X, Y, and Z, as is done for instance by J. G. Ramsay (Folding and fracturing of rocks, New York, 1967, p. xv).

a axis always corresponds to the direction of mass transport in the rock if its only description is that it lies at right angles to the most prominent lineation in the most conspicuous *s* plane or that it is, or is not, the axis of a girdle of fabric elements. Another misconception that has become firmly entrenched is the assumption that the *ac* plane of the fabric is the only plane in which effective movements have taken place (see pp. 436–41).

Fabric habit – One category of fabric elements deserves special mention. It concerns single crystals with an inequant habit quite different from their usual crystallographic habit or even bearing no fixed relation to the internal structure of the crystal at all. If it is found that such a habit is distinctly related to the fabric of the rock in which it occurs, it is referred to as *fabric habit* (*Gefügetracht*)[1] (Fig. XIII–1). The relation between fabric and habit is usually of a simple type, such as linear parallel to *b* or *a*, or planar parallel to *ab*, {*h o l*} or {*o k l*}; or combinations of these, often with a triaxial fabric habit. It may be due to processes of differential erosion and sedimentation, to plastic deformation of crystals, or perhaps most frequently to crystallization under non-hydrostatic stress in the presence of an intergranular liquid capable of dissolving and precipitating the pertinent mineral.[2] Among the common rock-forming minerals quartz, calcite, and olivine exhibit most noticeably such a habit conditioned for the greater part by the fabric. Triaxial fabric habit of quartz grains with long axis parallel *a*, median axis parallel *b*, and short axis parallel *c* is shown in Fig. XIII–11A and B, with weak lattice orientation of optic axes perpendicular to {*h o l*} and {*o k l*} appearing in Fig. XIII–12 (*left*).

Statistical treatment of fabric data

In the preceding paragraphs the statistical nature of preferred orientation of fabric elements has been mentioned. When dimensional orientation rigorously prefers one plane or one direction in space and the fabric elements are sufficiently large, the field-geologist may determine the frequency modus of orientation without the use of visual or mechanical aids. But dimensional fabric elements may be too small to allow the unaided eye to differentiate between long and short diameters. In this event visual or photometric averaging can be carried out in thin or polished sections under the petrological or the binocular microscope. To detect even the strongest optical orientation, however, requires the use of a petrological microscope with standard accessories such as the gypsum plate. Qualitative analyses of optically visible lattice orientation and techniques for quantitative measurement of optical orientation in uniaxial crystals involving comparatively little time and

---

[1] B. Sander, *Einführung in die Gefügekunde der Geologischen Körper*, Vol. II, Vienna, 1950, pp. 66–7.

[2] P. Hartman and E. den Tex, 'Piezocristalline fabrics of olivine in theory and nature', 22nd Int. Geol. Congress, New Delhi, Proc. Pt IV, pp. 84–114, 1964.

requiring the use of standard petrological equipment only, are available. However, adequate study of optical orientation requires the use of a universal stage which instrument permits ready identification of the most important crystal-optical vectors, as well as sufficiently accurate measurement of their orientation in random sections. More recently metallurgists and petrophysicists have developed techniques for the measurement of bulk orientation in crystalline aggregates by means of X-ray diffraction, thermal expansion, seismic wave propagation, intensity of magnetization, and other derived properties of the crystal lattices concerned. These methods have a stronger averaging effect and are much less time-consuming, which is desirable in statistical work, but orientation data on individual crystals, such as only the optic and X-ray universal stage can yield, are often indispensable for petrotectonic purposes.[1]

The statistical presentation of the accumulated orientation data presents its own problems. Schmidt[2] pioneered the use of Lambert's azimuthal equal-area projection. Its plotting device (now known to all structural petrologists as the Schmidt-net) is operated in the same manner as the stereographic Wulff-net, as has been explained in Chapter VI.

Both pole (or π) diagrams, and cyclographic (or β) diagrams may be used for fabric study, and if the density contours of π-diagrams are suitably shaded in a tone suggestive of the density class they represent (Fig. XIII–2), a reasonably accurate and self-explanatory picture of the geometry of the fabric emerges. A β-diagram is equally amenable to density analysis if the mutual intersections of the great circles are counted (Fig. XIII–2C). In petrofabric analysis all plotting is done on the lower hemisphere of reference. This permits ready comparison and makes the observer look down the dip and plunge of structures as he normally does in the field.

Types of orientation maxima

The fabric may be resolved into components according to the shape of the areas of maximum density of orientation. There are three fundamental types of maximum concentration in a density diagram:

(i) the *circular* or *point-maximum*, in which the orientation data cluster around a point (i.e. directions tend to be parallel to a line (Fig. XIII–2A, C, D).

[1] R. Brinkman, W. Giesel, and R. Hoeppener, 'Über Versuche zur Bestimmung der Gesteinsanisotropie', *N. Jb. Geol. Pal.*, Mh. 1, 1961, pp. 22–33. R. Lautenbach, 'Geomagnetische Gefügeforschung in der Nordöstlichen Heide Mecklenburgs', *Geophys. u. Geol.*, Bd. F.1, 1959, pp. 89–96. K. von Gehlen, 'Die röntgenographische und optische Gefügeanalyse von Erzen, ins besondere mit dem Zahlröhr-Texturgoniometer', *Beitr. Miner. Petr.*, Bd. 7, 1960, pp. 340–88. N. I. Christensen and R. S. Crosson, 'Seismic anisotropy in the upper mantle', *Tectonophys.*, Vol. 6, 1968, pp. 93–107.

[2] W. Schmidt, 'Gefügestatistik', *Min. Pet. Mitt.*, Vol. 38, 1925, pp. 392–423.

(ii) the *girdle maximum*, where the data are concentrated along a great circle of the projection (i.e. directions tend to be confined to a plane (Fig. XIII–2G); and

(iii) the *cleft-girdle* or *annular maximum*, which has the data approximating to a pair of opposite small circles of the projection (i.e. directions tend to lie in a double conical surface) (Fig. XIII–2H).

A point maximum indicates the presence of either a planar or linear fabric component depending on whether the orientation data used are poles of planar, or penetration points of linear fabric elements respectively. Thus, if poles to the basal cleavage of mica have been measured, a point maximum stands for foliation due to planar parallelism of mica flakes. On the other hand, the same maximum would signify the presence of a lineation due to linear parallelism of hornblende needles, if their c axes had been plotted.

A similar ambiguity applies to the significance of the girdle maximum, but here the relations are the reverse, A lineation is represented by a girdle of $\{001\}$ poles of mica crystals, while a foliation figures as a girdle of hornblende c axes.

The cleft-girdle maximum is also connected with either a linear or a planar fabric component. This type of concentration means that the measured direction tends to make a constant angle other than 90° with a fixed line, the angle being the semi-vertical angle of the cone the axis of which constitutes the lineation or the pole to the foliation. The rhombohedral carbonates and quartz often exhibit cleft-girdle arrangements of their optic axes. Both the calcite and dolomite lattices contain important twin- and translation-glide lines in the rhombohedral lattice planes, which may have acted as a fabric control, while $α$-quartz has three directions of maximum elasticity inclined at approximately 70° to the optic axis.

Combinations of the three fundamental types of density maxima are common. A girdle of mica cleavage poles containing one or more subordinate point maxima is the expression of a linear structure modified by one or more co-axial planar components (Fig. XIII–2F, G), while a similar principal point maximum lying in one or more partial girdles indicates the presence of a planar structure modified by one or more co-planar linear components (Fig. XIII–2D). For linear fabric elements the structural significance of these density patterns is, of course, exactly the opposite. The more complex fabrics have point maxima situated outside girdles which do not have a common intersection. This means that the planar components of the fabric are not co-axial nor the linear components co-planar.

Obviously, the fabric has a statistical symmetry determined by the nature, number, and mutual orientation of the fabric components. We will later see how the kinematic interpretation of the fabric hinges on the proper recognition of its symmetry.

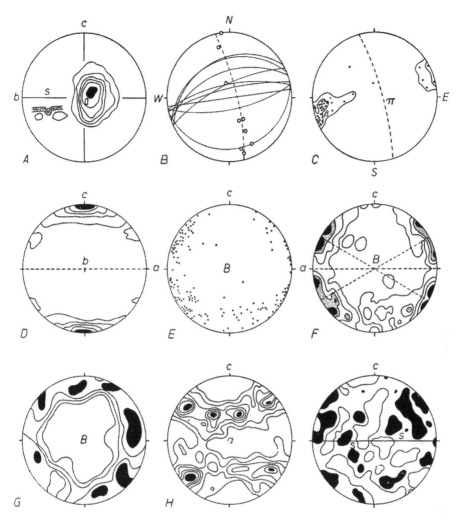

FIG. XIII-2. FUNDAMENTAL TYPES OF FABRIC DIAGRAMS

A. Point-maximum indicating lineation parallel to *a*. 180 optic axes of quartz in a pre-crystalline slickensided mylonite from the Meliboekus Granite, Odenwald, Germany. Contours: (25–20)–10–8–6–4–2–0 per cent per 1 per cent area. (*After B. Sander*, 1950)

B. *β*-diagram (or cyclographic projection) of bedding planes showing co-axial disposition with poles along a great circle (*π*) normal to the fold-axis (*B*). (*After B. Sander*, 1948)

C. Same as 3B showing only the intersections (*β*) with approximate density contours and the *π*-circle normal to the maximum of *β*-intersections.

D. Point-maximum situated within incipient girdle indicating planar structure slightly modified by linear component. 250 poles to {001} of muscovite grains from Moine Schist, Scotland. Contours: 5–4–3–2–1–0 per cent per 1 per cent area. (*After F. C. Phillips*, 1937)

E. Scatter diagram of 150 optic axes of quartz in a slightly gneissic granite from Lake Manapouri, New Zealand, projected upon the *ac* plane of the megascopic fabric.

F. Same as 2E but counted out and contoured revealing a girdle-maximum containing two double point-maxima symmetrically situated with regard to the *ab* plane. Linear structure modified by planar components probably due to late flattening. Cont. 6–4–2–0·67 per cent. (*After F. J. Turner*, 1948)

G. Girdle-maximum of 140 optic axes of quartz in a biotite schist from the Isergebirge, Silezia. Contours: 4–3–2–1–0. (*After B. Sander*, 1950)

H. Annular or cleft-girdle maximum of 284 optic axes of quartz in a granulite from Hartmannsdorf, Saxony. Cont. 5–4–3–2–1–0·5–0. (*After B. Sander*, 1950)

I. Fortuitous preferred orientation of 251 optic axes of quartz in a limestone from Roumania, bedded parallel to *s*. Isotropic fabric of the quartz lattice. Contours: 3–2–1–0. (*After B.*

Anisotropism of fabric

A fabric showing significant preferred orientation is usually referred to as *anisotropic*, whereas the *isotropic* fabric has its orientation data distributed in a random fashion over the sphere of reference showing chance concentrations only, because the sample we draw from it is neither infinitely large nor perfectly random (Fig. XIII−21). The question as to where fortuitous concentration ends and preferred orientation significant of anisotropism begins is clearly a most important one which should be answered for each maximum or minimum. Reproducibility of orientation maxima in diagrams made from variously orientated sections through different parts of a rock is usually regarded as sufficient evidence that the fabric is anisotropic, but maxima which cannot be reproduced in this manner do not constitute conclusive evidence to the contrary, since an anisotropic fabric may be *inhomogeneous*, in other words it may vary according to its location within a rock-unit studied without becoming isotropic as a whole.[1] However, single plots of orientation data (being finite samples drawn from an infinite population of isolated events in a continuum of area) may reveal their uniformity, or derived isotropism, by the fact that the frequency of their density distribution is governed by a Poisson-type law. This law states that the probability for a certain orientation to occur a specific number of times per unit area of the projection is an exponential function of the expectation of the sample or, in this case, of the average number of orientations per unit area. Inserting the expectation figure of our sample into the successive terms of the Poisson exponential function, we may find the frequency distribution of the sample to be expected from isotropic orientation in the parent population. The differences between the values thus found and the frequencies of observed densities per unit area may now be tested to see whether they could be accounted for by the laws of chance. For this purpose we employ any one of the standard tests for significance of deviation (or reality of association) specially adjusted to three dimensional orientation data.[2] Some of these make use of the value of

[1] Preferred orientation not significant of fabric anisotropism may also arise from the fact that the number of grains of a certain mineral, cut by a particular thin section of a rock, is dependent on the orientation of that section with regard to a possible preferred dimensional orientation of the mineral under consideration. For instance, mica flakes in a schist will turn up in greater numbers in sections at right angles, than in sections cut parallel to the schistosity. The two types of section may be used in combination, or a correction factor may be applied to results obtained from one type of section in order to eliminate this effect. See B. Sander, D. Kastler, and J. Ladurner, 'Zur Korrektur des Schnitteffektes in Gefügediagrammen heterometrischer Körner', *Sitz. Ber. Oester. Akad. Wiss., Math.-Nat. Wiss. Kl., Abt. I*, Vol. 163, 1954, pp. 401–23. Correction tables in *Ann. Univ. Sarav., Nat. Wiss., Scientia*, Vol. 6, 1957, pp. 337–58.

[2] H. Winchell, 'A new method of interpreting fabric diagrams', *Amer. Miner.*, Vol. 22, 1937, pp. 15–37. H. J. Pincus, 'The analysis of aggregates of orientation data in the earth Sciences', *Journ. Geol.*, Vol. 61, 1953, pp. 482–509. O. Braitsch, 'Quantitative Auswertung einfacher Gefügediagramme', *Heidelb. Beitr. Miner. Petr.*, Vol. 5, 1956, pp. 210–26. D. Flinn, 'On tests

$\chi^2 = \pm \sum \dfrac{(O-E)^2}{E}$ (Chi-square test) in which O stands for observed and E for expected frequency of orientation per unit area. Obviously if the parent population of the fabric is isotropic its χ^2, or that of an infinitely large random sample thereof, should be zero. Finite samples of any size will yield a χ^2 whose deviation from zero is a function of the number of degrees of freedom of distribution of the data over the unit areas (number of effective theoretical frequencies minus two) and of the number of trials run (significance level expressed in per cent of total number of samples drawn). If the observed χ^2 of one sample exceeds the expected value in one out of twenty similar samples that could be drawn from an isotropic fabric (5 per cent significance level) we may be reasonably sure that our parent population was anisotropic. Should it exceed the χ^2 value in one out of a hundred isotropic samples (1 per cent significance level), the significance of its preferred orientation may be regarded as beyond doubt.

However, having rejected the *null hypothesis* of its deviation from an isotropic parent population, does not mean that we have gained insight into the geometry of the anisotropic population present. To obtain this we should test the probability of deviation of our data from density distributions other than the uniform or isotropic distribution. Comparatively simple examples of such a model are: the *circular normal distribution* representing a point-maximum, the *equatorial distribution* underlying a great circle girdle, and the *elliptical distribution* as model of incomplete girdles and elongate point-maxima.[1] Thus an elongate point-maximum of fabric elements may be tested against a contour diagram constructed on the basis of Bingham's probability density function for the elliptical distribution using the same real shape parameters as were derived by maximum likelihood estimation from the orientation pattern under scrutiny[2] (Fig. XIII−3A).

Instead of the one per cent square grid commonly used as testing cells, the diagram may be divided into more or less deformed hexagons of one per cent area,[3] or into any number of zones and sectors of equal area (Fig. XIII−3B). The *ten-cell zone test* was claimed by Winchell[4] to be appropriate for the

---

of significance of preferred orientation in three-dimensional fabric diagrams', *Journ. Geol.*, Vol. 66, 1958, pp. 526–39. E. Dimroth, 'Eine Theorie der Korngefügestatistik', *N. Jb. Miner.*, Mh. 10, 1962, pp. 218–29.

[1] C. Bingham, 'Distributions on the sphere and on the projective plane', Ph.D. thesis Yale Univ., 1964, VII + 93 pp. G. S. Watson, 'The statistics of orientation data', *Journ. Geol.*, Vol. 74, 1966, pp. 786–97. M. R. Stauffer, 'An empirical statistical study of three-dimensional fabric diagrams as used in structural analysis', *Can. J. Earth Sci.*, Vol. 3, 1966, pp. 473–98.

[2] A. van Zuuren, 'Structural petrology of an area near Santiago de Compostela (N.W. Spain)', *Leidse. Geol. Meded.*, Vol. 45, 1969, p. 48, fig. IV−2.

[3] F. Kalsbeek, 'A hexagonal net for the counting-out and testing of fabric diagrams', *N. Jb. Miner.*, Mh. 7, 1963, pp. 173–6.

[4] H. Winchell, 'A new method of interpretation of petrofabric diagrams', *Amer. Miner.*, Vol. 22, 1937, pp. 15–36.

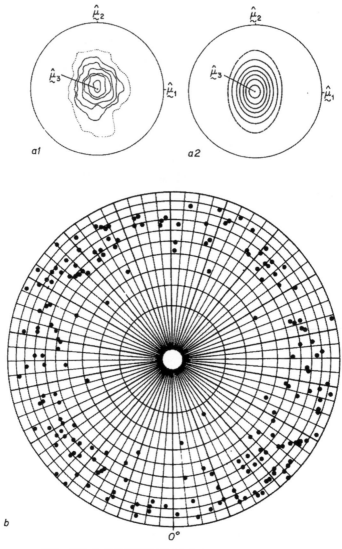

FIG. XIII-3. SIGNIFICANCE TESTS OF PREFERRED
ORIENTATION

A. 1. Fabric diagram of 200 β-axes of clinopyroxene from an eclogite at
Cabo Ortegal (N.W. Spain). 2. Distribution according to Bingham's
probability density function with real shape parameters derived from
the data of diagram 1. (*After van Zuuren, 1969*)

B. Scatter diagram of 198 optic axes of quartz forming a cleft-girdle of
apical angle ~70°. Plot divided into ten zones and seventy-two
sectors of equal area. A ten-cell zone-test yields $P < 0.001$, whereas
twenty-two out of a hundred samples drawn at random from a uniform population would show a similar distribution of points over the
seventy-two sectors within the outer five zones: no significant point-
maxima.

significance of a point-maximum, a small- and a great-circle girdle simultaneously. Indeed, if used in conjunction with a probability density function for the circular normal or equatorial distribution, it may serve a better purpose than the directionally unrelated cells of the square and hexagonal

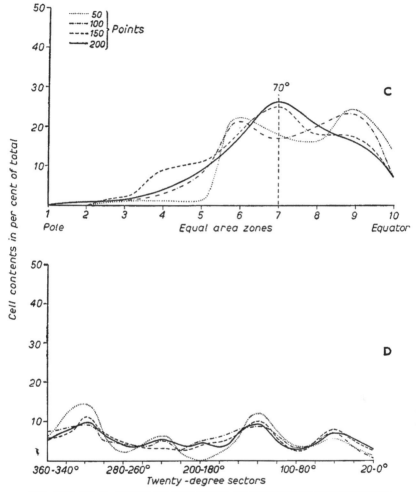

FIG. XIII-3C, D. SIGNIFICANCE TESTS OF PREFERRED ORIENTATION

c. Graph of the development of preferred orientation in ten equal area zones of scatter diagram 3B. Skewed frequency curve with peak at 70° from central pole. Minimum sample size: 150 points. Cleft-girdle significant.

D. Similar graph for eighteen 20° sectors of the same diagram. Normal frequency curve of composite nature. Minimum sample size 100 points. Some significant preferred orientation within cleft-girdle?

testing grids. Since, in this case, we would be testing a surmised orientation pattern against the corresponding anisotropic model distribution, we should dispose the zones or their axis as closely as possible parallel to the girdle or the point-maximum to be tested.

The development of preferred orientation is also worthy of watching. Chance concentrations decrease as more data are plotted, while those due to anisotropism increase until an optimum sample size is reached. If measurements are to be kept to a minimum it is advisable to keep a graphical record of the cell-contents as plotting proceeds.[1] If a single girdle or point-maximum is expected, the cells of the Winchell zone test are obviously most appropriate, while any number of *equal area sectors* may be superimposed to form additional loci of registration for the development of subsidiary point-maxima within a principal concentration of girdle-type (Fig. XIII–3B, C, D).

The *degree of preferred orientation* is another characteristic of the fabric. Whereas in an ideally uniform distribution n points occupy each $1/n$ of the total area of the hemisphere of reference, preferred orientation causes increasingly smaller parts of it to be occupied by an increasingly larger proportion of points. This is expressed by the ratio of the percentages of total points to the percentage of total area occupied by each density class in the fabric diagram. The standard density contour diagram gives an approximation of both these quantities, but Barnick has shown that the scatter diagram, when contoured on a *free counter* basis, using a counter of a size proportional to $1/n$ of the total area, yields an orientation factor that is independent of the orientation of a grid and of the number of points per density class.[2]

Inhomogeneity of the fabric

Like a body or a field of matter a fabric is said to be *homogeneous* if any two parts of the same size are identical when considered under the same circumstances and within the limits of a given *domain*. The domain under consideration may be of the size of a thin section, a hand-specimen, an outcrop, or an even larger unit of rock.

Once anisotropism of the fabric has been established on the strength of its positive correlation or its deviation from Poisson expectancy, failure to reproduce significant maxima in diagrams made from different portions of, say, a microfold, a minor fold, or a major fold, indicates *inhomogeneity* of the fabric on the scale employed.

Consider, for instance, the case of a bed laid down under uniform conditions of sedimentation. Suppose that a depositional fabric develops which is homogeneous throughout the bed. Now let us assume that the bed is warped or bent into a fold and that the fabric is flexed along with it, then its orientation in space is no longer identical in opposite limbs and in the crest or

[1] I. Larsson, 'A graphic testing procedure for point diagrams', *Amer. Journ. Sci.*, Vol. 250, 1952, pp. 586–93.

[2] F. K. Drescher-Kaden, 'Zur Darstellung des Regelungsgrades eines Gefüges', *Tscherm. Min. Pet. Mitt.*, Vol. 4, 1954, pp. 159–77. G. Fischer, 'Über die Auswertung von Gefügediagrammen', *Abh. dtsch. Akad. Wiss., Berlin, Kl. III*, Vol. 19, 1960, pp. 283–99. H. Barnick, 'Zur Frage der Normung und Darstellung des Regelungsgrades von Gefügediagrammen', *Tscherm. Min. Pet. Mitt.*, Vol. 11, 1966, pp. 1–28.

trough of the fold (Fig. XIII—4A). This is a fabric rendered inhomogeneous by a deformation which itself is inhomogeneous or *non-affine*. However, there is no regular relation between homogeneity of deformation and homogeneity of fabric, as much depends on the scale of the domain for which homogeneity of fabric is considered. A homogeneous deformation by pure shear on two equivalent sets of slip planes may produce an apparently homogeneous fabric

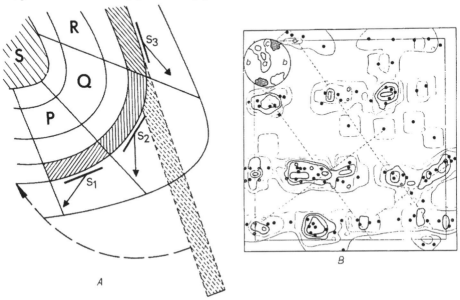

FIG. XIII—4. INHOMOGENEITY OF FABRIC

A. Inhomogeneous flexure fold in quartzite from the Brenner, Tyrol. Sectors P, Q, and R are not homogeneous as regards orientation of the microfabric *s* plane defined by maxima of quartz optic axes ($s_{1, 2, 3}$). The fabric may be unrolled into a homogeneous flat bed as shown by the dashed extension of s_3. (*Modified from B. Sander, 1930*)

B. Axial distribution analysis of part of a thin section cut parallel to the *ac* plane of a quartzite from the Rensenspitze, Austria. Inset: contoured fabric diagram of quartz optic axes. Only grains contributing to the shaded portion of the inset have been plotted as points (106) and contoured: (8–5)–4–3–2–1 per cent per 1 per cent area analysed. Distance between counting centres: 16 mm. Inhomogeneous fabric with regard to optic axes orientation in adjoining laminae. (*After H. Ramsauer, 1941, and B. Sander, 1950*)

on all scales larger than a thin section. The slip planes may be spaced at several millimetres and crystals adjoining them could be of a different orientation from similar crystals in between. Inhomogeneities of this kind require extensive testing. For instance, the optic axes of quartz crystals in general may appear to be oriented more or less at random in such a rock, but *selective* fabric analyses using quartz grains of different size, shape, association, or location could well yield very distinct patterns of preferred orientation, some of which are highly significant of anisotropism and others less so or not at all. These patterns are best shown individually as so-called *partial* diagrams of quartz optic axes, since their summation in a *collective* diagram may weaken or eliminate significant partial maxima so as to render the fabric apparently isotropic. In some such cases of apparent isotropism and suspected inhomogeneity of fabric it may be very difficult if not impossible to differentiate between groups of grains of a mineral on which to carry out selective analyses. Here the *axial distribution analysis* (A.D.A.) comes to the rescue.[1] This most exhaustive of fabric analyses records both the orientation (*Drehlage*) and the relative position (*Ortslage*) of each grain of quartz, calcite, or olivine. The result is a series of contoured plots showing the areal distribution of each group of differently oriented grains (Fig. XIII–4B) or a sketch of a thin section in which the grains contributing to different maxima of preferred orientation are marked by a distinctive colour or shade. Statistics requiring some hundred grains per orientation group, an A.D.A. involves much measuring and computing time. Flinn has devised a test for the significance of clustering of similarly oriented grains in axial distribution diagrams.[2]

PETROTECTONICS

The descriptive results of petrofabric analysis can be used for a variety of purposes. They may serve to characterize a rock more fully than does its compositional petrology, its lithology, or its fossil content. The comparing and contrasting of fabric diagrams on a purely geometrical basis has obvious applications in sedimentary stratigraphy, but the igneous and metamorphic rocks afford equally promising fields for comparative petrofabric analysis especially where inclusions of metamorphic rocks occur in igneous terrains or *vice versa*. The sequence of geological events, and the mode of emplacement in particular, may be established from fabric analogies between various types of inclusions on the one hand and a plutonic host rock or a metamorphic country rock on the other. In this way it has been shown, for instance, that some alpine-type peridotites are metamorphic tectonites that were deformed and emplaced in an essentially solid state.[3]

However, the principal use of the fabric is the correlation it allows to be made between the visible strains of the rock, the individual movements of the fabric elements and the systems of stresses and forces causing both. This is the field of *petrotectonics*, which is not merely a tectonic study on a microscopic scale but also one leading to fundamental conclusions regarding

[1] H. Ramsauer, 'Achsenverteilungs analysen an Quarztektoniten', Dissertation, Min.-Pet. Institut, Alte Univ., Innsbruck, No. 304, 1941, pp. 1–26. B. Sander, *Einführung in die Gefügekunde der Geologischen Körper*, Vol. II, 1950, Vienna, pp. 39–42, 178–88, 404–5. F. J. Turner, D. T. Griggs, R. H. Clark, and R. H. Dixon, 'Deformation of Yule marble', Pt VII, *Geol. Soc. Amer. Bull. 67*, 1956, pp. 1259–1293. A. L. G. Collée, 'A fabric Study of Lherzolites', *Leidse Geol. Meded. 28*, 1963, pp. 45–63.

[2] D. Flinn, 'On the statistical analysis of axial distribution diagrams', *N. Jb. Miner.*, Mh. 2, 1965, pp. 54–64.

[3] E. den Tex, 'Origin of Ultramafic rocks, their tectonic setting and history', *Tectonophys.*, Vol. 7, 1969, pp. 457–88.

mechanisms of deformation and internal movements of rocks. Although there is still some contradiction and ambiguity in the hypotheses of petrotectonics, as shown by the various textbooks [1] on the subject, its fundamental concepts are generally accepted.

Development of preferred orientations by componental movements

The kinematic significance of fabric depends on the participation of the fabric elements in the relative movements that have affected rocks, in such a way that the differential displacements of material points are reflected in the orientation of fabric elements. This may come about through the influence of currents or other factors in the settling of grains of sedimentary rocks, through movements in partially crystallized magmas or through the orientation and reorientation of fabric elements during clastic and plastic deformation in igneous and metamorphic rocks. All such differential adjustments or readjustments involving fabric elements are termed *componental movements* (*Teilbewegungen*) provided always that they do not disrupt the continuity of the rock so that a line, continuous before deformation, remains so on the scale employed throughout the process. They are termed *penetrative movements* (*Durchbewegungen*) since they affect every single fabric element of the rock.

Componental movement between and within crystal grains

Since most rocks are polycrystalline aggregates, the movements involved are firstly such as may take place between grains and groups of grains. These *intergranular movements* are rotations, slidings, and diffusions in the liquid or solid state which tend to separate the grains while forces existing across grain boundaries and chemical precipitations of dissolved material tend to maintain or re-establish their adherence. In unconsolidated sedimentary and partially crystallized igneous rocks intergranular movement is the principal mechanism of orientation. Water or magma, whether stagnant or flowing, may for the present purpose be considered to maintain the continuity of the rock and to control the movements of grains in two-phase systems. Thus flakes of mica, tablets of feldspar, prisms of hornblende, and flat or elongate pebbles are arranged in a subparallel manner according to grain shape so as to occupy the most stable position in relation to the forces of gravity and laminar flow, and the resulting preferred orientation is essentially dimensional. In metamorphic deformation the componental movements take place mainly within the lattices of the constituent crystal grains. Mechanical twinning, translation-gliding,

---

[1] B. Sander, *Einführung in die Gefügekunde der Geologischen Körper*, Vol. I and II, Vienna, 1948–50. H. W. Fairbairn, *Structural Petrology of Deformed Rocks*, Addison-Wesley, Cambridge, Mass., 1949. F. J. Turner and L. E. Weiss, *Structural Analysis of Metamorphic Tectonites*, McGraw-Hill, New York, 1963. F. Karl, *Anwendung der Gefügekunde in der Petrotektonik, Tl. 1*, Grundbegriffe, Tektonische Hefte 5, Pilger Verl., Clausthal, 1964.

bend-gliding, kinking and torsional twisting are *intragranular movements* giving rise to preferred orientations of the crystal lattices involved since they are fixed in relation to specific planes and directions in these lattices, permitting plastic deformation of each lattice according to a uniform plan. It is obvious that such movements can occur only when the particular planes and directions of twinning, gliding, etc., are suitably oriented with regard to the forces causing them. If they do not happen to lie subparallel to planes and directions of potential shearing movement or of maximum flow and elongation in the rock, they may be brought into such parallelism by external rotation of individual crystals.

Crystals such as the micas, in which one glide plane is dominant, may be expected to undergo appreciable external rotation in order that gliding may become possible on it in the required direction. Others, such as the cubic minerals garnet and magnetite, are merely rotated externally because their lattices lack planes and directions of easy gliding, twinning, or other intragranular movement.[1] Again the cubic crystals of galena and fluorite have so many possible glide planes and directions that they may yield plastically without notable external rotation.[2] In general, experiments with marble and ductile metals show that the greater the deformation the more strongly pronounced the preferred orientation of twin and glide planes becomes.[3]

As twinning or translation-gliding proceeds, any pre-existing crystal structures, such as twin-lamellae or cleavages, not parallel to the active glide plane or glide line, are rotated with reference to them or to some other internal co-ordinate system like the crystallographic axes (see pp. 117–18). Since the axis and the maximum angle of this *internal rotation* are fixed by the active twin or glide plane and the glide line, this leads to further preferred orientation of elements of the crystal lattice. *External rotation* (with regard to co-ordinates outside the crystal) does not necessarily cause preferred orientation of the crystal lattice, since the axis of rotation and the maximum angle of rotation are not fixed in relation to it.

Physico-chemical aspects of componental movement

So far the emphasis has been on componental movements of a purely mechanical nature. Rotation, sliding, twinning, gliding, bending, kinking, and twisting of crystals constitute what are called the *direct componental movements* of a rock.

[1] O. Mügge, 'Bewegungen von Porphyroblasten in Phylliten und ihre Messung', *Neues Jahrb. Min.*, Beil. Bd. 61A, 1930, pp. 469–510.

[2] M. J. Buerger, 'The plastic deformation of ore minerals', Pt I (Galena), *Amer. Miner.*, Vol. 13, 1928, pp. 6–14. D. Korn, 'Ein deformiertes Gefüge Flüszspat-Quartz-Kupferkies aus einer Mittel Schwedischen Sulfidlagerstätte', Neues Jahrb. Min., Beil. Bd. 66A, 1933, pp. 433–59.

[3] W. Boas, 'Deformation of a Duplex Brass by cold working', *Proc. Aust. Inst. Min. Met.*, N.S., No. 123, 1941, pp. 123–34. D. Griggs, F. J. Turner, I. Borg, and J. Sosoka, 'Deformation of Yule marble: Pt V', *Geol. Soc. Amer. Bull.*, Vol. 64, 1953, pp. 1327–42.

Being of a penetrative nature these mechanical movements are not only the immediate cause of preferred orientation, but they also have a stirring effect on the poly-phase system of rock and pore fluid much like the one a chemist uses to speed up the solution of crystals. Thus, the minerals present may be unstable under the conditions prevailing at the time of deformation, in which case the reactivity of these minerals to form stable associations is greatly stimulated by penetrative mechanical movements.

Piezo-crystallization – Of more immediate importance for the fabric of igneous and metamorphic rocks is the fact that elastically anisotropic minerals, when nucleating or growing in a non-hydrostatic stress field, tend to occupy a thermodynamic equilibrium orientation, which may be a crystal-lattice orientation and – in the presence of a liquid phase capable of dissolving and precipitating the minerals concerned – also a dimensional orientation leading to a schistosity entirely due to recrystallization under directed pressure (*crystallization-schistosity* of Becke and Wright).[1] The term *piezocrystallization*, coined by Weinschenk[2] for protoclastic textures in plutonic rocks and virtually obsolete as such, was redefined by Hartman and den Tex[3] to denote the principle of crystallization under non-hydrostatic stress leading to preferred mineral orientations. Despite its early recognition the influence of piezocrystallization on the development of rock fabrics has generally been neglected until a sounder basis was laid with the aid of appropriate thermodynamic models.[4] Piezocrystalline fabrics are now being recognized in rapidly increasing numbers, especially in deepseated metamorphic and plutonic rocks. Thus, certain cleft-girdle fabrics of quartz optic axes in granitic rocks[5] are held to be of a piezocrystalline nature as well as the principal γ'-olivine fabrics perpendicular to a primary schistosity in garnet peridotites.[6] Evidence has recently been forthcoming that the oceanic upper mantle is seismically anisotropic owing to recrystallization under steady-state creep.

[1] F. Becke, 'Über Kristallisationsschieferung und Piezokristallisation', *Int. Geol. Congress*, *C.R. Session 10*, Vol. 2, Mexico, 1906, pp. 1179–85. F. E. Wright, 'Schistosity by crystallization', *Amer. J. Sci.*, 4th ser., Vol. 22. 1906, pp. 224–230.

[2] E. Weinschenk, 'Dynamométamorphisme et Piézocristallisation', *Int. Geol. Congress*, *Mem. Session 8*, Paris, 1900, pp. 326–41.

[3] P. Hartman and E. den Tex, 'Piezocrystalline fabrics of olivine in theory and nature', *Int. Geol. Congress, Proc. 22nd Session*, New Delhi, 1964, Pt IV, pp. 84–114.

[4] W. B. Kamb, 'The thermodynamic theory of non-hydrostatically stressed solids', *J. Geophys. Res.*, Vol. 66, 1961, pp. 259–71. M. Kumazawa, 'A fundamental thermodynamic theory on non-hydrostatic field and on the stability of mineral orientation and phase equilibrium', *J. Earth Sci.*, Nagoya Univ., Vol. 11, 1963, pp. 145–217. K. Ito, 'Thermodynamics of non-hydrostatically stressed solids with geologic applications', *J. Fac. Sci.*, Univ. Tokyo, Sec. II, Vol. 16, 1966, pp. 347–79. A. G. McLellan, 'A thermodynamical theory of systems under non-hydrostatic stresses', *J. Geophys. Res.*, Vol. 71, 1966, pp. 4341–7.

[5] H.-J. Behr, 'Zur tektonischen Analyse magmatischer Körper unter besonderer Berücksichtigung des Quarzkorngefüges' I, Freib. Forschungshefte Bd. C 215, 1967, pp. 9–60.

[6] E. den Tex, 'Principal γ-olivine fabrics: their tectonic and metamorphic significance', in *Experimental and natural rock deformation* (ed. by P. Paulitsch), Springer, Berlin, 1970, pp. 486–95.

Mimetic crystallization – Equally important for the development of preferred orientation are the movements of material in solution and the preferential growth of crystals along planes and directions of easy penetration (*Wegsamkeit*) in rocks. Most of the common rock-forming minerals are anisotropic as regards rate of growth, and these will orient themselves in such a manner as to suffer least resistance from a solid environment. Thus mica flakes tend to grow with their basal cleavage parallel to planar components of the fabric that constitute planes of easy penetration, such as bedding, cleavage, slip planes, and joints. Stellate, sheaf- and pencil-like patterns result from the preferential growth of prismatic minerals, like hornblende and tourmaline, with their long axis parallel to a plane of least resistance or to a direction of easy penetration such as a fold axis, a common intersection of bedding, cleavage and joints, or a system of slip-striae on a slickensided fault plane. Sander coined the term '*Abbildungskristallisation*' (*mimetic crystallization*) for this process of developing preferred orientations that form a *belteroporic fabric*, i.e. a cast of a pre-existing fabric which was anisotropic as regards ease of penetration by growing crystals.[1] Relaxation of stress involving hysteresis of the elastic portion of the strain of a rock is deemed to be responsible for most types of mimetic crystallization.

All these movements involving chemical activity are termed *indirect componental movements*. Their principal function is to simplify or emphasize patterns of preferred orientation, to cause segregation of minerals and to maintain or re-establish the continuity of the deformed rock.

Time-relationships of deformation and crystallization

The relation of cause and effect which appears to exist between mechanical and chemical movement, or between deformation and recrystallization, is not a simple one since the natural conditions, including T- and P-gradients, may retard or accelerate either process almost to the exclusion of the other. This is brought out by the threefold division of time-relationships proposed by Sander.[2]

1. *Post-crystalline deformation* or *pre-tectonic crystallization* in which the componental movement is mainly direct (i.e. purely mechanical). The process represents finite deformation in high levels of the crust and may be compared with the metallurgical technique of cold working. Given mineral associations which are stable under these conditions, it causes rupture and granulation, deformation lamellae in quartz, twinning in feldspars and rhombohedral carbonates, bending and kinking of micas and olivines, but no recrystallization of

---

[1] B. Sander, *Gefügekunde der Gesteine*, Springer, Vienna, 1930, pp. 156–73. It should be noted that the belteroporic fabric is not necessarily a complete cast of the pre-existing fabric.

[2] B. Sander and M. Pernt, 'Zur petrographisch-tektonischen Analyse, I', *Jahrb. Geol. Bundesanst., Vienna*, Vol. 73, 1923, pp. 183–253.

any consequence. In other words, it is the domain of mechanical disruption as found in cataclasites, mylonites, phyllonites and other deformed but virtually non-metamorphic rocks.

2. *Pre-crystalline deformation* or *post-tectonic crystallization* in which the visible componental movements are indirect, although considerable direct movement may have taken place before or during the early stages of re-crystallization and neomineralization. This is the case when geoïsotherms rise and geoïsobars fall upon relaxation of orogenic stresses. Apart from the influence of phase equilibria, the process is comparable to that of annealing whereby recrystallization is obtained at high temperatures in metals pre-viously subjected to cold-working. Porphyroblastesis and metamorphic dif-ferentiation may produce new structures at the expense of early mechanically formed structures, but mimetic crystallization is usually more effective in preserving or modifying the latter, for instance as trains of inclusions in porphyroblasts.[1] Typical representatives may be found among the products of regional and dynamothermal metamorphism such as schists, gneisses, amphibolites and blastomylonites, and among contact-metamorphic rocks in which preferred orientations have been preserved.

3. *Para-crystalline deformation* or *syntectonic crystallization* in which piezo-crystallization dominates although effects of both direct and indirect compon-ental movement may be visible.

This occurs for instance in semi-solid igneous intrusives during or shortly after their emplacement, when mineral associations stable at the high tem-peratures still prevailing, are deformed *in situ*. It also operates in the deforma-tion of rocks simultaneously adjusted to conditions of high temperature and is clearly demonstrated in alpine-type peridotites and granulites,[2] and in certain schists and gneisses by the sigmoid attitude of trains of inclusions in porphy-roblasts that have been rotated during growth (e.g. 'snowball' and 'pin-wheel' garnets),[3] and by microfolds outlined by polygonal as well as smoothly

[1] Helicitic structure as defined by E. Weinschenk, *Allgemeine und Spezielle Gesteinskunde*, Freiburg i/Br., 1900. See also J. B. Howkins, 'Helicitic textures in garnets from the Moine rocks of Moidart'. *Trans. Edinb. Geol. Soc.*, Vol. 18, 1961, pp. 315–24.

[2] E. den Tex, 'Origin of ultramafic rocks, their tectonic setting and history', *Tectonophys.*, Vol. 7, 1969, pp. 457–88. H.-J. Behr, 'Beiträge zur petrographischen und tektonischen Analyse des sächsischen Granulitgebirges', *Freib. Forschungshefte*, Vol. C 119, 1961, pp. 5–118.

[3] J. S. Flett, in 'The Geology of Ben Wyvis, etc.', *Mem. Geol. Surv. Scotland*, 1912, p. 111. P. Niggli, 'Die Chloritoidschiefer des Nordostrandes des Gotthard-massives', *Beitr. Geol. Karte der Schweiz*, N.F., Vol. 36, Bern, 1912, pp. 1–94, L. J. Krige, 'Petrographische Untersuchungen im Val Piora und Umgebung', *Eclog. geol. Helv.*, Vol. 14, 1916, pp. 519–654. W. Schmidt, 'Bewegungsspuren in Porphyroblasten kristalliner Schiefer', *Sitz. Ber. Kais. Akad. Wiss.*, Vienna, Math. Nat. Kl. Abt. *I*, Vol. 127, 1918, pp. 293–310. F. Becke, 'Struktur und Kluftung', *Fortschr. Min. Krist. Petr.*, Vol. 9, 1924, pp. 185–218. O. Mügge, 'Bewegungen von Porphyroblasten in Phylliten und ihre Messung', *Neues Jahrb. Min.*, Vol. 6, Beil. Bd. A, 1930, pp. 469–510. H. H. Read, 'A Contemplation of Time in Plutonism', *Quart. Journ. Geol. Soc. Lond.*, Vol. 105, 1949, pp. 101–56.

curved trains of mica flakes.[1] It is the domain of deformation at low levels in the earth's crust and upper mantle. Similarly hot-working of metals reduces their work-hardening and promotes simultaneous recrystallization.

More recently the threefold division was refined and extended, when it was realized that both orogenic deformation and regional metamorphism are rhythmic processes that oscillate while waxing and waning on a larger scale thus giving rise to several *kinematic and metamorphic phases* per orogenic cycle. Among the many attempts to establish criteria for the correlation of metamorphic mineral growth with successive phases of deformation and relaxation, those by Zwart and Johnson,[2] based respectively on the Hercynian orogeny in the Central Pyrenees and the Caledonian orogeny of the Scottish Highlands, appear to be most successful and exhaustive. In applying such correlation schemes it has to be borne in mind that deformation and metamorphism not only wax and wane in time but that, while doing so, they both migrate through orogenic space at different rates and following different patterns.[3] Thus a specific time relationship between the growth of say biotite and the second kinematic phase established for the outer regions of an orogene may not be applicable to the inner portion of it.

Mechanics of orientation of fabric elements

It would be beyond the scope of this book to enumerate the actual patterns of preferred orientation observed even for common fabric elements such as optic axes of quartz and calcite, and poles to the basal cleavage of mica.

Numerous hypotheses have been advanced to explain the mechanism by which each of these patterns is developed,[4] but in fact the behaviour of most of the common rock-forming minerals under conditions of natural stress is as yet insufficiently known to provide the greater part of these hypotheses with a reasonably sound foundation. Critics of petrotectonics have often seized on this previously weak link in the chain of its thought, but the body of evidence showing relations of cause and effect between specific componental movements and certain patterns of preferred orientation is expanding rapidly.

[1] B. Sander, *Einführung in die Gefügekunde der geologischen Körper. Vol. II. Die Korngefüge*, Springer, Vienna, 1950, pp. 279–306.

[2] H. J. Zwart, 'On the determination of polymetamorphic mineral associations, and its application to the Bosost area (Central Pyrenees)', *Geol. Rundsch.*, Bd. 52, 1962, pp. 38–65. M. R. W. Johnson, 'Some time relations of movement and metamorphism in the Scottish Highlands', *Geol. en Mijnb.*, Vol. 42, 1963, pp. 121–42.

[3] E. den Tex, 'A commentary on the correlation of metamorphism and deformation in space and time,' *Geol. en Mijnb.*, Vol. 42, 1963, pp. 170–6.

[4] For full references in English, see E. B. Knopf and E. Ingerson, 'Structural Petrology', *Geol. Soc. Amer.*, *Mem. 6*, Chapters 10, 11, and 12, 1938. H. W. Fairbairn, *Structural Petrology of Deformed Rocks*, Addison-Wesley, Cambridge, Mass., Chapters 2, 3, 9, and 10, 1949, and F. J. Turner and L. E. Weiss, *Structural analysis of metamorphic tectonites*, McGraw-Hill, New York, 1963.

Fabric analyses have been carried out on sediments, glaciers, and lava flows, the movements of which can be observed in their natural course of action,[1] and experimental deformation of suitable rocks and minerals under controlled laboratory conditions is yielding further information. Thus, some naturally occurring patterns of preferred orientation of calcite are adequately explained by mechanical twinning in a single sense of a specific direction along the flat rhombohedron $\{01\bar{1}2\}$ and by translation-gliding in the opposite sense of a different direction along the unit rhombohedron $\{10\bar{1}1\}$, so that the principal stress axes may be located by analogy with their location in experimentally produced fabrics of this type.[2] Similar results have been obtained with dolomite, quartz, olivine, pyroxene, and ice.[3]

Movement picture

Notwithstanding incomplete knowledge of the specific orienting mechanism, the componental movements may be summed up in terms of such visible effects in the fabric as can be integrated with an overall tectonic plan. Thus a *movement picture* (*Bewegungsbild*) arises which may be taken to represent the field of differential displacements or strains of material points in a rock subject to deformation. This is possible where discrete surfaces and bodies, such as bedding planes, idiomorphic crystals, oöliths, and fossils are caught in the process of deformation (cf. Fig. XIII–7D and pp. 124–32). It constitutes an important corollary of the kinematic picture derived from symmetry considerations regarding the fabric as is done in the following pages.

[1] E. Ingerson and J. L. Ramisch, 'Studies of unconsolidated sediments. I. Quartz fabric of current and wind ripple marks', *Min. Pet. Mitt.*, Vol. 4, 1954, pp. 117–24. H. Bader, 'Introduction to ice petrofabrics', *Journ. Geol.*, Vol. 59, 1951, pp. 519–36. G. F. Rigsby, 'Crystal fabric studies on Emmons Glacier, Mount Rainier, Washington', ibid., pp. 590–8; H. Philipp, 'Bewegung und Textur in magmatischen Schmelzflüssen', *Geol. Rundsch.*, Vol. 27, 1936, pp. 321–65; R. Brinkmann, 'Kluft- und Korngefügeregelung in Vulkaniten', *Geol. Rundsch.*, Vol. 46, 1957, pp. 526–45. A. C. Waters, 'Determining direction of flow in basalts', *Amer. J. Sci.*, Vol. 258A, 1960, pp. 350–66.

[2] D. T. Griggs, F. J. Turner et al., 'Deformation of Yule marble, I–VII', *Bull. Geol. Soc. Amer.*, Vol. 62, 1951, pp. 853–905, 1385–416; Vol. 64, 1953, pp. 1327–52; Vol. 67, 1956, pp. 1259–94. F. J. Turner, 'Nature and dynamic interpretation of deformation lamellae in calcite of three marbles', *Amer. Journ. Sci.*, Vol. 251, 1953, pp. 276–98.

[3] J. Handin and H. W. Fairbairn, 'Experimental Deformation of Hasmark Dolomite', *Bull. Geol. Soc. Amer.*, Vol. 66, 1955, pp. 1257–73. N. L. Carter, J. M. Christie, and D. T. Griggs, 'Experimental deformation and recrystallization of quartz', *J. Geol.*, Vol. 72, 1964, pp. 687–733. H. C. Heard and N. L. Carter, 'Experimentally induced "natural" intragranular flow in quartz and quartzite', *Amer. J. Sci.*, Vol. 266, 1968, pp. 1–42. C. B. Raleigh, 'Experimental deformation of ultramafic rocks and minerals' in P. J. Wyllie (ed.), *Ultramafic and Related rocks*, Wiley and Sons, New York, 1967, pp. 191–9. D. T. Griggs, F. J. Turner, and H. C. Heard, 'Deformation of rocks at 500 to 800°C', *Geol. Soc. Amer.*, Mem. 79, 1960, pp. 39–104. H. Bader et al., 'Der Schnee und seine Metamorphose', *Beitr. Z. Geol. d. Schweiz-Geotechn. Ser.-Hydrol.*, Lief. 3, 1939, pp. 55–61. S. Steinemann, 'Experimentelle Untersuchungen zur Plastizität von Eis', ibid., Lief. 10, 1958, pp. 46–50.

Tectonites and their classification

Following Sander[1] all rocks affected by interrelated componental movements are to be classed as *tectonites*. This definition comprises a great diversity of rocks ranging from mylonites to glaciers and from fluxion-gneisses to gravity-settled shales,[2] the only requirement being that their componental movements, when added up, present an integral picture of the movement of the rock as a whole. In so far as this notion of componental movement is acceptable, there is no reason to restrict the tectonites to the deformed rocks as has been done by some authors. On the other hand, it reduces the class of non-tectonites to a rather insignificant group of rocks with isotropic fabric.

A first division can be made between *primary* and *secondary tectonites*. In the primary tectonites the componental movement takes place mainly by fluid flow in a medium such as air, water, or molten rock. They include *fusion tectonites* (igneous rocks showing flow structure) as well as *deposition* and *apposition* tectonites (sediments deposited from still and moving water or wind). The secondary tectonites comprise all deformed and metamorphic rocks which have piezocrystallized or yielded by solid flow, i.e. by componental movement in an essentially solid medium.

Another subdivision of the tectonites, on a kinematic basis, was introduced by Sander,[3] but has not been given much prominence by recent authors. Here the main groups are:

S-tectonites, in which slip in one direction on a single set of planes should account for the single point-maximum of poles of planar fabric elements, such as mica $\{001\}$ in shear-folds, or of penetration points of linear fabric elements, such as quartz optic axes in slickensides (Fig. XIII−2A and 7D).

R-tectonites, in which external rotation of fabric elements is supposed to be the cause of girdle-maxima of poles to planar fabric elements about the axis of rotation and of point-maxima of linear fabric elements coinciding with this axis (e.g. flexure-folds and rolled porphyroblasts) (Fig. XIII−2G).

B-tectonites, in which the *B*-axis is defined by the intersection of two or more sets of *s* planes with componental movements in the directions at right angles to the common intersection *B*. The fabrics of phyllonites and other rocks flattened on two complementary sets of slip planes are typical of the *B*-tectonite group, while that of the flexural-slip folds may be included on the strength of their incomplete girdles, elongate point-maxima and other transitional features between *S*- and *R*-tectonites (Fig. XIII−2E and F).

$B \wedge B'$ and $B \perp B'$ *tectonites* are special cases of *B*-tectonites showing

[1] B. Sander and M. Pernt, 'Zur petrographisch-tektonischen Analyse, I', *Jahrb. Geol. Bundesanst., Vienna*, Vol. 73, 1923, pp. 183–253.

[2] Sander does not include the purely gravitative depositional rocks in his tectonite concept but it seems unreasonable not to do so, when rocks formed by simple upward movement in a fluid medium are so classified (volcanic necks, salt domes, etc.).

[3] B. Sander, *Gefügekunde der Gesteine, Vienna*, 1930, pp. 220–2.

more than one intersection of at least four sets of slip planes or more than one axis of rotation, both of which are *B*-axes as defined in the preceding paragraph. The angle between *B* and *B'* may be of any value but is very often approximately 90°, suggesting that the *B*-axes were formed penecontemporaneously (Fig. XIII–10, 11, and 12).

This classification has its use as a broad geometrical grouping. Thus, *S*-tectonites typically possess *planar fabrics* and tend to split into slabs or foliae, *R*-tectonites have principally *linear fabrics* yielding pencils or rods, while *B*-tectonites are *plano-linear* inasmuch as their fabric is equally constituted by planar and linear components, such that they split into lenses and corrugated slabs. Flinn[1] has proposed to relabel these three geometrical classes as *S*-, *L*-, and *L–S*-tectonites respectively, and to relate them to the oblate, prolate, and triaxial deformation ellipsoids. However, even in the simply planolinear (*L–S*) fabric system, the relations implied by Sander to exist between fabric and movement are not always confirmed by the results of experimental investigations.

Symmetry of fabric, movement, strain, and stress

The *symmetry concept* of Curie[2] has proved of far more consequence in petrotectonics. It entails the hypothesis that the symmetry elements of stress, strain, and movement, which are related mutually and to the fabric as cause to effect, should constitute the minimum content in symmetry elements of the fabric. *Subfabrics*, representing one fabric element only, may again have a higher symmetry than the total fabric, depending on the crystal symmetry of the fabric element concerned and on the mechanism of its orientation. As in geometrical crystallography the symmetry of the fabric, considered as a body, can be expressed in terms of the following *symmetry elements*: rotation axes, reflection planes, translation lines and centres. Moreover, some of the crystal systems (i.e. triclinic, monoclinic, and orthorhombic) provide a suitable framework for a broad classification of fabric symmetries. *Axial symmetry* refers to that of a cylinder, and has been invoked by Sander as a comprehensive system for all fabrics of a symmetry higher than orthorhombic and lower than spherical. Of course the symmetry of a fabric, resulting from the *statistical* parallelism of fabric elements, is by no means approaching perfection so closely as does the symmetry of a crystal.

The symmetry of stress, strain, and movement can be made visible with the aid of the stress tensor, the strain ellipsoid, and a similar reference body for preferred directional movements. It has been common usage to employ the strain or deformation ellipsoid as reference body for the symmetry of move-

[1] D. Flinn, 'On the symmetry principle and the deformation ellipsoid', *Geol. Mag.*, Vol. 102, 1965, pp. 36–45.
[2] P. Curie, 'Sur la Symétrie', *Bull. Soc. Minér.*, France, Vol. 7, 1884, pp. 417–57.

ment, but spheres, cylinders, cones, prisms, pyramids, rhombohedra, and less symmetrical polyhedra serve a more specific purpose when the polarity of movements is to be accounted for.[1] In the following the main symmetry systems of fabric and movement, strain or stress are briefly reviewed.

Spherical symmetry (Schönfliess point groups: K ∞ h and K ∞). The symmetry of an isotropic fabric is *spherical* and the movements, strains, or stresses causing it are random, or otherwise cancelling out. Each direction is an infinitive-fold symmetry axis, with or without a reflection plane at right angles to it, depending on its homo- or heteropolarity. Geological examples of the *homopolar class* (K ∞ h) are: spherulites, geodes, oöliths, armoured mud balls and randomly oriented hornfels fabrics. In the *heteropolar class* (K ∞) no geological examples can be cited.

Axial symmetry (Schönfliess point groups: D ∞ h and C ∞ v). Having one infinite-fold axis with an infinite number of reflection planes parallel to it, the axial symmetry system is referable to an ellipsoid of revolution or a cylinder or to a cone, depending on the homo- or heteropolarity of the axis. If rotational movement may be neglected for the present purpose, the presence or absence of twofold symmetry-axes normal to the main axis is immaterial. However, when the movement in the axial direction is irrotational and homopolar, the circular section normal to the cylinder axis is a plane of symmetry as well. Movements exhibiting this *homopolar type of axial symmetry* (K ∞ h) are the rolling of clay or dough into rods, the drawing of metal wire, or the 'spreading' of a ball of clay into a flat disc by compression between a pair of parallel plates. 'Extrusion' and 'spreading' are processes widely used in the ceramic industry and Williamson[2] has shown that rutile needles of random orientation contained in a ball of clay are preferentially oriented parallel to the circumference of a cylinder formed by spreading, and parallel to the axis of an extruded cylinder. Geological fabrics of homopolar axial symmetry are rare but they include convolute folds, magmatic rolls, boudins, discoid lenses, and cigar-shaped pebbles in strongly stretched or compressed rocks.

However, if the movement in the axial direction is heteropolar, the circular section normal to it is no longer a plane of symmetry. Movements exhibiting this *heteropolar type of axial symmetry* are: the industrial process of extrusion, the hydraulic mechanism of pipe-flow, and geological events such as the intrusion of salt domes and igneous stocks as well as the settling of particles

---

[1] B. Sander, 'Zur petrographisch-tektonischen Analyse, II', *Jahrb. Geol. Bundesanst.*, Vienna, Vol. 75, 1925, pp. 181–236. W. Schmidt, 'Gefügesymmetrie und Tektonik', ibid., Vol. 76, 1926, pp. 407–30. M. S. Paterson and L. E. Weiss, 'Symmetry concepts in the structural analysis of deformed rocks', *Geol. Soc. Amer. Bull.*, Vol. 72, 1961, pp. 841–82. M. Kirchmayer, 'Das Symmetrie-konzept von CURIE 1884 in der Makrogefügekunde', *N. Jahrb. Geol. Pal. Abh.*, Vol. 122, 1965, pp. 343–50. D. Flinn, 'On the symmetry principle and the deformation ellipsoid', *Geol. Mag.*, Vol. 102, 1965, pp. 36–45.

[2] W. O. Williamson, 'Lineations in Three Artificial Tectonites', *Geol. Mag.*, Vol. 92, 1955, pp. 53–62.

from still water or magma in a sedimentary or igneous basin.[1] Essentially the symmetry of basins, domes, stocks, and plugs is heteropolar as compared with that of cylinders and discs, and the corresponding fabric suffers a similar loss of symmetry, though local subfabrics may be homopolar (Fig. XIII–5; 2A, G).

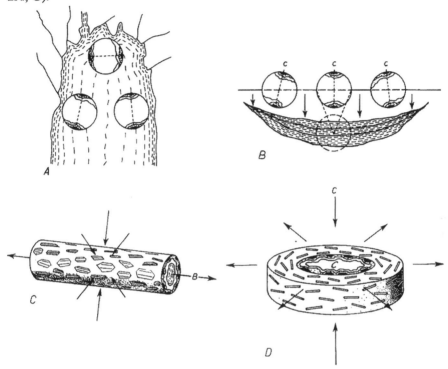

FIG. XIII–5. AXIAL SYMMETRY OF FABRIC IN DIAGRAMMATIC FORM

A. Vertical section of an igneous stock or cupola showing linear structure of hornblende prisms. Insets: local fabric diagrams of hornblende c_v axes. Heteropolar axial symmetry. (*Modified from R. Balk, 1936*)

B. Vertical section of lacustrine sedimentary basin showing direction of settling and ultimate orientation of flaky particles. Insets: local fabric diagrams of poles to clay flakes. Heteropolar axial symmetry.

C. Homopolar axial symmetry as expressed by the fabric of gypsum tablets in a rod produced by kneading (*left*) and of rutile needles in a disc obtained by 'spreading' of a ball of clay (*right*). Insets: integrated fabric diagrams. (*Modified from W. O. Williamson, 1955*)

Orthorhombic symmetry (Schönfliess point groups: D_{2h} and C_{2v}). Fabrics, strains, and stresses exhibiting this type of symmetry are referable to a triaxial ellipsoid or a prism when homopolar, and to a pyramid when heteropolar. The minimum symmetry requirements are two orthogonal reflection planes intersecting in a twofold axis. Movement in the direction of the single twofold axis is heteropolar, whereas movements at right angles to it remain

---

[1] On a small scale this polarity may not be visible, except when the sedimentary bedding is graded or the igneous banding rhythmic.

homopolar though not of uniform magnitude. Intrusions of magma or salt in a conduit of elliptical or rectangular cross-section, and the filling process of a clastic dyke answer this description of movement, while the same applies to the formation of graded bedding or rhythmic banding in a sedimentary or igneous basin of oblong shape. All these movements produce *heteropolar*

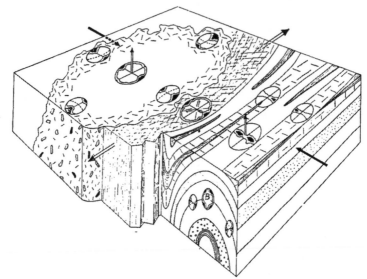

FIG. XIII-6. ORTHORHOMBIC SYMMETRY OF FABRIC IN DIAGRAMMATIC FORM

Block diagram showing, in the foreground, a symmetrical fold of the flexural slip type with fabric diagrams of mica cleavage poles as insets (monoclinic on the limbs, orthorhombic in the hinges). The fabric is inhomogeneous and the overall symmetry as revealed by integrated diagrams is orthorhombic. In the far distance an igneous intrusive body of oval cross-section is shown containing a flow structure with linear and planar components. Insets are fabric diagrams of mica cleavage poles. The fabric is inhomogeneous and the overall symmetry as revealed by integrated fabric diagrams is orthorhombic. The middle distance exhibits a strip of country rock flattened by slip on two symmetrical equivalent sets of planes intersecting in the vertical *B*-axis. Fabric diagram (inset) of mica cleavage poles reveals total symmetry: three symmetry planes at right angles to each other, three twofold axes at their intersections: homopolar orthorhombic. Arrows represent major ($\Longleftarrow\!\!=\!\!=\!\!=\!\!\Longrightarrow$) and minor ($\longrightarrow \longleftarrow$) deformation axes of the geological bodies concerned. Relative compression parallel to $\longrightarrow \longleftarrow$ is consistent with all three fabrics. *B* is the intersection of prominent sets of slip planes or the fold-axis.

structures frequently containing a linear component as is seen in oscillation ripple marks.

In *homopolar symmetry* of this type there are three orthogonal reflection planes and three twofold axes parallel to their mutual intersections. The most appropriate example of homopolar orthorhombic symmetry (D_{2h}) is afforded by the tectonic process of flattening, whereby the rock material is spread bilaterally in one of the reflection planes by compression perpendicular or tension parallel to it. The flattening movement takes place mainly on two or more symmetrical and equivalent sets of slip planes, which are more or less obliquely disposed towards the effective flattening or spreading plane that contains their common intersection and mean deformation axis (*B*). Being irrotational this strain may be referred to a triaxial ellipsoid[1] which is the shape actually assumed by spherical pebbles that have been subject to a flattening deformation. The periodic structure of a series of symmetrical folds exhibits orthorhombic symmetry, which may also be described as homopolar, if translation over half a period is allowed for in the horizontal reflection plane and in the direction of the horizontal twofold axis that runs perpendicular to the fold axis (*B*).

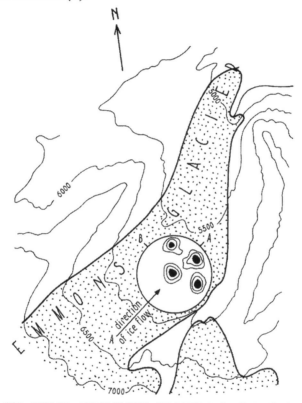

FIG. XIII–7A. MONOCLINIC SYMMETRY OF FABRIC IN DIAGRAMMATIC FORM

Contour map of the Emmons Glacier, Mount Ranier, Washington. Showing direction of ice flow at location of fabric diagram made from 167 optic axes of ice crystals. The ice subfabric has full monoclinic symmetry with *B* as a twofold axis of symmetry. (*After Rigsby, 1951*)

Monoclinic symmetry (Schönfliess point groups: C_{2h}, $C_{1h} = C_s$). Here the only remaining symmetry element is: one reflection plane (C_s), which may be supplemented by a twofold axis at right angles to it (C_{2h}). Generally speaking

[1] The equilibrium stress tensor, causing this type of strain, also has orthorhombic symmetry.

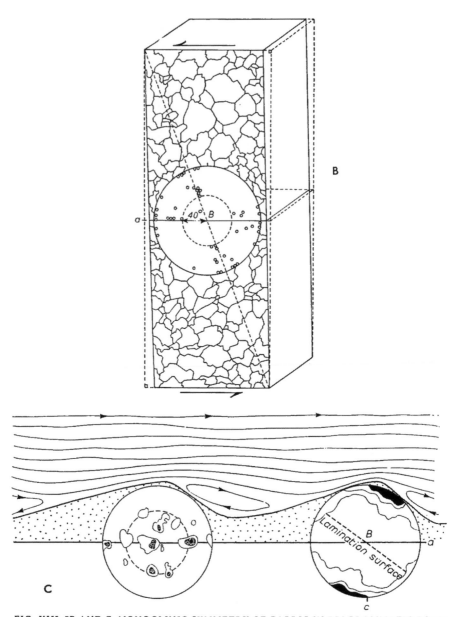

FIG. XIII–7B AND C. MONOCLINIC SYMMETRY OF FABRIC IN DIAGRAMMATIC FORM

B. Prism of glacier ice with fabric of random orientation experimentally deformed by a shearing couple acting in a direction *a* of a plane *aB*. Front face showing somewhat serrated mosaic texture of ice crystals after deformation, and preferred orientation of thirty-eight optic axes in *ac* girdle with point maxima near *a*, and the long diagonal of the *ac* face. Homopolar monoclinic symmetry. (*After Bader and Haefeli, 1939*)

C. Cross-section of current ripple-marks showing shape of ripples, stream-lines of water flow and fabric diagrams of quartz optic axes (*left inset*) and of poles to clay flakes or flat pebbles (*right inset*). The symmetry of both subfabrics is homopolar, but that of the ripples as a whole is heteropolar. (*After Twenhofel, 1926, and Ingerson and Ramisch, 1954*)

movements of monoclinic symmetry, such as simple shear and laminar or turbulent flow, take place either on a single set or on two or more sets of asymmetrical, non-equivalent slip planes intersecting in B (Fig. XIII–7B, D). The resulting strain is rotational in the AB plane and its stress tensor has monoclinic symmetry. If the relative movements on regularly spaced, parallel slip planes are equal and interchangeable, the strain is homogeneous, and its symmetry is *homopolar* possessing a twofold axis (B) in the slip plane at right angles to the slip direction (Fig. XIII–7B). This is demonstrably so in those asymmetrical shear folds in which the amount of slip remains constant on equally spaced slip planes (Fig. XIII–7D, lower part).

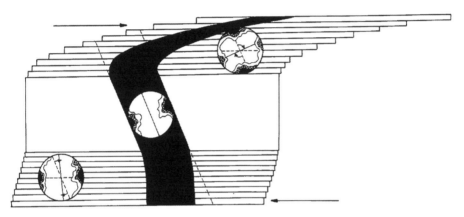

FIG. XIII–7D. MONOCLINIC SYMMETRY OF FABRIC IN DIAGRAMMATIC FORM

Cross-section of 'Gleitbrett' fold caused by slip on one set of planes. The planes are irregularly spaced and the slip does not remain uniform throughout the deformed area. A bed obliquely cutting the slip plane is shown in black. The undeformed central slice is a 'Gleitbrett'; the upper portion is deformed by variable, the lower by uniform slip on equally spaced slip planes. Insets: fabric diagrams of mica cleavage poles. Overall symmetry: heteropolar monoclinic; B is the normal to the only symmetry plane but is not, itself, a twofold axis of symmetry.

More often, however, the mechanism of slip is variable, the planes being spaced irregularly or the relative amounts changing from plane to plane, such that the resulting strain is inhomogeneous and its symmetry *heteropolar* retaining only the reflection plane normal to the slip planes and to B and parallel to the direction of slip (Fig. XIII–7D, upper part). Most dunes, current-ripple marks, lava flows, and glaciers exhibit this type of symmetry, their movement being contained only by a floor and by two opposite banks exerting equal pressure or friction (Figs. XIII–7A, C). The same is true for any single asymmetrical fold and for a series of irregular *Gleitbrett*-folds (Fig. XIII–7D, upper part).

Planar fabric elements are brought into near-parallelism with the effective movement plane, as in an edgewise conglomerate. Linear fabric elements tend to be oriented parallel to B which is an axis of external rotation in simple shear.

Monoclinic fabric permits the ready identification of the mean deformation axis B as the normal to the only plane of symmetry, but proper recognition of the major and minor deformation axes is not such a simple matter as in most orthorhombic fabrics.

Triclinic symmetry (Schönfliess point group: $C_1 = S_2$). This lowermost form of symmetry has only a onefold rotation axis (C_1) and a twofold rotation–reflection axis or inversion centre (S_2) as elements. Triclinic movements are like monoclinic ones (simple shear and flow), except that they take place on irregular surfaces or along curved lines being contained by pairs of opposite banks exerting unequal pressure or friction.[1] The corresponding strain is rotational in more than one plane and its stress tensor has triclinic symmetry.

Arcuate fold belts of asymmetrical cross-section and curvature exhibit triclinic symmetry of movement and fabric. The arc of the Jura Fold Belt with its drawn-out and southerly deflected western flank is a case in point as Wegmann[2] has rightly argued. Moreover, the prevalence of sinistral over dextral wrench faults suggests a forward movement of this belt, suffering more friction along its S.W. than along its N.E. bank (see Fig. XIII–8A).

Schmidegg[3] suggested that the triclinic symmetry of some of the steeply plunging folds in the lower East Alpine fold nappes of the Central Tyrol may be the expression of their unilateral retardation whereby the axes have been steepened towards the source of friction. This would cause trends to become strongly curvilinear on a regional scale. The term *Schlingentektonik* is used for variably trending folds of triclinic symmetry found in formations ranging from the crystalline basement to the high-level calcareous nappes of the North Tyrol[4] (Fig. XIII–8C).

Symmetry reduction by overprinting of imposed on inherited fabrics. Few tectonite fabrics have been formed in one act. Frequently one or more deformations and their fabrics are imposed or overprinted on a pre-existing or inherited fabric. The symmetry of such a *composite fabric* is the more reduced the fewer symmetry elements are shared by the *inherited* and the *imposed* fabrics. Simple relations between inherited, imposed, and composite fabric are, for instance, visible in a bedded sequence of rocks on which a horizontally directed uniaxial compression has imposed a vertical axial plane cleav-

[1] K. A. Howard, 'Flow direction in triclinic folded rocks', *Amer. J. Sci.*, Vol. 266, 1968, pp. 758–765.

[2] E. Wegmann, 'Ueber einige Züge von unter geringer Bedeckung entstandenen Falten', *Min. Pet. Mitt.*, Vol. 4, 1954, pp. 187–92.

[3] O. Schmidegg, 'Steilachsige Tektonik und Schlingenbau auf der Südseite der Tiroler Zentralalpen', *Jahrb. Geol. Bundesanst., Vienna*, Vol. 86, 1936, pp. 115–49.

[4] A. Fuchs, 'Untersuchungen am tektonischen Gefüge der Tiroler Zentralalpen: II. Kalkalpen-Achensee-Kaisergebirge', *Neues Jahrb. Min.*, Abh. 88B, 1943, pp. 337–73.

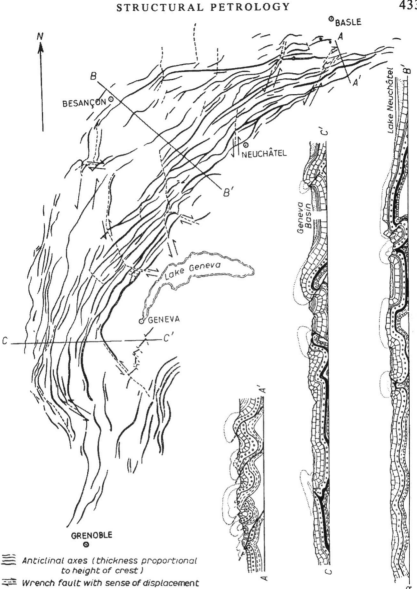

FIG. XIII–8A. TRICLINIC SYMMETRY OF FABRIC IN DIAGRAMMATIC FORM
Structural map and cross-sections of the Jura Fold Arc in France, Switzerland, and
Germany. Asymmetrical folds with irregularly curved axial trends. A meridional set of
sinistral wrench faults prevails. No reflection planes of symmetry. (*After Alb. Heim,
1919*)

age resulting in the composite structure of folds or, on smaller scales,
crenulations and lineations.

A single thin section yielding monoclinic subfabrics of quartz optic
axes and mica cleavage poles not sharing any symmetry element between

themselves has been reported by Weiss[1] in a phillite from Anglesey. He points out that triclinic fabrics should not be taken *a priori* as indicative of two separate acts of deformation, since slip on a single set of shear planes may cause stratified mica flakes to rotate bodily around their intersections with a slip plane, whereas the optic axis of quartz may prefer orientation in a girdle normal to B of the imposed fabric, a direction that does not

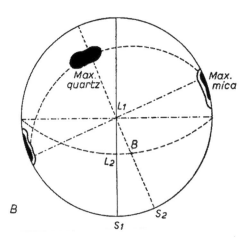

FIG. XIII–8B AND C. TRICLINIC SYMMETRY OF FABRIC IN DIAGRAMMATIC FORM

B. Synoptic diagram of quartz optic axes and mica cleavage poles in a phyllite from Anglesey (England), showing principal maxima only. L_1 and L_2 are visible lineations, B is the (invisible) normal to the plane of monoclinic symmetry in the quartz fabric. s_1: original foliation; s_2: plane of transposition of mica flakes. (*After L. E. Weiss, 1955*)

C. 466 optic axes of quartz from a vertically plunging fold in gneiss from the Cima Leinert, South Tyrol. Inset: oriented specimen from which section was made showing generalized trend of mica flakes. Contours: 5–4–3–2–1–0 per cent per 1 per cent area. Equal area projection. Lower hemisphere. (*After O. Schmidegg, 1936*)

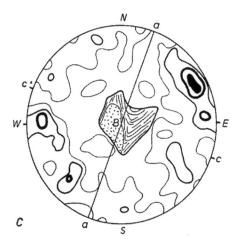

necessarily coincide with the intersection between previously tilted bedding and the imposed set of slip planes (Fig. XIII–8B).

The experiments carried out on Yule Marble have clearly shown the influence of the inherited fabric on the symmetry of the composite fabric.

[1] L. E. Weiss, 'Fabric Analysis of a Triclinic Tectonite and its bearing on the geometry of flow in rocks', *Amer. Journ. Sci.*, Vol. 253, 1955, pp. 225–36.

Turner[1] has pointed out that control of the inherited fabric over the componental movements causes the symmetry of the strains and hence of the composite fabric, to be determined not only by the symmetry of the stress tensor, but also by the symmetry of the inherited fabric and by the number of

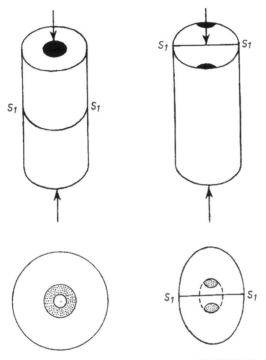

FIG. XIII-9. INFLUENCE OF THE SYMMETRY OF STRESS AND OF THE INHERITED FABRIC ON THE SYMMETRY OF THE COMPOSITE FABRIC

Test cylinders of marble with optic axes concentration of calcite shown as insets. Homopolar axial symmetry of inherited fabric.

A. One component compression normal to foliation. Symmetry of composite fabric remains axial: calcite optic axes stippled in cross-section which remains circular.

B. One component compression parallel to foliation. Symmetry of composite fabric becomes orthorhombic. Calcite optic axes orientation stippled in cross-section which has become oval in outline. (*After F. J. Turner, 1957*)

symmetry elements it has in common with the stress tensor. Thus, if a foliated marble of axial symmetry is subjected to a one-component compression normal to the foliation it will retain its axial symmetry because all symmetry elements are shared between the stress tensor and the inherited fabric. How-

[1] F. J. Turner, 'Lineation, symmetry and internal movement in monoclinic tectonite fabrics', *Bull. Geol. Soc. Amer.*, Vol. 68, 1957, pp. 1–18.

ever, if such a compression is applied parallel to the foliation, the fabric of the deformed marble will be orthorhombic since only three planes and three twofold axes of symmetry are common to two ellipsoids of revolution oriented with their short axes at right angles to each other (Fig. XIII–9).

LINEATION, SYMMETRY, AND STRAIN

Along with the development of its special techniques structural petrology has seen many arguments over the kinematic significance of fabric and of related macrostructures such as cleavage, deformed pebbles, and lineation. The fabric interpretation of lineation has at one stage developed into a controversy of such proportion, that it requires special mention in a survey of petrotectonics however brief.

Geometry of lineation

All linear structures in rocks, visible with the unaided eye, are termed lineations. They arise from statistical linear parallelism of either planar or linear fabric elements as seen in slate 'pencils' and mineral elongation. On geometrical grounds lineations may be grouped as follows:

(i) fold-axes on all scales due to rotation of bedding planes around a common axis (Fig. XIII–6; 2B, C).

(ii) lineations due to common intersection of s planes such as schistosity, cleavage, bedding, foliation, slip planes, etc. (Fig. XIII–6, 7).

(iii) lineations due to linear parallelism of lenses, disc-shaped pebbles, platy inclusions, mica flakes, etc. (Fig. XIII–5, 6).

(iv) lineations due to parallelism of mineral streaks, veins, rods, and boudins, of elongate pebbles and ooliths, of quartz spindles, tourmaline prisms, etc. (Fig. XIII–1, 5, 6).

(v) lineations due to parallelism of grooves and striae on surfaces of slip such as slickensided faults and glacial pavements.[1]

Fabric of lineation

Some of the fabric elements constituting linear structures are usually amenable to petrofabric analysis. In the first section of this chapter it has been shown how lineations may be reflected in the fabric by girdles of mica cleavage poles and by point-maxima of quartz optic axes *inter alia*. The lineation is either parallel to the axis of the girdle or it plots in the point-

[1] Linear structures sub i, ii, iii, and v have at least one planar component. For a full discussion of lineation, see E. Cloos, 'Lineation. A critical review and annotated bibliography', *Geol. Soc. Amer.*, Mem. *18*, 1946, 122 pp., and *Supplement: Review of literature 1942–52*, 1953, 14 pp.

maximum of the projection. However, the relation is not always so simple. Frequently fold axes, common intersections of shear planes, flow lines, and even slip-striae on major fault planes are surrounded by girdles of quartz optic axes.[1] Now shear plane intersections are parallel to the B-axis of the strain ellipsoid, while slip-striae are usually at right angles to B. This means that there is no simple rule relating lineation with fabric and movement on the basis of preferred orientation especially of quartz optic axes.

On this score structural petrologists have in the recent past split up into two rather dogmatic schools of thought. The so-called orthodox or normal interpretation of linear fabric has its roots in Schmidt's early work and claims that a lineation parallel to the axis of a girdle of fabric elements is always a B lineation with movement restricted to the plane of the girdle (plane deformation in ac). It is based on the assumption that all girdle-fabrics are due to rotation of fabric elements around B or to shear planes intersecting along B.[2]

Other workers adhere to what is known as the revised or parallel interpretation of linear fabric. They maintain that tectonic movement is always parallel to the lineation when coinciding with a girdle axis of fabric elements.[3] In some cases, such as tapering salt pipes, the mechanism of obstructed canal flow, invoked by Anderson to explain the movement parallel to fold axes and girdle axes of fabric elements, is applicable, but in such three-dimensional strains distinction should be made between principal and subordinate movement directions, and it is manifestly misleading to designate, e.g. the axes of fold nappes and their lineations as directions of tectonic movement merely because they are normal to girdles of fabric elements.

Symmetry of lineation

Since the symmetry of fabric, strain, and stress are interrelated, we have better criteria for the kinematic significance of lineation than are afforded by

[1] B. Sander, *Gefügekunde der Gesteine*, Springer, Vienna, 1930, pp. 239–43. T. G. Sahama, 'Struktur und Bewegungen in der Granulitformation des Finnischen Lapplands', *Bull. Comm. Géol. Finl.*, No. 101, 1933, pp. 82–90. H. Closs, 'Tektonische Folgerungen aus Graubündner Quarzgefügediagrammen', *Min. Pet. Mitt.*, Vol. 46, 1935, pp. 403–15. H. Martin, 'Ueber Striemung, Transport und Gefüge', *Geol. Rundsch.*, Vol. 26, 1935, pp. 103–8. A. Kvale, 'Petrofabric analysis of a quartzite from the Bergsdalen Quadrangle, W. Norway', *Norsk Geol. Tidsskr.*, Vol. 25, 1945, pp. 193–215. R. Balk, 'Fabric of quartzites near thrust faults', *Journ. Geol.*, Vol. 60, 1952, pp. 415–35.

[2] W. Schmidt, 'Gefügestatistik', *Min. Pet. Mitt.*, Vol. 38, 1925, pp. 392–423. H. W. Fairbairn, 'Structural Petrology of the Claire River Syncline, Tweed, Ontario', *Trans. Roy. Soc. Canada, Sec. 4*, Vol. 29, 1935, pp. 21–5. F. C. Philips, 'A fabric study of some Moine Schists and associated rocks', *Quart. Journ. Geol. Soc. Lond.*, Vol. 93, 1937, pp. 581–620. F. J. Turner, 'The metamorphic and plutonic rocks of Lake Manapouri, Fiordland, New Zealand, Pt III', *Trans. Roy. Soc. New Zealand*, Vol. 68, 1938, pp. 122–40.

[3] E. M. Anderson, 'On lineation and petrofabric structure and the shearing movement by which they have been produced', *Quart. Journ. Geol. Soc. Lond.*, Vol. 104, 1948, pp. 99–125. T. Strand, 'Structural Petrology of the Bygdin Conglomerate', *Norsk Geol. Tidsskr.*, Vol. 24, 1945, pp. 14–31. Chr. Oftedahl, 'Deformation of quartz conglomerates in Central Norway', *Journ. Geol.*, Vol. 56, 1948, pp. 476–87.

girdles and point-maxima of fabric elements. Sander has long insisted that in monoclinic fabrics a B lineation must be normal to the only plane of symmetry and may be the only twofold axis of symmetry.[1]

Recently, an imposing array of orientation data from naturally and experimentally deformed rocks, in which the lineation conforms to Sander's symmetry rule, has been presented by Turner.[2] In orthorhombic fabrics of heteropolar type lineation parallel to B may be distinguished from lineation parallel to A by the fact that the former is normal to a plane of symmetry (see p. 427, and Fig. XIII–6). The same rule may be applied to axial fabrics of heteropolar type if it is borne in mind that lineations at right angles to the heteropolar direction of movement are normals to planes of symmetry and all others not (see p. 427 and Fig. XIII–5). The argument from symmetry fails in triclinic fabrics when curved and oblique lineations occur that are never normal to a reflection plane. Only those straight lineations appertaining to subfabrics of higher symmetry not sharing any elements with other subfabrics may be used for symmetry considerations (see p. 431 and Fig. XIII–8B).

Lineation and crossed strains

It is perhaps not a mere coincidence that the present conflict over the kinematic significance of lineation is focused on a limited number of structures: on flattened and arcuate folds, on orthogonal cross-folds, on conglomerates containing both flattened and elongate pebbles, and on constricted or flattened intrusive bodies such as salt domes and syntectonic plutons. These structures are most conveniently referred to a triaxial form-ellipsoid in which the mean axis (B) is not of the same length as the diameter of the sphere before deformation.[3] Extension or compression parallel to B produces fold axes, shear plane intersections and other girdle axes in directions at right angles to B. Thus a fabric containing two or more crossed girdles develops, and its triaxial strain may be resolved into crossed two-dimensional strains each with its axis of no deformation $(B, B^1,$ etc.) normal to a girdle of fabric elements, as suggested by Sander.[4] This widely adopted practice has led to the misconception that crossed strains and their B lineations are separate in time as well as in space, though it has repeatedly been pointed out that their rectangular relation $(B \perp B')$ is too frequent to be accidental and indicates a single cause

[1] B. Sander, *Gefügekunde der Gesteine*, Springer, Vienna, 1930, p. 108.

[2] F. J. Turner, 'Lineation, symmetry and internal movement in monoclinic tectonite fabrics', *Bull. Geol. Soc. Amer.*, Vol. 68, 1957, pp. 1–18.

[3] W. Schmidt, *Tektonik und Verformungslehre*, Borntraeger, Berlin, 1932, pp. 60–72. D. Flinn, 'On folding during three-dimensional progressive deformation', *Quart. J. geol. Soc., London*, Vol. 118, 1962, pp. 385–428.

[4] B. Sander, *Gefügekunde der Gesteine*, Springer, Vienna, 1930, pp. 225–6, 238–9; idem, 'Ueber Striemung, Transport und Gefüge', *Geol. Rundsch.*, Vol. 27, 1936, pp. 298–302.

such as flow of layered salt into a tapering conduit or arcuation of folds by lateral friction.[1]

Figs. XIII–10, 11, and 12 illustrate on various scales the $B \perp B'$ fabric of an isoclinally folded and flattened phyllite of Lower Palaeozoic age from the Snowy Mountains of New South Wales. The major fold axis (B) is parallel to the common intersection of bedding (s_1) and shear planes (s_2, s_3). It is also normal to a partial girdle of mica cleavage poles. B' is marked by a lineation due to common intersection of bedding (s_1) and shear planes (s_4, s_5) which are

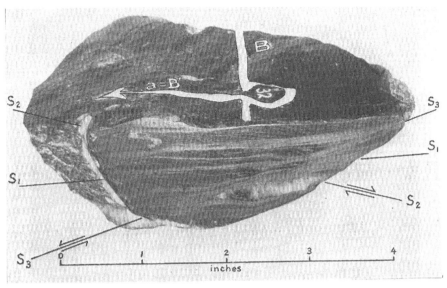

FIG. XIII–10. $B \perp B'$ TECTONITE DUE TO ISOCLINAL FOLDING AND PENE-CONTEMPORANEOUS FLATTENING OF LOWER PALAEOZOIC QUARTZ-PHYLLITE FROM MT. KOSCIUSKO, NEW SOUTH WALES

Front face cut at right angles to B showing traces of bedding s_1 and shear planes s_2, s_3. Striation parallel to B' on top surface.

responsible for the flattening movement with elongation parallel to B. B' is normal to a complete girdle of mica cleavage poles. Yet the B lineation is normal to the most prominent and persistent symmetry plane of the total fabric which has acquired orthorhombic tendencies since the folds became isoclinal and were eventually flattened (compare Fig. XIII–6).

When folded beds are thus stretched and broken up into lenses and rods, inhomogeneities of fabric are likely to accompany those of composition so that lineations and girdle axes parallel to B may be prominent in one locality, and those parallel to B' in another. It is therefore of the utmost importance to

[1] E. Cloos, 'Lineation und Bewegung, eine Diskussionsbemerkung', *Geologie*, Vol. 7, 1958 (von Bubnoff Gedenkschrift), pp. 307–12. R. Balk, 'Structure of Grand Saline Salt Dome, Van Zandt County, Texas', *Bull. Amer. Assoc. Petrol. Geol.*, Vol. 33, 1949, pp. 1791–829.

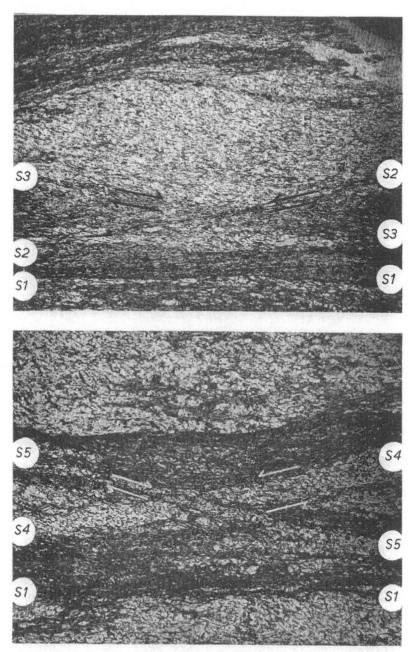

FIG. XIII–11. PHOTOMICROGRAPHS OF SECTIONS CUT FROM SPECIMEN
SHOWN IN FIG. XIII–10

A. Photomicrograph of section at right angles to B. Shows stretching of the bedding
plane s_1 on the slip planes s_2 and s_3. Plane polarized light. Lin. magn.: $\times 12 \cdot 5$.

B. Photomicrograph of section at right angles to B'. Shows stretching of the bedding
plane s_1 on the slip planes s_4 and s_5 which intersect in B'. Plane polarized light.
Lin. magn.: $\times 12 \cdot 5$.

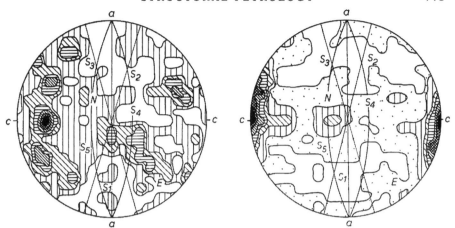

FIG. XIII-12. 300 QUARTZ OPTIC AXES (*left*) AND 900 MICA CLEAVAGE POLES (*right*) FROM SECTIONS SHOWN IN FIG. XIII-11 SHOWING THE VARIOUS *s*-PLANES AND THE ORTHORHOMBIC SYMMETRY (WITH MONOCLINIC AND TRICLINIC ASPECTS) OF THE FABRIC

Contours: (7)–6–5–4–3–2–1–0 per cent per 1 per cent area. Equal area projection at right angles to *B*. Lower hemisphere.

determine the principal plane of regional symmetry of a tectonic unit, and not to generalize unduly the fabric of alleged key areas.

Kinematic syntheses of homogeneous fabric domains are essential to acquire coherent movement pictures in areas of inhomogeneous structure and lithology. This approach has served well in settling the fabric argument in the Caledonides of Scotland and Norway where orthogonal and oblique lineations, each with a girdle of fabric elements, have been gradually stripped off their isolated, antagonist interpretations.[1]

---

[1] F. C. Philips, 'Lineation in the N.W. Highlands of Scotland', *Geol. Mag.*, Vol. 84, 1947, pp. 58–9; idem, 'Lineation in Moinian and Lewisian rocks of the Northern Highlands of Scotland', ibid., Vol. 86, 1949, pp. 279–87. E. M. Anderson, 'Lineation in Highland Schists', ibid., Vol. 88, 1951, p. 219. D. B. McIntyre, 'Lineation in Highland Schists', ibid., pp. 150–1. A. Kvale, 'Linear structures and their relation to movement in the Caledonides of Scandinavia and Scotland', *Quart. Journ. Geol. Soc. Lond.*, Vol. 109, 1953, pp. 51–74. G. Wilson, 'Mullion and Rodding Structures in the Moine Series of Scotland', *Proc. Geol. Assoc. Lond.*, Vol. 64, 1953, pp. 118–51. G. R. McLachlan, 'The bearing of rolled garnets on the concept of *b*-lineation in Moine rocks', *Geol. Mag.*, Vol. 90, 1953, pp. 172–6. P. Clifford, M. J. Fleuty, J. G. Ramsay, J. Sutton, and J. Watson, 'The development of lineation in complex fold systems', *Geol. Mag.*, Vol. 94, 1957, pp. 1–24. M. R. W. Johnson, 'The structural history of the Moine thrust zone at Lochcarron, Wester Ross', *Trans. Roy. Soc. Edinburgh*, Vol. 64, 1960, pp. 139–68, A. L. Harris and N. Rast, 'The Evolution of Quartz fabrics in the metamorphic rocks of Central Perthshire', *Trans. Edinburgh Geol. Soc.*, Vol. 18, 1960, pp. 51–78, D. M. Ramsay, 'Microfabric studies from the Dalradian rocks of Glen Lyon, Perthshire', ibid., Vol. 19, 1962, pp. 166–200. J. M. Christie, 'The Moine thrust zone in the Assynt region Northwest Scotland', *Univ. Cal. Publ. Geol. Sci.*, Vol. 40, 1963, pp. 345–440. J. G. Ramsay, 'Structure and metamorphism of the Moine and Lewisian rocks of the North-west Caledonides', in *The British Caledonides*, ed. by M. R. W. Johnson and F. H. Stewart, Oliver & Boyd, London, 1963, Chapter 7, pp. 143–75, N. Rast, 'Structure and metamorphism of the Dalradian rocks of Scotland', ibid., Chapter 6, pp. 123–42. M. Lindström, 'Tectonic fabric of a sequence of areas in the Scandinavian Caledonides', *Geol. Fören. Forhandl.*, Vol. 83, 1961, pp. 15–64.

CHAPTER XIV

Geomorphology and structure – Morphotectonics

The influence of structure on geomorphology is profound. Not only are the great features in the morphology of the globe – mountain chains, plateaux, rift valleys, continental margins, and the like – predominantly of structural origin and formed by folding, faulting, or warping, but many of the minutiae of land forms, the shape of hills or mountains, and the trends and patterns of streams, are controlled by structure through the action of agents of weathering and erosion on complex rock masses. As the zoologist's scalpel dissects the parts of an animal, so do the physiographic agents probe and dissect the earth. This is a fundamental tenet of physiographic lore, but has been strangely neglected as a method of study by which inferences may be drawn from geomorphology as to geological structure. The topic has recently received some attention as *structural geomorphology*, but *morphotectonics* is preferred as having a wider connotation.[1]

Use of photo-geology

Until aerial photography became possible, geologists had to rely on observations on the ground and on geographic maps for the study of morphotectonics, but photo-interpretation has now become one of the most potent of geological methods. The techniques of map-making from air photographs belong to the science of photogrammetry, on which standard works should be consulted, but geological data are inserted by the geologist rather than by the cartographer, and skill in photo-geology is founded largely on the geologist's ability to interpret land forms. This chapter outlines some of the more important aspects of the subject, especially from the structural viewpoint, but for sound work a full appreciation of all the principles of geomorphology is required, which involves a vast field, beyond the scope of this book. Photo-geology is, however, now so widely used that a few introductory remarks may be made, and illustrations from aerial photographs will be given where possible.

[1] F. A. Melton, 'Aerial photographs and structural geomorphology', *Journ. Geol.*, Vol. 67, 1959, pp. 355–70. E. S. Hills, 'A Contribution to the Morphotectonics of Australia', *Journ. Geol. Soc. Aust.*, Vol. 3, 1956, pp. 1–15.

Although much use can be made of single prints, it is normal practice to examine pairs of adjacent photographs stereoscopically, marking geological data either on the prints or on a sheet of transparent material placed over them. The dip of beds exposed on dip-slopes or in gorges may be estimated, but with purely visual inspection it must be noted that the stereoscope enlarges the vertical scale so that both the topography and dips appear exaggerated. Photographs are usually most revealing in arid or semi-arid

FIG. XIV–1A. AERIAL PHOTOGRAPH OF FOLDED
AND FAULTED PRE-CAMBRIAN BASALT, COOLGARDIE,
WESTERN AUSTRALIA

regions, but even in densely vegetated tropical terrains photo-geological work may still be carried out. In New Guinea, the approximately constant heights of tree tops above the ground permits the recognition of scarps which are faithfully reproduced in the foliage, and again the characteristics of formations with respect to the texture of drainage, a tendency to form low or high topographic features and the like, may yield useful information. Joints and faults are often revealed by differences in vegetation, and igneous rock masses whether dykes or other forms, may, because of the chemical characteristics of the rock, afford either a sparse or a dense vegetation-cover (Fig. XIV–1).

FIG. XIV–1B. BASIC DYKE, SOUTH OF MARBLE BAR,
WESTERN AUSTRALIA
(*R.A.A.F. photo: reproduced with permission*)

The use of special photographic emulsions and of radar, and increasing experience and skills in photo-interpretation, have made *remote sensing* as such methods are in general termed, an extremely useful adjunct in geological reconnaissance.[1] Previous field experience in a region is generally desirable for a proper appraisal of aerial photographs, and it is as an adjunct to field work that they find their greatest use.

Principles of geomorphology

Landforms result from the interaction of surface agents and earth forces (hypogene forces) on rocks. Diastrophism produces tectonic landforms, which are subsequently modified by weathering, erosion, and deposition. If, for simplicity, it is assumed that a period of diastrophism has produced new tectonic landforms which subsequently remain as passive subjects for erosion, the new land surface is regarded as representing the initial stage of a cycle of erosion. Although not all the processes are fully understood for all climates, it will be clear that the features found at various stages of a cycle will be fundamentally affected by the physiographic agents dominant in the region. Thus a normal (pluvial or humid) cycle is recognized where running water is the chief agent; an arid cycle for desert conditions where wind action is very important and running water relatively less so; glacial and marine or shore-line conditions also produce their own characteristic landforms, but the cyclic development is not so clearly understood for glaciers, and for shorelines it is of a different type from land surfaces. In this chapter we shall not discuss

---

[1] 'Manual of Photographic Interpretation', 1960, *Amer. Soc. Photogramm.*; 'Manual of Colour Aerial Photography', 1968, ibid.; *Proceedings of Symposia on Remote Sensing of Environment, 1962–9*, Vols I–VI, Univ. Michigan.

shorelines or glacial or aeolian action, for to do so would take us too far into geomorphology, and stream action is so important even in deserts that its effects in moulding the landscape are dominant in most arid regions other than sandy deserts.

Commencing with the tectonically produced initial land surface, the normal cycle progresses with increasing dissection, through young and mature stages to old age, when a landscape of very low relief will have been formed. During youth and early maturity of the cycle there will still be relics of the initial surface remaining as plateaux, but there are many other ways in which plateau-levels may be formed at various elevations, and adequate guidance for their recognition as ancient peneplains, pediments, terraces, or other similar truncated surfaces must be sought from the field data.

Although most of the features of landscapes are due to tectonic action followed by erosion and deposition, constructional landforms such as volcanoes, lava flows, sand dunes, and uplifted marine features such as coral reefs and bars are also to be recognized, and climatic, volcanic, and tectonic 'accidents', relatively minor events which disturb the physiographic regime, may all produce features of the landscape.

Tectonic landforms

Fold and block mountains, fault troughs, broad basins, and domes, and indeed all landforms that result directly from diastrophism reveal important structural information simply from their contours and topography. Although, it is true, the morphology of major tectonic landforms is complicated by the interaction of erosion, deposition, and diastrophism, the realization that the key to the understanding of a region may lie in the direct relationship of topography to structure carries the force of a revelation, and is indeed insufficiently stressed in teaching geomorphology.[1]

Equally, however, the realization that structures such as the great East African Rift Valleys, for long thought to be entirely young grabens, in reality originated in Mesozoic movements or perhaps even earlier, has opened new vistas for tectonic interpretation especially with regard to the antiquity of major earth-fractures, which may have been reactivated many times before finally producing a tectonic landform of recent age[2] (see also p. 341).

[1] This arises because the chief interest of many geomorphologists lies in the sequential landforms developed from an initial surface, the description of which is often highly generalized and unrelated to actual examples. This attitude in turn derives from the separation of geology and geomorphology – which latter is often regarded as appropriate in schools of geography, so that the geological structure is regarded as 'given' – a notion that is basically unsound. The works of de Martonne and C. A. Cotton are, however, notable for the excellence of their treatment of the actualities of initial tectonic landforms – see especially E. de Martonne, *Traité de Géographie Physique*, Paris, 1932–5 (3 vols), C. A. Cotton, *Geomorphology*, Christchurch, 1942.

[2] F. Dixey, 'The East African Rift System', *Colonial Geol. and Min. Res., Supp. Series*, Bull. Suppt. No. 1, 1956, 77 pp.

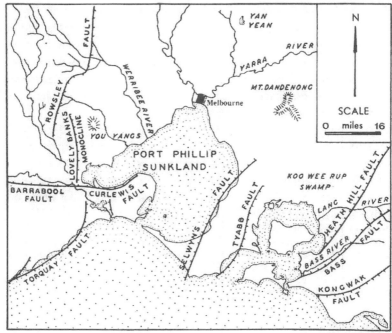

FIG. XIV–2. PHOTOGRAPH OF A RELIEF MODEL OF SOUTH-CENTRAL VICTORIA, AND
DIAGRAM SHOWING THE DIRECT EFFECTS OF FAULTING ON TOPOGRAPHY

A glance at a relief map of a large area such as is given in any good atlas will reveal major tectonic landforms in the regional distribution of high and low land, of basins and plateaux and the margins of the continental blocks at the edge of the shelf. Fig. XIV – 2 shows an area around Melbourne, Victoria, that is virtually a direct expression of tectonic landforms, and many other examples may be found.

Character profiles

Hobbs introduced the term *character profile* into geomorphology, for section views or profiles.[1] In field reconnaissance, profiles afford valuable guidance on structure, and although the concept is implied in the standard principles of geomorphology it is worth-while drawing particular attention to it.

Davis' trilogy of structure, process, and stage affords all that is required for geomorphological understanding, if under 'structure' is included all the characteristics of rocks including their lithology, and further if the concept of 'stage' is broadened to permit earth movements during the currency of a cycle. For physiographic purposes three structural types may be recognized: (*a*) massive rocks, (*b*) stratified rocks, (*c*) complex or heterogeneous rocks.

In *massive rocks* structural control of physiographic processes is virtually lacking. Few rock masses are massive in this sense, for even in granites strong jointing is generally present, but thick formations composed of beds only a few feet or inches thick and not differing greatly in resistance to erosion, approximate to this condition. Where deeply dissected, profiles generally exhibit a complex of 'spurs', that is, of residual ridges which bifurcate when followed down their crests (Fig. XIV – 3).

With *stratified structures*, in which well-defined strata of differing resistance to weathering and erosion are interbedded, important differences arise according to whether the rocks are flat-lying, tilted, or folded.

Table-mountains, mesas, and buttes, typically with vertical or near-vertical sides due to jointing in the hard capping, and a catenary profile in the underlying soft rock, are formed where a flat-lying resistant formation or lava-flow overlies softer sediments. *Cambering* of the plateau edges (see p. 72) may result if the underlying beds yield and flow to the valleys, and in rocks which readily form mud-flows the profile is modified into a hummocky land-slide topography. With a series of alternating hard and soft beds, the profile shows a step-like repetition of ledges with vertical edges and catenary slopes, beautifully represented in Holmes' drawings of the Colorado Plateau and the Grand Canyon (Fig. XIV – 4) which should be examined by every student of geology.[2]

[1] W. H. Hobbs, *Earth Features and their Meaning*, New York, 1912.
[2] Atlas to accompany the 'Tertiary History of the Grand Canyon District', by C. E. Dutton, *U.S. Geol. Surv.*, 1882.

FIG. XIV–3. A SYSTEM OF SPURS DUE TO DEEP DISSECTION OF PHYSIOGRAPHICALLY
MASSIVE ROCKS, IN THE VICTORIAN HIGHLANDS
(*Photo: B. Coghlan*)

FIG. XIV–4. TOPOGRAPHIC EXPRESSION IN A DISSECTED REGION OF FLAT-BEDDED,
ALTERNATELY SOFT AND RESISTANT BEDS
Vishnu's Temple, Grand Canyon of the Colorado. (*After Holmes*, U.S. Geol. Surv.)

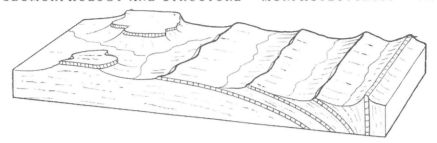

FIG. XIV–5. THE EFFECT OF VARIATION IN DIP ON TOPOGRAPHIC EXPRESSION
IN A REGION OF STRATIFIED ROCKS

In tilted beds the character of the profile depends largely on the angle of dip and the degree of induration and type of jointing in the resistant beds. In general, erosion produces *dip-slopes* and *escarpments* in tilted beds, and with low dips the escarpments may be several miles apart with intervening broad valleys, each flanked on one side by an escarpment, on the other by a long dip-slope. This is the *cuesta* type of topography. As the dip steepens so do the dip-slopes become shorter (Fig. XIV–5): *strike-ridges* follow the basset-edges of the strata. Similar profiles are also found on the flanks of folds in stratified rocks and also with block-faulting, and although the local profile may give little assistance in distinguishing tilted beds due to faulting from those tilted by other mechanisms, other evidence may serve. In a thick tilted sequence of beds successive ridges are developed from different strata,

FIG. XIV–6. SINUOUS STRIKE-RIDGE FOLLOWING BASSET-EDGE OF RESISTANT BED
IN A STRUCTURAL BASIN, WEST PAKISTAN
(Photo: E. S. H.)

whereas with faulting, repetition of the same bed may be the cause of a series of ridges. Again, the trend of strike-ridges in folded rocks follows swings in the strike, and is in general more or less sinuous in plan (Fig. XIV − 6). *Fault-ridges*, on the other hand, are either straight or sweep in simple curves. Among the common profiles of strike-ridges may be mentioned *homoclinal ridges* with well-defined dip-slopes, generally fairly steep; *razorbacks*, with a sharp crest and steep flanks, usually indicating near-vertical dips, and *hogbacks* with rounded profiles, an indication of massive resistant beds (or dykes) with steep dip.

In regions of complex structure, due to the absence of persistent dipping strata no long ridges are found, but the topography is variable and there is no dominant type of character profile. However, much can be made out in such regions from the pattern of dissection, as seen in photographs or good maps.

Drainage patterns

Perhaps the greatest aid that geomorphology offers to structural geology is to be got from drainage patterns. In the process of drainage development on a land surface undergoing dissection, the streams, in general, become adjusted to structures through the influence of hard and soft rocks, and of faults and joints in determining stream courses. As we shall see, in certain circumstances there is a notable lack of adjustment, which in itself, if properly understood, may afford valuable information.

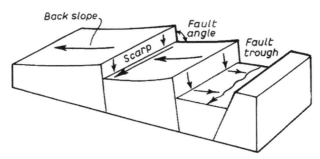

FIG. XIV–7. CONSEQUENT STREAMS IN A BLOCK-FAULTED
REGION

Before deep dissection, consequent stream patterns are often simple, with few tributaries because of the poor development of integrated slopes leading to the trunk streams. Deeper dissection also invariably leads to complexity, with numerous short tributaries to the main streams. As these cut down, the easier erosion of soft rock, either in soft beds or in weak zones along faults and major joints, causes the channels to shift and eventually a close adjustment of streams to structure takes place. After considerable erosion, when the initial surface has largely disappeared, it may be difficult to determine which, if any, of the streams are consequents, but this is immaterial as our interest now shifts from the *surface* and the geological events that have affected it, to the structures *within* the rocks of the region, whatever their age. Thus dissection of an uplifted peneplain of Palaeozoic rocks will reveal the structures, whether Palaeozoic or younger, that are present in those rocks, and we may learn little about the streams that flowed when the peneplain was first uplifted, because their courses have long since been modified. Streams that follow belts of soft rocks, or faults or joints in a dissected rock mass are *subsequent streams*.[1]

A *trellis* or *grapevine* pattern is formed where trunk streams have a dominant parallel trend, with right-angled bends linking parallel reaches, and

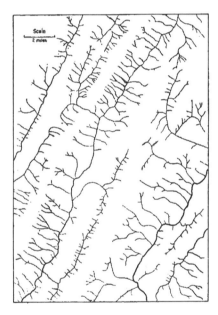

FIG. XIV–8. GRAPEVINE STREAM-PATTERN IN DISSECTED HARD AND SOFT ROCKS WITH STRIKE RIDGES BETWEEN THE VALLEYS

(*After Nevin, 1942*)

numerous minor tributaries at right angles to the major streams. The grapevine pattern is generally an indication of stratified structure with strike-ridges of hard rocks forming the interfluves and the trunk streams located in the softer beds (Fig. XIV−8). Because of cross-joints, the numerous minor tributaries are generally at right angles to the major streams, and the few right-angled bends in these are also controlled by joints or cross faults. The same pattern is however also found in igneous rocks where there is a strong dominant set of faults or joints, as shown in Fig. XIV−9.

Where the strike is sinuous, corresponding flexures are found in the stream pattern and in the dissection of a dome the subsequent streams may take on a circumferential, circular, or elliptical pattern. Many salt domes have been

[1] The author prefers not to complicate this simple concept so that no reference is made here to obsequent or resequent streams.

discovered by the dissection of the dome formed above the salt plug, commonly with an inward-facing escarpment in hard beds. Again, a similar pattern may be found in igneous rocks, especially in granite stocks (Fig. XIV–10) and in ring-complexes.

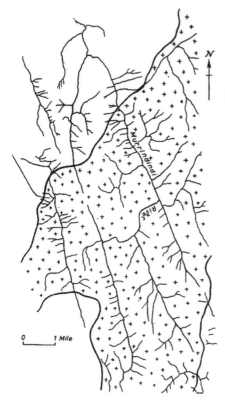

FIG. XIV–9. GRAPEVINE STREAM-PATTERN DETERMINED BY MASTER JOINTS IN GRANITE, MURRINDINDI AREA, VICTORIA

In the dissection of rocks in which no structures are strong enough to determine definite drainage lines, the streams generally take on sinuous courses giving a *dendritic* drainage pattern (Fig. XIV–11), and are termed *insequent streams.*[1]

Even in a region where structures are prominent and many streams follow them, there may be exceptions. Large streams by virtue of their greater power and their width show a less intimate relationship than smaller to joints and

[1] The author is one of a number who have illustrated supposedly insequent drainage by a pattern which on more confident analysis may be recognized as a network. See E. S. Hills, *Physiography of Victoria*, Melbourne and Christchurch, 3rd edn, 1951, Fig. 112; but see also C. M. Nevin, *Principles of Structural Geology*, New York, 3rd edn, 1942, Fig. 134, p. 206, which purports to be insequent but in fact illustrates an excellent network. Although the absence of any one obvious trend might be taken to conform with W. M. Davis' definition of insequent, it seems proper to use the term more strictly in the sense indicated above, since truly insequent drainage patterns do, indeed, exist.

FIG. XIV–10A. ANNULAR DRAINAGE PATTERN IN DISSECTED DOME
Middle Dome near Harlowton, Montana
(*Photo: Dr Barnum Brown*)

faults, and indeed many major streams transgress important structural features that might normally have been expected to influence them. Apart from the effects of glaciers and ice-caps, lack of adjustment to structure may imply rapid dissection in a young stage of the cycle of erosion, or the *superimposition* of the streams onto structures in an undermass, after a covering series of rocks is dissected through. Lack of adjustment reveals itself most strikingly in river gorges. Where a large stream abruptly enters a high block of country

FIG. XIV–10B. ANNULAR DRAINAGE
PATTERN IN GRANITE STOCK,
HARCOURT, VICTORIA

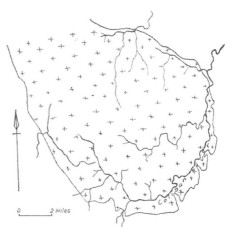

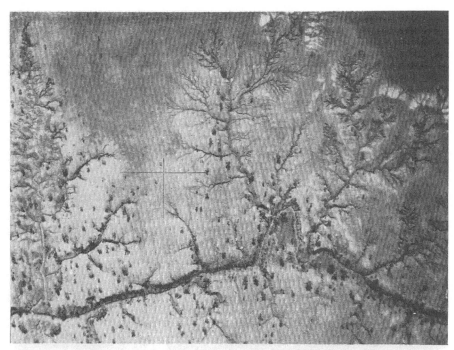

FIG. XIV–11. DENDRITIC DRAINAGE PATTERN (WITH INSEQUENT STREAMS)
NOTABLY IN THE TRIBUTARIES, NORTHERN AUSTRALIA

(*R.A.A.F. photo: reproduced with permission*)

FIG. XIV–12A. ANTECEDENT STREAM CROSSING A SLOWLY UPLIFTED WARPED
AREA. THE FINKE RIVER, MACDONNELL RANGES, CENTRAL AUSTRALIA

(*After Madigan*)

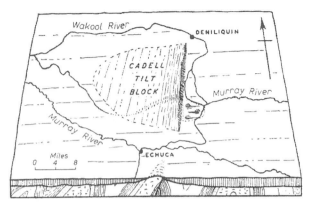

FIG. XIV–12B. DEFEATED STREAM BLOCKED BY RECENT
FAULT–THE MURRAY RIVER NEAR ECHUCA, VICTORIA
(*From Hills,* Physiography of Victoria, *Melbourne, 1967*)

FIG. XIV–13. STREAM TRANSVERSE TO RIDGE OF RESISTANT
ROCKS, CENTRAL AUSTRALIA

through which it flows in a gorge the possible causes should be considered *seriatim* until the most likely is found. Evidence for superimposition from a cover may lie in the presence of remnants of the overlying beds on nearby hills; again, however, faulting or warping might be postulated. With slow uplift of the high block, a large stream may be able to erode its bed with sufficient rapidity to maintain its course despite uplift across it. The stream is then said to be *antecedent*, but such streams are rare and more often the stream is blocked or *defeated* (Fig. XIV–12B), forming a lake or swamp which drains around the ends of the uplift. *Inherited* stream courses are those which originated on an old-land and were later rejuvenated. On a peneplain they may have left their structurally controlled subsequent courses and wandered during the old stage of the cycle. Warping may then rejuvenate these streams, but again the slopes formed by warping may be so strong as to cause a new set of consequent streams to develop. In either eventuality, the streams

FIG. XIV–14. A BRAIDED STREAM – THE NEALE RIVER IN FLOOD, CENTRAL AUSTRALIA

(*Photo by courtesy Dept. of National Mapping*)

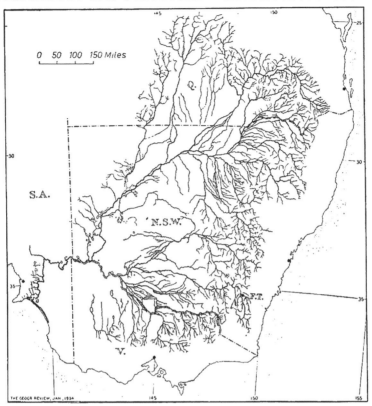

FIG. XIV–15. THE MURRAY RIVER SYSTEM

Note anastomosing patterns in regions of aggradation, especially in N.S.W., and numerous lineaments followed by major streams. (*Cf.* Fig. XI–21)

will erode new valleys, irrespective of the major structures of the basement. This is thought to be the explanation of the remarkable transverse streams in the Macdonnell Ranges of Central Australia (Fig. XIV–12A), which cross a series of quartzite ridges through chasms, many of which are only a few feet wide. The chasms are structurally determined by major joints, but the transverse rivers *inherited* their courses from an older land surface, remnants of which may be seen as plateau-residuals.

Although we are generally more interested in regions where rock outcrops, or is covered only by a sedentary soil, the physiographic features characteristic of alluviated regions may afford data of structural value.

Anastomosing and braided stream patterns (Fig. XIV–14, 15) are formed by *aggrading* streams, that is, by streams which are depositing alluvium. The anastomosing pattern is developed over large areas by slow-flowing streams whereas the braided is more typical of faster-flowing rivers. In great semi-desert plains the valleys of braided streams may be several miles wide and confined only between low banks, but braided courses are also typical of the narrow flood plains of heavily laden streams, for example, those fed by glacial melt-water. The plains of the Murray River basin in Australia exhibit very extensive anastomosing and braided stream systems, yet the courses of the trunk streams show a clear relationship to geological structure, quite unexpected in a region of such low topographic relief. Fig. XIV–15 shows the stream pattern, major trends representing faults, and broad folds being shown by the long straight stretches. Some of the folds are due to compaction of the sediments in the basin over buried bedrock ridges, others to diastrophism.

Texture of dissection

The spacing of stream courses, whether of perennial or of intermittent streams, in any area of country determines the *texture* of dissection. The term is generally used in a broad sense to cover the whole area, but in fact each main rock formation within the area has its own characteristic texture of dissection which, under uniform conditions of precipitation and aspect, is determined by the weathering properties of the substance of the rock, and its joint pattern. Indeed, in air photographs the texture of dissection is one of the distinguishing characteristics of rock formations that assists in their recognition (Fig. XIV–16).

Folds

Systematic treatment of the topographic expression of folding should naturally include all types of folds, but sufficient will have been gathered from the principles outlined above to make a full treatment of very gentle folds in domes and basins, or very large folds in regional uplifts unnecessary. Discussion may most profitably be limited to folds of moderate size, having clearly defined anticlinal and synclinal axes, in well-stratified rocks. Excellent descriptions of folds in an early stage of dissection are available concerning the Juras and the Persian folded zone. For folds thoroughly truncated by erosion the Appalachians are a classic example, and the arid parts of Australia also afford many fine studies of topographic expression in truncated fold belts.

Stages in the dissection of folds are shown in Fig. XIV–17, 18. Longitudinal consequent streams will occupy the synclinal valleys, and lateral consequents flow down the flanks of the anticlines. These lateral consequents dissect the flanks of the anticlines as a fault scarp is dissected, maximum erosion, however, being midway down the fold where volume is adequate and gradient is steepest. As the gulches widen, second-order tributaries begin to work at the foot of scarps where a hard stratum is broken through, and eventually these, subparallel to the consequents, develop into major tributaries, bearing scarps facing the anticlinal crest on both sides. If the anticline has a soft core, this may be eroded away and a major anticlinal valley will

FIG. XIV–16. STRATA WITH DIFFERENT TEXTURE OF DISSECTION, FLINDERS
RANGES, SOUTH AUSTRALIA

(R.A.A.F. photo: reproduced with permission)

then develop, but often another hard stratum is revealed on dissection, which again reproduces the form of the anticline as an anticlinal mountain. In the meantime, accumulation of alluvium will generally go on in the synclines.

The topography of folded beds that have been peneplained and subsequently subjected to renewed erosion is generally notable for intimate differential erosion of hard and soft beds, each of which may form its own strike ridge or valley. The former peneplain level for a time is recognizable in accordant summit levels, so that, especially in aerial views, the countryside has the appearance of having been combed (Fig. XIV–18). The geometry of the folded beds is virtually the same as would appear on a map.

FIG. XIV–17. DIFFERENTIAL EROSION OF AN ANTICLINE – KUH-I-KIALAN, IRAN
(By courtesy British Petroleum)
The central arch is capped with Asmari Limestone overlying Eocene marls in the core of the
fold and flanked with soft Lower Fars evaporites (Miocene).

With deeper erosion, harder beds again reproduce anticlinal mountains and some resemblance to the dissection of young folds is to be seen.

Faults

Many of the features of fault topography will be evident from the previous discussion of faulting in Chapter VII but we may refer here more particularly to the erosion of fault blocks and to the effects of strike-slip faults, which latter have been studied particularly in New Zealand and in California (Fig.

FIG. XIV–18. AERIAL VIEW SHOWING PATTERN OF DISSECTION IN A REGION OF
ALTERNATING STRONGLY FOLDED HARD AND SOFT ROCKS IN FLINDERS RANGES,
SOUTH AUSTRALIA

(R.A.A.F. photo: reproduced with permission)

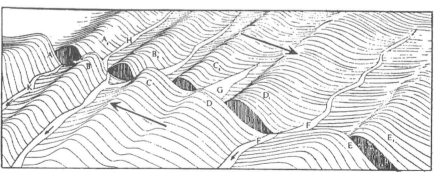

Shutter-ridges A B C D E   Offset stream F   Faultpond G   Diverted stream H   Old bed of diverted stream K

FIG. XIV–19. PHYSIOGRAPHIC EFFECTS OF STRIKE-SLIP FAULTING

(After Lensen, 1958)

XIV−19).[1] The diagram shows *shutter-ridges* blocking valleys, offset streams deviated around shutter-ridges and ponding of streams with the formation of dry valleys downstream from the blockage.

The dissection of fault scarps normally occurs by numerous short consequents, which because of their steep gradient, actively erode notches in the scarp and deposit alluvial cones, which may coalesce to form a *fault apron* along the base of the scarp. An important effect of this deposition is to force streams in the fault angle or on the graben-floor away from the scarp. In an early stage of dissection the scarp, or triangular facets of it, may be clearly visible, but after prolonged erosion and weathering the scarp front retreats,

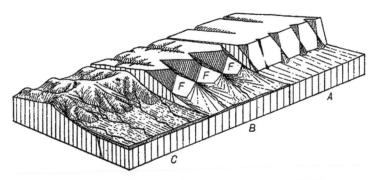

FIG. XIV–20. STAGES IN THE DISSECTION OF A FAULT SCARP

A, initial stage with dissection commencing on fault scarp and back slope; B, further dissection with growth of alluvial cones along foot of fault scarp, and facets (F) of the scarp remaining; C, maturely dissected fault scarp.

the fault facets disappear and become round-ended spurs (Fig. XIV−20). At this stage and later, the position of the fault line may be difficult to infer, as it will not longer coincide with the mountain-front. The scarp is then a maturely dissected fault scarp.[2] Most faults attain their ultimate displacement by repeated small movements rather than one grand event, so that new bases to the fault facets may continue to be formed, revealing the fault plane. Such renewed movements may effect piedmont or fault apron alluvia, which may previously have built backwards and encroached on the upthrown block. Or again, due to secondary tension in the piedmont deposits, scarplets parallel to the main fault, and facing either in the same direction, or the opposite direction, or both (so as to form a small graben) are commonly formed, although they are not likely to persist for long and are therefore more characteristic of active regions. Many such scarplets in North America, New Zealand, and Japan have been formed in historic times during earthquakes.

Where stratified rocks have been affected only by faulting a close relation-

---

[1] G. J. Lensen, 'The Wellington Fault from Cook Strait to Manawatu Gorge', *N.Z. Journ. Geol. Geophys.*, Vol. 1, 1958, pp. 178–96.
[2] Not, however, a fault-line scarp, q.v. below.

ship between dip and the tilt of fault blocks is to be expected, and the structure of the region may thus be obvious. An additional effect indicating the location of fault planes or strong monoclines, which are physiographically similar to faults, is shown where beds are strongly upturned by drag. This may be shown by soft rocks on young faults, or, after erosion, by hard strata projecting as a razorback, such as is well shown in the Flatirons along the Rocky Mountain front in Colorado (Fig. XIV—21).

Strong evidence of faulting is available where differences in topography and rock structure occur along a well-defined line (Fig. XIV—22). In many such examples we have to deal, however, with a *fault-line scarp* rather than a

FIG. XIV–21. THE FLATIRONS, ROCKY MOUNTAIN FRONT, COLORADO
(*Photo: W. C. Bowen*)

fault scarp. A fault-line scarp is one resulting exclusively from erosion along a fault, and containing no inheritance from the original fault-topography. Fault-line scarps may face either in the same direction as the original scarp or in the opposite direction, in which latter case the scarp is described as an obsequent fault-line scarp. The formation of fault-line scarps is due to the effects of differential erosion on the more and less resistant beds brought into juxtaposition by faulting. Although many such scarps are formed after the original topography has first been obliterated during the old stages of an erosion cycle, and later rejuvenation permits the active erosion of the softer rocks on one side of the scarp as shown in Fig. XIV—23, the intervention of planation is not a pre-requisite, for an obsequent fault-line scarp may develop on one side of a fault while the maturely eroded fault scarp still persists on the other.

Again, many old faults with thousands of feet of throw may have no

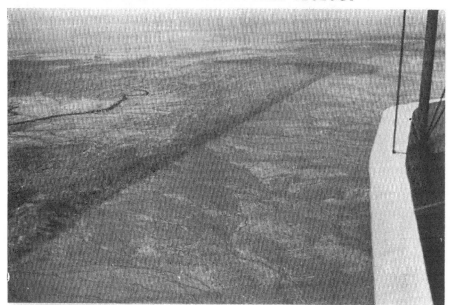

FIG. XIV–22. THE WARRATTA FAULT, WEST OF TIBOOBURRA, NEW SOUTH WALES
(*Photo: Haddon King*)

topographic expression whatsoever, if the beds on either side are of similar lithology. This is the case with the Whitelaw Fault at Bendigo, Victoria, along which Lower and Upper Ordovician graptolitic slates are brought into contact. The fault is not even marked by a topographic lineament at this locality. On the other hand, in northwest Australia low mountain ranges several miles long, once thought to represent residual strata, have been shown to be silicified breccias and crush-conglomerates formed along faults; and again, many great lineaments marked by streams and lakes, especially in the Pre-Cambrian shields, are fault lines.

Fault gaps and fault-line gaps – A gap formed by a displacement in the line of a ridge, caused directly by faulting, is a *fault gap*: a similar gap due to faulting followed by erosion is a *fault-line gap*. Fault gaps are caused chiefly by the lateral displacements that go on along transcurrent or wrench faults; fault-line gaps are often formed by the differential erosion of more and less resistant beds affected by dip-faults.

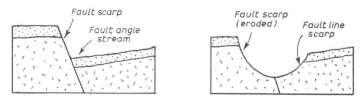

FIG. XIV–23. FAULT-LINE SCARP AND ERODED FAULT SCARP

Joints

Major joints which may be followed for miles and which cut several beds are of considerable importance in determining drainage patterns, indeed more so than might be expected with structures that are simply cracks, or faults with

FIG. XIV–24. WEATHERING AND EROSION TO FORM VALLEYS ALONG MASTER JOINTS IN FLAT-LYING SANDSTONE, THE RUINED CITY, ARNHEM LAND, NORTH AUSTRALIA
(Photo by Australian Army Public Relations)
The topography greatly resembles 'tower karst' in limestones. The scale is given by trees in the valleys.

minute displacement. Prominent joints are common in brittle, flat-lying sediments, but igneous and metamorphic rocks may also have strong regional jointing (Fig. XIV – 9).

Lineaments

We have, so far, considered morphotectonics as if a good deal were known about the geology of a region, but in the early stages of an investigation this knowledge may, in fact, be meagre. On the other hand, topographic maps and air photographs may be available, on which it will immediately be obvious that physiographic trends shown by streams, ridges, and shorelines exhibit a certain geometrical regularity, notably by the parallelism of straight features.

FIG. XIV–25. PART OF THE CANADIAN SHIELD IN THE ICE-SCOURED REGION,
SHOWING INTERSECTING LINEAMENTS IN A REGION OF COMPLEX GEOLOGY.
STAPLE LAKE AREA, NORTH OF YELLOWKNIFE, N.W. TERRITORY, CANADA

(*Royal Canadian Air Force photo: reproduced with permission*)

Although the geological data may at first be quite inadequate to decide what
are faults, joints, or folds, the study of the *lineaments* as each straight physiographic element is termed,[1] affords very valuable information not only by
pin-pointing localities to which ground surveys should devote particular
attention but also by providing a tectonic network, the chief directions of
which must represent important fractures and fold trends that afford essential
data for structural analysis (Fig. XIV – 25).

It is, however, incorrect to regard a tectonic network as exclusively a
fracture pattern, for it may represent joints, faults, strike of beds, or even
intimate fabric elements of schists and gneisses, such as foliation and lineation. Sander has referred to this latter as *Gefügerelief.*[2] An example in Otago,

[1] W. H. Hobbs, 'Repeating Patterns in the Relief and Structure of the Land', *Bull. Geol. Soc.
Amer.*, Vol. 22, 1911, pp. 123–76, used *lineament* in this sense, and the recent substitution of
linear robs the concept of much of its symbolism.

[2] B. Sander, *Einführung in die Gefügekunde der geologischen Körper*, Vol. I, 1948, pp. 206–
8, Vienna.

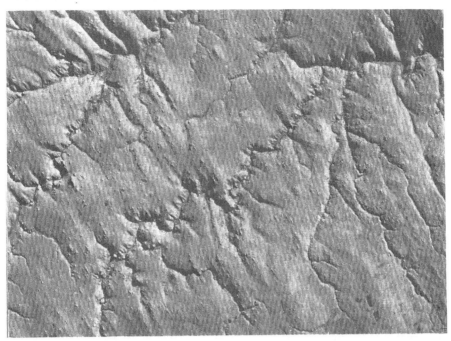

FIG. XIV–26. *Gefügerelief* IN SCHISTS, CENTRAL OTAGO, NEW ZEALAND
(*After F. J. Turner*)

New Zealand, has been described by Turner[1] (Fig. XIV–26), and Larsson has made a detailed study of the relationship between geomorphology and rock structures in the Blekinge region, southeast Sweden.[2]

No geological trend is absolutely straight, for local diversion of streams may readily occur and there are minor complications associated with faults, joints, and the like. It is therefore legitimate in examining tectonic patterns to neglect small and local deviations, as in 'smoothing' a curve on a graph. On the other hand a tendency to overlook curved trends and also to neglect those that deviate slightly from pronounced trends in the area must be consciously guarded against. Above all the lineaments must be drawn purely objectively. Statistical analysis of the results may follow, for instance by constructing 'roses' (azimuth frequency diagrams),[3] but continual reference to geology and

[1] F. J. Turner, '*Gefügerelief* illustrated by "Schist Tor" Topography in Central Otago, New Zealand', *Amer. Journ. Sci.*, Vol. 250, 1952, pp. 802–7.

[2] I. Larsson, 'Structure and Landscape', *Lund Studies in Geogr., Ser. A. Physical Geogr.*, No. 7, 1954.

[3] P. H. Blanchet, 'Development of Fracture Analysis as Exploration Method', *Bull. Amer. Assoc. Petrol. Geol.*, Vol. 41, 1957, pp. 1748–59; E. S. Hills, 'Morphotectonics and the Geomorphological Sciences with Special Reference to Australia', *Quart. Journ. Geol. Soc. Lond.*, Vol. 117, 1961, pp. 77–89; D. U. Wise, 'Regional and Sub-Continental sized Fracture Systems Detectable by Topographic Shadow Techniques', *Proc. Conf. on Research in Tectonics, Geol. Surv. Canada*, Paper 68–52, pp. 175–99.

topography is equally as important as statistical treatment. This is particularly true since streams having remarkably straight courses may systematically deviate, during dissection, from the geological control that first determined them, or, again, in deeply dissected country a stream may, on a map or photograph, show as *apparent trend* due to the fact that it follows an inclined fault or joint on a steep gradient rather than horizontally (Fig. XIV–27).

Recognition of mega-lineaments – Although it is quite possible for different observers to reach substantial agreement in the recognition of lineaments in small areas, a glance at a number of continental or world maps prepared by various authors is sufficient to show that precision and agreement in the

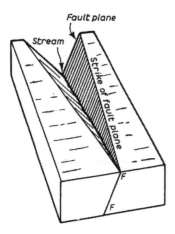

FIG. XIV–27. DEVIATION OF STREAM COURSE FROM THE STRIKE OF AN INCLINED FAULT PLANE

Based on examples in the Snowy Mountains area, New South Wales.

location of lineaments of great magnitude is rarely attained.[1] Owing to the formation of secondary fractures and folds related to large primary structures and to inhomogeneities in the crust, innumerable small lineaments exist and the recognition of major structures must be firmly based on sound topographical, geological, and geophysical evidence. Local lineaments may readily be studied on accurate topographic maps or aerial photos, but megalineaments are best revealed on relief models, which may be viewed from all angles and with varied lighting. It is not possible to view at one time the many detailed maps needed to cover a large region, even were these to be available on the same scale and projection.

Fig. XIV – 29 shows some mega-lineaments in the northeast Pacific Ocean and in Australia.[2]

[1] Compare for instance the maps of R. Staub, 'Grundsätzliches zur Anordnung und Entstehung der Kettengebirge', in *Skizzen zum Antlitz der Erde*. Festschrift for the Seventieth birthday of L. Kober, Vienna, 1953. R. A. Sonder, *Mechanik der Erde*, Stuttgart, 1956, and F. A. Vening Meinesz, 'Shear Patterns in the Earth's Crust', *Trans. Amer. Geophys. Union*, Vol. 28, 1947, pp. 1–61.

[2] In addition to the works cited below for certain of the continents, see H. W. Menard, 'Development of Median Elevations in Ocean Basins', *Bull. Geol. Soc. Amer.*, Vol. 69, 1958,

Continental and global lineaments

The existence of repeating trends that together form a pattern in areas of continental dimensions and perhaps even over the whole earth has long been recognized and is still a topic of lively discussion. Werner, the great teacher of the Freiberg mining school, held that the parallel strike of mineral veins in many regions affords evidence of their simultaneous origin, and Élie de Beaumont in 1852 developed this line of thought in his study of mountain chains, attempting to prove that systems with parallel strike are of the same age throughout the world. He proposed a pentagonal network representing

FIG. XIV–28. THE CHIEF TREND DIREC-
TIONS OF GREAT BRITAIN
(*After Lapworth from Wills*)

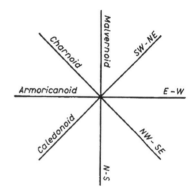

the chief mountain systems of the globe, and in Europe he indicated up to twenty-one trends thought to be characteristic of mountain chains of different ages. This hypothesis obviously could not be substantiated, but it led John Philips[1] to recognize certain characteristic trend directions in Great Britain, where four such are currently named,[2] as shown in Fig. XIV–28.

Although these trends have no real time significance, the notion that each is strongly developed in the fold belts and major faults formed at one period (e.g. Caledonian Folding, Caledonoid trends) still persists in some measure. A similar arrangement of major trends along meridional, latitudinal, and diagonal directions has been pointed out by almost all who have studied

---

pp. 1179–86; idem, 'Deformation of the Northeastern Pacific Basin and the West Coast of North America', ibid., Vol. 66, 1955, pp. 1149–98. M. Ewing and B. C. Heezen, 'Some Problems of Antarctic Submarine Geology', *Amer. Geophys. Union*, Geophysical Mon. No. 1, 1956, pp. 75–81. P. St. Amand, 'Geological and geophysical synthesis of the tectonics of portions of British Columbia, the Yukon Territory, and Alaska', *Bull. Geol. Soc. Amer.*, Vol. 68, 1957, pp. 1343–70. J. D. Moody and M. J. Hill, 'Wrench-Fault Tectonics'; ibid., Vol. 67, 1956, pp. 1207–46 (with numerous references). J. T. Kingma, 'The Tectonic History of New Zealand', *N. Z. Journ. Geol. Geophys.*, Vol. 2, 1959, pp. 1–55. J. H. Illies, 'An Intercontinental Belt of the World Rift System', *Tectonophysics*, Vol. 8, 1969, pp. 5–29.

[1] *Manual of Geology*, Part I, Ed. Seeley, London, 1885, pp. 343–8.

[2] L. J. Wills, *Physiographical Evolution of Britain*, London, 1926. A. Morley Davies, in *Handbook of the Geology of Great Britain* (C. J. Stubblefield, Ed.), London, 1929, Chapter 1, 'Morphology'.

lineaments over large areas.[1] This applies for instance to Europe,[2] Africa,[3] Australia,[4] and South America.[5]

Global tectonic networks – The existence of a pattern of megalineaments over areas as large as sub-continents or ocean basins is demonstrable; but the question of whether there is a global pattern to which the continental and oceanic megalineaments conform, has not been satisfactorily answered, although continental drift would suggest that this is unlikely. Nevertheless the search for such a pattern will continue and earlier syntheses are relevant to various aspects of mega-tectonics.

Prinz emphasized that it is the diagonal lineaments of the globe that are the more prominent excepting in high latitudes,[6] a generalization that is also borne out by later and more detailed studies. The existence of a structural network on the surface of the moon, governing the shape of the large hexagonal craters and the alignment of 'dykes' and other linear features, is discussed by Puiseaux.[7] Wasiutyński[8] accounts for this, as well as for the pattern of orogens in the earth, and for the canals and fields of Mars, by reference to convection cells.

Vening Meinesz has drawn a global network of two conjugate shear directions that fit closely many of the world's mega-lineaments. He first studied prominent topographic lineaments, almost exclusively in the ocean floors, on a statistical basis, and then erected an hypothesis that might account for the observed trends. This hypothesis assumes a displacement of the crust relative to the sub-crustal layers, through 90° of latitude, at the longitude of

[1] C. Lapworth, 'The Heights and Hollows of the Earth's Surface', *Proc. Roy. Geogr. Soc.*, Vol. 14, 1892, pp. 688–97. M. Bertrand, 'Sur les Déformations de l'Écorse Terrestre', *C.R. Acad. Sci., Paris*, Vol. 114, 1892, pp. 402–6. W. H. Hobbs, 'Repeating Patterns in the Relief and Structure of the Land', *Bull. Geol. Soc. Amer.*, Vol. 22, 1911, pp. 123–76.

[2] A. P. Karpinsky, 'Ocherki Geologicheskago Proshlago Evropeiskoi Rossyi', *Acad. Sci. U.S.S.R.*, Moscow, 1947 (Scientific Popular Series). R. A. Sonder, 'Die Lineamenttektonik und ihre Probleme', *Ecl. Geol. Helv.*, Vol. 31, 1938, pp. 199–238. R. Staub, 'Grundsätzliches zur Anordnung und Entstehung der Kettengebirge', in *Skizzen zum Antlitz der Erde*. Festschrift for the Seventieth birthday of L. Kober, Vienna, 1953. J. J. Sederholm, 'Weitere Bermerkungen über Bruchspalten mit besonderen Beziehung zur geomorphologie von Fennoskandia', *Bull. Comm. Géol. Finlande*, Vol. 6, No. 37, 1913.

[3] P. Fourmarier, 'Les Traits directeurs de l'Evolution Géologique du Continent Africain', *C.R. 14th Geol. Congr.*, Madrid, 1926. E. Krenkel, *Geologie Afrikas*, Berlin, Vol. 1, 1925, Fig. 4.

[4] E. S. Hills, 'Some Aspects of the Tectonics of Australia', *Proc. Roy. Soc. N.S.W.*, Vol. 79, 1946, pp. 67–91; idem, 'The Tectonic Style of Australia', in *Geotektonisches Symposium zu Ehren von Hans Stille*, Stuttgart, 1956; idem, 'A Contribution to the Morphotectonics of Australia', *Journ. Geol. Soc. Aust.*, Vol. 3, 1956, pp. 1–15.

[5] F. E. von Estorff, 'Tectonic Framework of North-Western South America', *Bull. Amer. Assoc. Petrol. Geol.*, Vol. 30, 1946, pp. 581–90.

[6] W. Prinz, 'Sur les Similitudes que presentent les Cartes terrestres et planetairs' *Ann. Obs. Roy. Bruxelles*, No. 58, 1891, pp. 304–37.

[7] P. Puiseaux, 'Les formes polygonales sur la Lune', *Bull. Soc. Astronom.*, France, Vol. 21, 1907.

[8] J. Wasiutyński, 'Studies in Hydrodynamics and Structure of Stars and Planets', *Astrophys. Norweg.*, Vol. 4, 1946, 497 pp.

Calcutta.[1] Vening Meinesz has pointed out that many major trends, especially those of the continental margins, conform rather with the bisectrices of his shear pattern than with the calculated shear directions; but if so many directions can be considered to belong to the network it becomes very difficult to check the perfection of 'fit', especially when dealing with small-scale world maps and permitting some deviation due to local crustal inhomogeneities. Boutakoff has shown that many lineaments conform with great circles of the globe, symmetrically distributed about the earth's axis of rotation and congeneric with the equator. He points out that variations such as have been observed in the rotational speeds of the earth[2] could cause a net of shear planes having this arrangement. Sonder[3] conceives the existence of a *rhegmatic pattern* due chiefly to tidal and other forces in the body of the earth, and *special* features connected with individual folds, faults, and other inhomogeneities.

Small-circles – The observation of Sollas[4] that island and mountain arcs closely approximate to small circles of the globe was interpreted by Lake[5] to imply that they represent the intersection of thrust planes with the earth's surface. Seismological data on the depth of origin of earthquakes indicate that the foci lie in zones, which, as is especially clear in the western Pacific, dip beneath the continental margin, and, at least in the crust if not in the substratum, are of the nature of underthrust faults.[6] In addition, the fracture zones that displace the mid-oceanic ridge system lie on small circular arcs.[7]

However, it must be said that there is as yet no agreement on a basic pattern of mega-lineaments, so that it is not possible thoroughly to test any theories that are propounded to account for them.[8]

Global expansion – One of the unresolved problems which is highly relevant to all theories of global tectonics, is the question as to whether the earth's radius has changed over time, and particularly, whether a large expansion may have occurred, with which the rifting and relative displacement of

[1] F. A. Vening Meinesz, 'Shear Patterns in the Earth's Crust', *Trans. Amer. Geophys. Union*, Vol. 28, 1947, pp. 1–61. On purely geological grounds this hypothesis has, so far as the author is aware, no factual support, but see J. D. Moody, 'Crustal Shear Patterns and Orogenesis', *Tectonophysics*, pp. 479–522.

[2] N. Boutakoff, 'The Great Circle Stress Pattern of the Earth', *Aust. Journ. Sci.*, Vol. 14, 1952, pp. 108–11. See also N. Thamm, 'Great Circles – the Leading Lines for Jointing and Mineralization in the Upper Earth's Crust', *Geol. Rundsch.*, Bd. 58, 1969, pp. 677–96.

[3] R. A. Sonder, *Mechanik der Erde*, Stuttgart, 1956.

[4] W. J. Sollas, 'The Figure of the Earth', *Quart. Journ. Geol. Soc. Lond.*, Vol. 59, 1903, pp. 180–8.

[5] P. Lake, 'Island Arcs and Mountain Building', *Geogr. Journ.*, Vol. 78, 1931, pp. 149–60.

[6] On the tectonic significance of the seismological data, see B. Gutenberg and C. F. Richter, *The Seismicity of the Earth and Associated Phenomena*, Princeton, 1949.

[7] W. J. Morgan, 'Rises, Trenches, Great Faults, and Crustal Blocks', *Journ. Geophys. Res.*, Vol. 73, 1968, pp. 1959–82.

[8] See also recent works of B. B. Brock, 'Structural Mosaics and Related Concepts', *Trans. Geol. Soc. S. Africa*, Vol. 59, 1956, pp. 149–97; 'World Patterns and Lineaments', ibid., Vol. 60, 1957, pp. 127–60 (discussion to p. 175).

continental blocks may have been associated. Theoretical considerations do not favour this hypothesis but the concept continues to attract attention.[1]

Global tectonics

A world-wide system of major structures is now known to exist in the ocean basins, including virtually continuous mid-ocean ridges, major fracture zones along which the ridge crests are displaced laterally, and a remarkable array of magnetic anomalies in parallel bands, which are parallel with the ridge crests and are also displaced along the transverse fracture zones. Interpretation of these structures has led to the concept of ocean-floor spreading, with which movements of the continents and of larger crustal plates are involved, these conforming well with palaeomagnetic and other data indicative of continental drift.[2] Thus for the first time in the history of geology and geophysics, a great many major phenomena are linked in a consistent model of global tectonics, including also the formation of oceanic trenches, earthquake distribution, and the origin of continental features such as orogenic belts and major fault zones such as the San Andreas. Certain igneous phenomena, and petrographic provinces, have also been interpreted in the light of the new global tectonics, but many fundamental aspects will require much further study, especially the role of convection currents in the mantle, and the involvement of the mantle in the production of crustal rocks.

The Mid-Ocean Ridge system – For long known chiefly from the Mid-Atlantic Ridge, the inter-connection of ridges in the oceans to form a world-wide system which, broadly regarded, bisects the major ocean basins, has been recognized.[3] The crests of most ridges are occupied by grabens and the rifts are often volcanic. Earthquakes are also associated. As opposed to Holmes' concept that the Mid Atlantic Ridge is composed of continental rocks, the ridges are formed of normal oceanic crust, mainly basaltic.

Fracture-zones – Following the discovery of megalineaments in the Pacific basin off the coast of North America,[4] which were interpreted as fracture-

---

[1] W. H. Bucher, *The Deformation of the Earth's Crust*, Princeton, 1933; idem, 'The Pattern of the Earth's Mobile Belts', *Journ. Geol.*, Vol. 32, 1924, pp. 265–90; N. Havre, *La Terre est un Astre pulsatile*, Paris and Liège, 1931; M. A. Ussov, 'Compression and Expansion in the History of the Earth', *Internat. Geol. Congr.*, 17th Session, 1937, Abstract papers, p. 144; M. J. Rickard, 'Relief of Curvature on Expansion – A Possible Mechanism of Geosynclinal Formation and Orogenesis' *Tectonophysics*, Vol. 8, 1969, pp. 129–44.

[2] X. Le Pichon, 'Sea-Floor Spreading and Continental Drift', *Journ. Geophys. Res.*, Vol. 73, 1968, pp. 3661–97; E. Irving, *Palaeomagnetism and its Application to Geological and Geophysical Problems*, New York, 1964; D. van Hilten, 'Evaluation of Some Geotectonic Hypotheses by Palaeomagnetism', *Tectonophysics*, Vol. 1, 1964, pp. 3–71; 'A Symposium on Continental Drift', *Phil. Trans. Roy. Soc. Lond.*, Ser. A, Vol. 258, 1965, pp. 1–323.

[3] H. W. Menard, 'Development of Median Elevations in Ocean Basins', *Bull. Geol. Soc. Amer.*, Vol. 69, 1958, pp. 1179–86.

[4] H. W. Menard, 'Deformation of the Northeastern Pacific Basin and the West Coast of North America', *Bull. Geol. Soc. Amer.*, Vol. 66, 1955, pp. 1149–98.

zones, it has been found that similar fractures, for the most part approximately latitudinal, exist in all the oceans and that the crests of the mid-ocean ridges are displaced by transcurrent strike-slip faulting, along the fractures (Fig. XIV − 29). It was in order to explain the apparent displacements on such fractures that Wilson conceived the transform fault mechanism which is now widely accepted for these and other major faults (see Chapter VII).

However, the geometrical effects associated with the fracture-zones and with the mid-oceanic ridges must, additionally, be related to the remarkable arrays of geomagnetic anomalies which also exist in the ocean floors.[1]

Geomagnetic anomalies − These comprise arrays of parallel straight bands of anomalies in the total magnetic field, among which alternate bands are polarized positively (corresponding with the present magnetic polarity of the earth) and negatively (that is with reverse polarization). Moreover, the intensity and reversal patterns are symmetrical about the mid-oceanic ridges. Although in some instances the spacing on one side differs from that on the other, matching pairs of ridges in series can readily be identified. The key to the understanding of this pattern derives from the study of reversals of the earth's magnetic field during geological time, the dating of such reversals having been achieved from palaeomagnetic studies on land in regions with a long history of Cainozoic vulcanism. What has been found is that a similar sequence of reversals exists in the magnetic lineaments of the ocean basins, and that these may be matched with the established time sequence, if it is assumed that the ocean floors are spreading away from oceanic ridge crests at a more or less constant rate. This rate, it is found, differs somewhat in different oceans and even on different sides of the same ridge and also has changed at times in rate and direction,[2] but these variations do not invalidate the general assumptions of ocean-floor spreading and, indeed, provide further support for it if local geological and other realistic details can be shown to require such variations. Rates of movement apart of the ridge flanks appear to range from 1 to 10 cm per year, while at times in the past, spreading may have ceased altogether. The bands of oceanic magnetic anomalies are displaced, like the ridge crests, by transcurrent strike-slip faulting along the fracture zones, and when the dated bands are used to indicate movements of the blocks in which they occur, it emerges that the movements of the earth's crust that are involved, may be accounted for by the relative displacement of large crustal plates.

Plate tectonics − Where two plates are moving towards each other the earth is

[1] H. W. Menard and T. Atwater, 'Origin of Fracture Zone Topography', *Nature*, Vol. 222, No. 5198, 1969, pp. 1037–40.

[2] A. Cox, G. B. Dalrymple, and R. R. Doell, 'Reversals of the Earth's Magnetic Field', *Sci. American*, Vol. 216, 1967, pp. 44–54; J. R. Heirtzler, 'Sea-Floor Spreading', ibid., Vol. 219, 1968, pp. 60–70; J. R. Heirtzler *et al.*, 'Marine Magnetic Anomalies, Geomagnetic Field Reversals, and Motions of the Ocean Floor and Continents', *Journ. Geophys. Res.*, Vol. 73, 1968, pp. 2119–36.

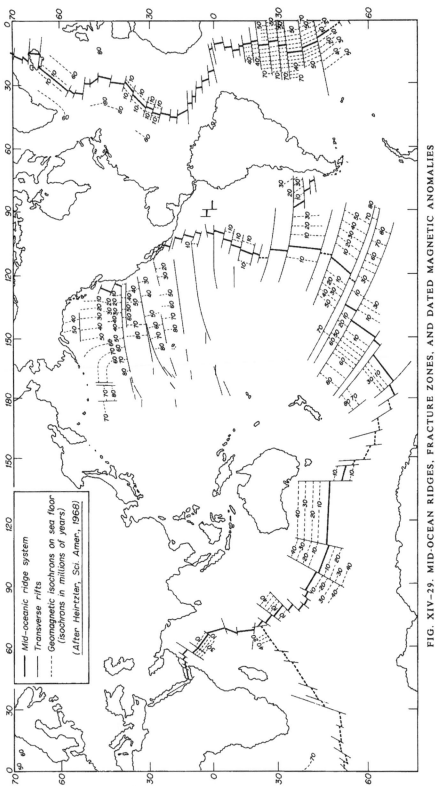

FIG. XIV-29. MID-OCEAN RIDGES, FRACTURE ZONES, AND DATED MAGNETIC ANOMALIES

Inset legend shows symbols for isochrons representing the age of the sea-floor, originating by spreading from the mid-oceanic ridges (heavy black lines). The isochrons are displaced along transverse fracture zones. (*After Heitzler, 1968*)

Legend:

Mid-oceanic ridge system
Transverse rifts
Geomagnetic isochrons on sea floor (isochrons in millions of years)
(After Heirtzler, Sci. Amer., 1968)

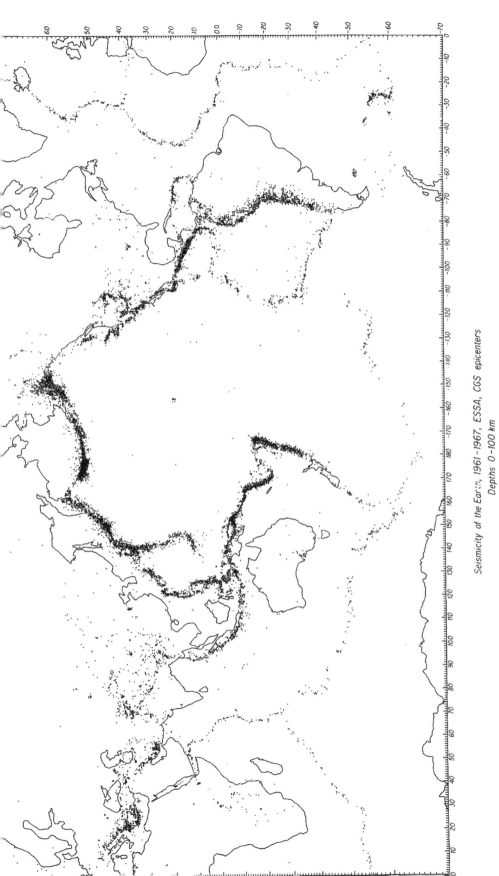

Seismicity of the Earth, 1961-1967, ESSA, CGS epicenters

Depths 0-100 km

FIG. XIV-30A. SEISMICITY OF THE EARTH, 1961-7.

Epicentres, depths 0-100 km (*From M. Barazangi and J. Dorman*, Bull. Seismol. Soc. Amer., *Vol. 59, No. 1, 1969*)

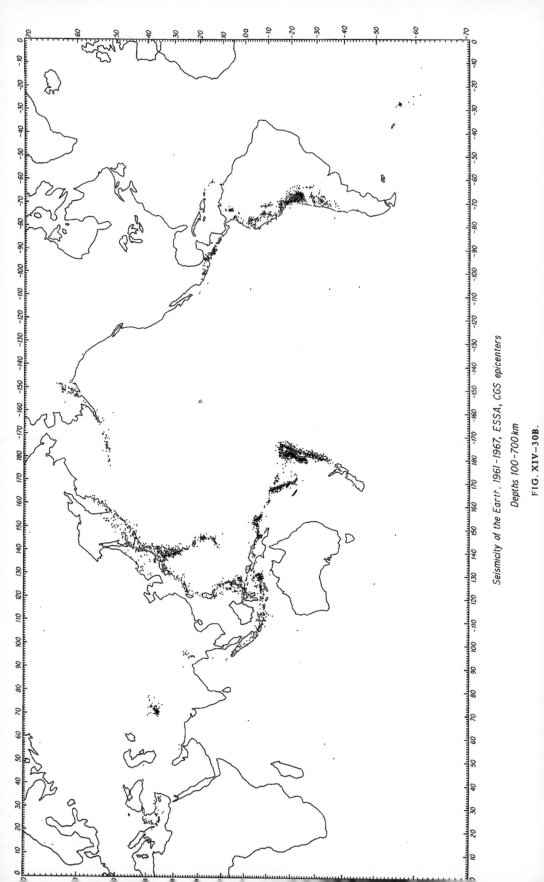

Seismicity of the Earth. 1961–1967, ESSA, CGS epicenters

Depths 100–700 km

FIG. XIV–30B.

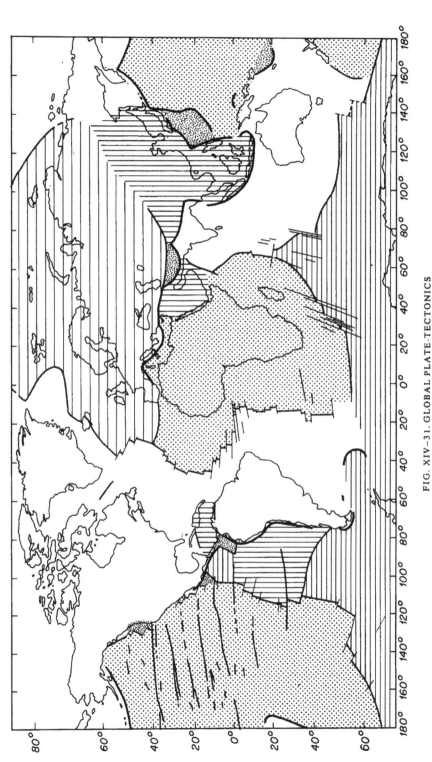

FIG. XIV-31. GLOBAL PLATE-TECTONICS

The crust is divided into units that move as rigid blocks. The boundaries between blocks are rises, trenches (or young fold mountains), and faults. The boundaries drawn in Asia are tentative, and additional sub-blocks may be required. (*After W. J. Morgan, 1968*)

most active seismically, although shallow earthquakes are also associated with the rifting of mid-ocean ridges and with the active portions of transform faults in the fracture zones (Figs XIV–29, 30).

Any regular displacement of a crustal plate over the globe may be referred to as rotation about an axis which emerges on the globe at a pole of rotation. The fracture zones are at right angles to the axis and the rate of spreading increases directly with the distance from the pole. Morgan[1] has proposed that the whole crust may be discussed by reference to six rigid plates (Fig. XIV–31).

From Figures XIV–29, 30A and 30B showing the mid-oceanic ridge system with its associated fracture zones, and the belts of shallow (0–100 Km) and deeper (100–700 Km) earthquakes, it will be seen how closely the mid-oceanic ridge system is connected with shallow-focus earthquakes. The theory of plate-tectonics, illustrated in Figure XIV–31 links the spreading of

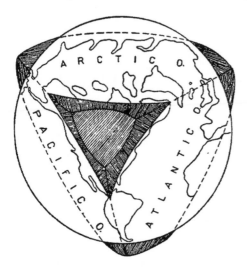

FIG. XIV–32. THE TETRAHEDRAL THEORY OF THE EARTH
(*After Hobbs*)

ocean floors with the impinging of plates against each other. Especially around the margin of the Pacific basin and in the East Indies, such impingements are associated with intermediate and deep-focus earthquakes, young mountain chains, island arcs, and submarine trenches lying on the oceanic side of sialic blocks, underthrust by the simatic ocean floors.

It is generally thought that streaming movements in the mantle, such as convection currents, cause the movements of plates by drag on the sub-crust. Orowan,[2] however, has pointed out certain difficulties in this regard, particu-

---

[1] W. J. Morgan, 'Rises, Trenches, Great Faults, and Crustal Blocks', *Journ. Geophys. Research*, Vol. 73, 1968, pp. 1959–82. See also J. F. Dewey and J. M. Bird, 'Mountain Belts and the New Global Tectonics', *Journ. Geophys. Research*, Vol. 75, 1970, pp. 2625–47.

[2] E. Orowan, 'The Origin of the Oceanic Ridges', *Sci. American*, Vol. 221, 1969, pp. 102–19.

larly in relation to the evidence for different rates of movement apart of opposed plates. He concludes from the persistence of ridges in the mid-oceanic position that the crustal plates cannot be rigid, and suggests the possibility that a global transvection might result from the existence of a hot pole in the earth, perhaps a result of meteoric impact as long ago as the end of the Pre-Cambrian. It is indeed difficult to conceive that the plates should be rigid, in the light of the magnitude and persistence of epeirogenic movements which have affected the continental parts of plates. Reconciliation of all the relevant geological and geophysical data has in fact still to be achieved despite the great advances recently made in global tectonics.[1]

It is well to recall that there is other evidence for axial symmetry in the earth, including the geodetic evidence that the earth is pear-shaped, which was long ago incorporated in the Tetrahedral Theory propounded by Lothian Green.

The Tetrahedral Theory – Lothian Green's Tetrahedral Theory, propounded in 1875,[2] accounted *ex hypothesi* for the preponderance of land in the northern hemisphere, the relative positions of the large continental masses (at the 'corners' of the tetrahedron) (Fig. XIV – 32), the existence of the great south-pointing peninsulas, and the land-encircled Arctic Sea, and it implied the existence of Antarctica. Green suggested that a contracting earth with a relatively rigid crust would tend to assume that regular geometrical form, the tetrahedron, which has the largest area per unit volume. The continental blocks, which are elevations on the body of the earth, correspond with the corners of the tetrahedron, the oceans with its faces.[3] Although this theory is not supported by precise data on the figure of the earth as it exists today, it is of interest in drawing attention to many major features of the globe that are otherwise unexplained.

The increased *tempo* of geological exploration in all continents, the application of new or improved methods of study, for example in age-determination and in palaeomagnetism, and our growing knowledge of the physics of the earth, aided by international collaboration in the dissemination of information, promise definitive results which, one hopes, will help to reveal to us the realities of the face and body of the earth.

[1] A. A. Meyerhoff, from an analysis of palaeoclimates, has argued that the major climatic zones have remained in the same relative position latitudinally throughout phanerozoic time, and that the continents may not, in fact, have been displaced by drift or by plate-tectonics. 'Continental Drift: Implications of Palaeomagnetic Studies. Meteorology. Physical Oceanography, and Climatology', *Journ. Geol.*, Vol. 78, 1970, pp. 1–51.

[2] L. Green, *Vestiges of the Molten Globe as exhibited in the Figure of the Earth's Volcanic Action and Physiography*, Part I, London, 1875; Part II, Honolulu, 1887. See also W. H. Hobbs, *Earth Evolution and its Facial Expression*, New York, 1921, Chapter 7.

[3] The theory is discussed by W. H. Bucher, *The Deformation of the Earth's Crust*, Princeton, 1933, pp. 464–8, by J. A. Steers, *The Unstable Earth*, London, 1932, and by S. W. Carey, 'The Asymmetry of the Earth', *Aust. Journ. Sci.*, Vol. 25, 1963, pp. 369–83 and 479–88.

Index of Authors

Index of Subjects

Index of Localities

Printed and bound by CPI Group (UK) Ltd, Croydon, CR0 4YY

17/10/2024

01775656-0018